# Electrical Power Distribution and Transmission

# Electrical Power Distribution and Transmission

Luces M. Faulkenberry
Walter Coffer
University of Houston

**PRENTICE HALL**
Upper Saddle River, New Jersey 07458    Columbus, Ohio

**Library of Congress Cataloging-in-Publication Data**

Faulkenberry, Luces M.
    Electrical power distribution and transmission / Luces M. Faulkenberry, Walter Coffer.
        p.      cm.
    Includes bibliographical references and index.
    ISBN 0-13-249947-9 (hc)
    1.  Electric power distribution.   I. Coffer, Walter.   II. Title.
TK3001.F38   1996
621.319—dc20

95-34051
CIP

Cover photo: © R. Kord / H. Armstrong Roberts
Editor: Charles E. Stewart, Jr.
Production Editor: Christine M. Harrington
Design Coordinator: Jill E. Bonar
Text Designer: Ruttle Graphics, Inc.
Production Manager: Pamela D. Bennett
Marketing Manager: Debbie Yarnell
Illustrations: Diphrent Strokes

© 1996 by Prentice-Hall, Inc.

**A Pearson Education Company**
**Upper Saddle River, NJ 07458**

Printed in the United States of America

ISBN: 0-13-249947-9

Prentice-Hall International (UK) Limited,London
Prentice-Hall of Australia Pty. Limited, Sydney
Prentice-Hall Canada Inc., Toronto
Prentice-Hall Hispanoamericana, S.A., Mexico
Prentice-Hall of India Private Limited, New Delhi
Prentice-Hall of Japan, Inc., Tokyo
Pearson Education Asia Pte. Ltd., Singapore
Editora Prentice-Hall do Brasil, Ltda., Rio de Janeiro

To Luke's children
**Laura and Matt Faulkenberry**
and
Walter's wife
**Syble Coffer**
In appreciation of their patience and understanding

# Contents

**CHAPTER 7    DISTRIBUTION LINE CONSTRUCTION    259**

**SECTION III — ELECTRICAL POWER TRANSMISSION**

**CHAPTER 8    TRANSMISSION SYSTEM OVERVIEW    297**

**CHAPTER 9    TRANSMISSION LINE PARAMETERS    356**

**CHAPTER 10    TRANSMISSION LINE FAULT CURRENT CALCULATION, PROTECTION, AND BULK POWER SUBSTATIONS    420**

# Preface

This book is a practical introduction to electrical power distribution and transmission. The basic concepts of the subject are introduced in a practical, down-to-earth manner, but the subject concepts are covered in considerable depth and detail. The book is designed to be easily read by students, we hope with great interest. Although designed as a textbook for electrical power technology programs, it should also be a good reference for people already working in the power field as technicians, technologists, and engineers.

This text book is designed for a two-term course in electrical power distribution (first semester) and transmission (second semester). The order can be reversed but some information from the distribution section, such as basic transformers and circuit breakers, will have to be presented in the transmission semester. The book can also be used as a one-term course in distribution and transmission if the depth of the course or the coverage of topics is limited. The student is assumed to have had courses in ac and dc circuits, algebra, and trigonometry as minimum prerequisites to a course using this text. Basic three-phase concepts are reviewed in the text but a previous course in which basic polyphase circuits are covered is helpful, as is a course on rotating electrical machines. Some very basic calculus is used in a couple of places in the book, but a course in calculus, while nice, is not needed because the calculus elements are briefly explained where they are first used.

The book is organized in three sections. The first section is a general section that presents a brief history of the electrical power industry, the relationship of the industry with other institutions, and basic concepts including three-phase phasors and per unit. The second section, Chapters 3 through 7, focuses on electrical power distribution. This section covers basic distribution concepts and layouts, transformers, equipment, substations, distribution and substation protection, and distribution system construction. The third section, Chapters 8 through 11, covers electrical power transmission. It includes: basic system concepts including power flow, stability, and economic dispatch; types of transmission systems; transmission line parameters and equivalent circuits; fault calculations with symmetrical components; transmission line protection; grounding; and transmission line construction.

# Introduction to the Electrical Power System

# CHAPTER 1

# Introduction to Electrical Power

This chapter presents a brief introductory discussion, a short history of electrical power, a discussion of the relationship of the electrical power utilities with regulatory agencies and customers, and an overview of the electrical power system.

## 1.1

### ELECTRICAL POWER IN THE MODERN WORLD

The electrical power industry is a mature, established industry. Electrical power is the prime source of energy that supports almost all of our other technologies. Electricity is the most convenient and omnipresent energy available today. Imagine a day without it.

Electrical power is an energy transportation system. It is a safe, convenient, efficient way to transport large amounts of power long distances. The high efficiency of electrical machines—generators are over 98% efficient with efficiencies reaching over 99%, transformer efficiencies routinely reach 98% and can reach over 99.5%, and electric motors have efficiencies that are routinely over 80% and many are over 90%—makes the conversion of energy to electricity for transportation and reconversion to heat, light, and mechanical power cost effective.

The alternative to electrical energy conversion for mechanical power is thermodynamic conversion of fossil fuel, hydro, and wind power. These all have serious constraints with regard to mobility or efficiency. Thermal engines, because of temperature and pressure differences currently obtainable, are limited to thermodynamic efficiencies of 40% or less, with higher efficiencies favoring large steam turbine installations powered by fossil or nuclear fuel. Higher efficiencies can be obtained by using the exhaust of gas turbines to provide the heat to produce steam for steam turbines. This is called *combined cycle generation,* which can provide efficiencies to about 60%. Although other sources of energy can be transported, electricity is only equaled by natural gas for efficiency of transport, and it must be converted to mechanical energy at a relatively low efficiency to obtain mechanical power. Only relatively small amounts of power, ideal for remote locations, can currently be obtained from solar and wind sources. Much research is being done in pursuit of alternatives for fossil fuels for the production of electric power. Nuclear fusion appears to be the most promising long-range alternate energy source for electric power generation, but much work remains before commercial power generation can begin with nuclear fusion as the thermal energy source.

One measure of the wealth of a nation is the total production of the society of that nation. Electrical energy conveniently provides light that lengthens both work and leisure hours. Power for work, from electric motors in tools, extends human physical work output from approximately 2.8 kW to many times that amount. The human effort guides the tools that do the work. Lately, much human effort guides computers that then guide the tools. Electricity provides power for both the computers and the tools. Thus, electricity is an integral part of the daily life of the citizens of any developed country. Electricity provides power for light, tools, and entertainment. It provides power for both work and play.

## 1.2

## A SHORT HISTORY OF ELECTRICAL POWER

In 600 B.C. Thales discovered that amber, when rubbed, attracted small objects. He had discovered static electricity. Centuries before Thales the Chinese had constructed magnetic compasses, and during the twelfth and thirteenth centuries the magnetic compass was the subject of experiments in Europe. It was not until 1600 that the first systematic discussion of electromagnetism based on experimentation was published by Englishman William Gilbert, who is called the father of modern electromagnetism. In 1660 Otto van Guericke discovered how to generate sparks of static electricity by rotating a sulphur ball on a wooden shaft. Allessandro Volta introduced the first storage battery in 1800. In 1825 Andre Ampere discovered the attractive and repulsive force, depending on current direction, between two conductors carrying a steady current. Note that the rate of discovery started to increase rapidly around 1800.

George Ohm published his discovery that current flow in a conductor is directly proportional to the applied potential and inversely proportional to its resistance in 1827. This relationship, called "Ohm's law," is still used. Only 4 years later Michael Faraday demonstrated the basic principles of the induction generator, transformer, and electric motor.

In 1858 the South Foreland lighthouse was equipped with an arc lamp, invented in 1808 by Sir Humphry Davies, using a steam generator for power. In 1864 James Clerk Maxwell published his electromagnetic equations, which were a great leap forward in electromagnetic theory.

Around 1873 a Belgian born engineer, Zenobe Theophile Gramme, developed the first practical electric generator and motor. Earlier (1835) American Thomas Davenport had invented the electric motor using the electromagnet developed by Joseph Henry in 1823. The Davenport motor was never developed into a practical machine.

The year 1876 is known primarily for the invention of the telephone by Alexander Graham Bell. A less well known development of that year was the first practical arc lighting system by Charles Brush.

Thomas Edison perfected the incandescent lamp in 1879. That same year the first company opened for the purpose of selling electricity. The California Electric Company of San Francisco opened to sell electric service for arc lamps to commercial establishments and factories. The year 1879 was also the year that Thompson, Westinghouse, and Stanley developed a practical transformer.

In 1882 Thomas Edison's Pearl Street station opened in Manhattan. It provided direct current (dc) for incandescent lighting to an area of about one sixth square mile. Figure 1.1 shows the Gable Street power plant in Houston, Texas in 1900. This plant replaced Houston's first electric generating plant, which was opened in December 1882, just a few months after the Pearl Street plant, and destroyed by a boiler explosion in 1898. For the next 20 years small electric generating stations sprang up in every city of any size, and by 1900 there were over 3000 stations in the United States.

The quest for efficiency has been a driving force throughout the history of electrical power. The steam turbine, invented by Parsons in 1884, was to play a big role in making the generation of electricity more efficient.

In 1881 Frenchman Lucien Gaulard and Englishman John D. Gibbs patented an alternating current (ac) transmission system in England. In 1885 George Westinghouse, the American mechanical engineer who invented the air brake, bought the American patents to the system from Gaulard and Gibbs. That same year Westinghouse and his associate William Stanley perfected the transformer and developed the constant potential ac generator. The next year, 1886, Stanley

*One-hundred years ago, in the summer of 1882, a group of investors formed the Houston Electric Light & Power Company? Construction began in September on a new plant and on the night of December 15, 1882, Houston entered the electrical age—as thirty crude carbon arc lights sputtered with brilliant white light.*

*At the turn of the century, a new plant called Gable Street, shown above, replaced the previous one which was demolished by a boiler explosion in 1898.*

*Since that first year, Houston Lighting & Power Company and electricity have played an indispensable role in the prosperity and growth of the Houston–Gulf Coast area.*

**FIGURE 1.1**   The Gable Street plant in Houston, TX began operation December 15, 1902 (Courtesy of Houston Lighting and Power).

demonstrated the practicability of ac transmission in Great Barrington, Massachusetts by transmitting power 4000 feet. He used a transformer to step the generated voltage up to 3000 V for transmission and another to step the transmitted voltage down to 500 V for use. The first commercial transmission of ac in the United States was from Willamette Falls to Portland, Oregon, 13 miles at 3300 V, in 1890.

In 1888 Nicola Tesla presented a paper that was to prove extremely important to the budding electrical power industry. He introduced the polyphase ac system, as well as the polyphase induction and synchronous motor. George Westinghouse recognized the importance of Tesla's work and bought the patents from him that same year. In 1891 the first three-phase system was installed between Lauffen and Frankfurt am Main in Germany. By 1894 five polyphase generating plants were operating in the United States. Four were three-phase systems and one was a two-phase system. Three-phase electricity quickly became the ac transmission and generation standard because three-phase power was provided smoothly, rather than in the pulsating manner of one or two phases. The first hydroelectric plant was installed in Niagara Falls, New York in 1886, but ac transmission, from Niagara Falls to Buffalo 22 miles away, did not begin until 1896. Alternating current, preferably three phase, was now firmly established as the preferred method of generating and transmitting electricity.

The 1902 World Fair in St. Louis, Missouri had as its theme electric lighting, which popularized electric lighting. The first steam turbine driven generator was installed in the United States in 1903. The 5 MW unit ushered in a new, more efficient era in electric power production.

Transmission voltages rose rapidly to about 40 kV, but the pin type insulators then available limited further increases in voltage. In 1907 two U.S. electrical engineers, Edward M. Hewlett and Harold W. Buck, invented the suspension insulator that allowed higher transmission voltages. By 1920 voltages of 132 kV were common, and some systems used 150 kV. In 1934 Boulder Dam (later called Hoover) on the Colorado River sent power to Los Angeles, 270 miles away, at 287 kV. Large, hollow, copper conductors were used to prevent *corona* loss.

Corona is the discharge of static electricity from a power line. It is minimized with large diameter lines because static electricity discharges best at sharp points, corners, and along small conductors. Corona became the limiting factor in transmission voltages. Very large diameter conductors were being used, even at low currents, to avoid corona. During World War II the Germans discovered that multiconductor bundles (two or more conductors held apart by periodic spacers) acted like one very large diameter conductor as far as corona is concerned. The first multiconductor installation was after the war in Sweden using double conductor bundles (two spaced conductors per phase) at 380 kV. Transmission voltages continued to rise rapidly. In the United States the first 345 kV line began operating in 1953, the first 500 kV line in 1965, and the first 765 kV line in 1969.

To improve further operating efficiency by increasing the immediately available reserve power (called spinning reserve), reduce duplication of capital expenditures, and reduce outages the electric utilities began to interconnect their systems for mutual benefit. This allowed a utility to supply power to another utility whose load was high at times when its own load was low, allowing both utilities to utilize their equipment more efficiently. Interconnection has been in use since the 1930s and is almost universally used today.

Although dc transmission was replaced by ac early in the history of electrical power because of the ease and efficiency with which ac voltage levels can be changed using transformers, *high voltage* transmission with dc has some advantages. There are no reactance losses on dc transmission lines, and there is no need to synchronize one user with another when using dc sources. Because the equipment required to convert the generated ac voltage to dc at the source and back to ac at the destination is expensive, dc is normally used to transmit power long distances (200 miles or more), to tie together systems for interconnection with very short lines so that the phases of the systems need not be synchronized, and when ac transmission is too lossy as with under water power cables. The Germans were the first to transmit high voltage dc. In World War II they transmitted 60 MW of dc power from the Elbe power station to Berlin at 400 kV ( + and − 200 kV). In 1954 Sweden connected the first commercial dc line, a 100 kV 20 MW submarine cable from the Swedish mainland to the Island of Gotland 60 miles away. The first U.S. dc line was finished in 1970. It was the Pacific Intertie, an overhead line that transmitted 1440 MW 825 miles from Celilo, Oregon to Sylmar, California near Los Angeles. Several dc systems were already in service in Canada, Europe, and Russia at that time. High voltage dc transmission is gaining in popularity as the conversion equipment becomes more efficient, reliable, and economical.

It is impossible in the available space to list all of the contributors to the electrical power industry. Some of the most important contributions were: Edison's dc distribution system, the ac transmission system, the large steam turbine, interconnection, and the regulated monopoly system of business for the electrical power generation and distribution companies. The last item is discussed in the next section.

By 1939 virtually all U.S. industries were powered, most by electricity. By 1945 75% of all U.S. farms had electricity (96% by 1960), and virtually all urban homes had electricity. In only a little over a century since Edison opened his Pearl Street Station in Manhattan electrical power has become an important part of the technology and the daily lives of the citizens of all developed nations.

**1.3**

## RELATIONSHIP BETWEEN ELECTRIC UTILITIES, CUSTOMERS, AND REGULATORY AUTHORITIES

Electric generating and distribution facilities were originally owned by individuals or small groups of investors. There were thousands of very small electric companies springing up in every large or medium sized town in the late 1800s. Some people generated electric power for their own use and sold to their neighbors, some set up small generating stations and sold to anyone who would buy (like "Mom and Pop" grocery stores), while businessmen seeking a profitable investment owned other generating stations. Each generator had to be connected to each customer it served. The many duplicate electrical distribution systems installed in many localities caused three problems. The first was danger to life and property caused by many crisscrossed, closely spaced, distribution systems. The second was the unacceptable crowding of public right of ways caused by the many distribution systems. The third was the extremely high capital cost for the many duplicate systems, which led to high rates.

The corrective step taken was to grant a franchise for an area to a single supplier of electricity. This cured all three problems but caused another one. Since the rates were no longer controlled by competition, another method of rate control had to be found. That method was regulation. By 1900 the consolidation of small, competing, electric companies into larger, investor owned, franchised systems was underway. In 1905 New York established the first public utility regulating commission. This began the transition of the electric generating and distribution industry into a set of state regulated monopolies. By 1924, 42 states had established public utility regulating commissions.

In 1920 the Water Power Act established the Federal Power Commission (FPC) to license the building and operation of hydroelectric facilities. Later its function expanded to regulating interstate commerce of power companies. It is now called the Federal Energy Regulatory Commission (FERC). During the Roosevelt administration a major push occurred to control the public utilities (investor owned utilities) through the FPC. An intense legal battle ensued. Eventually it was realized that both investor owned and government owned utilities are useful, and have a stabilizing influence on each other.

Samuel Insull, whose first job in the United States was private secretary to Thomas Edison but later became chairman of the Consolidated Edison Company that had substantial holdings in the Central United States, was very influential in shaping the electrical power industry in its early days. In two speeches, one in 1914 and another in 1916, he stated the reasons that electric utilities should be regulated monopolies. They are still valid.

The capital investment required for each dollar of revenue received is much higher in the electric generation and distribution business than it is in other industries. (Currently it is $4 or more for electric utilities, about $2 for railroads, and 30 to 50 cents for manufacturing.) Duplication of electric service in an area is costly and wasteful of both human and natural resources. Monopoly status, however, requires regulation to assure accountability to the public since the utilities have no direct competition. The regulators must assure the public of proper rate structures and business practices by the utilities, and the utilities of sufficient return on investment to make needed capital improvements and attract investors. Most cities do not have sufficient expertise and funds available to regulate large utilities effectively so the states have to shoulder the job of regulating public utilities. Insull warned of two problems associated with regulation. One is that the people appointed to regulatory commissions may not have sufficient knowledge of large business practices to understand the power industry. Another is politicization. Regulators and other officials might be tempted to be short-term public advocates concerning power rates for political gain at the expense of long-term availability of power at reasonable prices.

Besides the high capitalization cost there are some other aspects of the electric utility business that are unique, and many are also common to other utilities.

1. The product is intangible in the common sense of the word. It cannot be driven, displayed, or handled directly (at least it shouldn't be).
2. Customers are charged for service and amount used. Service is defined as making available the correct voltage and frequency without interruption. Often good service and low rates are in conflict because of the equipment investments necessary for continued good service.
3. The product cannot be stored in large quantities. Enough equipment must be installed and ready to operate on short notice to meet the peak demand. If, as expected, the storage of electrical power in superconducting rings becomes economical, then the storage of large quantities of electrical power for use at peak loads may become practical.
4. The utility is responsible for delivering the product so the transmission and distribution systems must operate properly at all times.
5. The utility must use public right of ways for delivering the product. Thus the utility must have the right to buy, lease, and in extreme cases condemn property for right of way to assure the delivery of electricity to the public.
6. The utility is responsible for reasonable precautions to ensure the safety of its workers and the public. It must assure there is no injury to people or property because of inadequate equipment. The electric utilities take their responsibility for public safety very seriously.

The U.S. government has always considered electrical power important for its citizens. Three examples of this concern are: the Tennessee Valley Authority, the Rural Electrification Administration, and the Bonneville Power Administration.

The Tennessee Valley Authority (TVA) is a U.S. government agency formed in 1933 to provide flood control, navigation improvement, power development, and recreational area development for the Tennessee River watershed. It was also to establish power rates to compare with the rates around the country. Originally power was a secondary goal of the TVA, but it quickly became its primary purpose. Since the TVA did not have to pay interest on loans for capital investment or dividends to stock holders, its rates soon assured it no competition in the area it serves.

Concern for the farmer and the desire to "reduce farm drudgery" caused the U.S. government to form the Rural Electrification Administration (REA) in 1935. The agency encouraged the formation of "electric membership cooperatives" and provided low interest loans to build distribution systems in areas too sparsely populated to attract investor owned electric companies at that time. The REA was very successful in bringing electricity to the farm, and helping to improve agricultural productivity and rural quality of life.

The Bonneville Power Administration (BPA) is a federal agency under the Department of Interior. It markets the power produced at the federally owned hydroelectric projects in the Columbia River Basin in the Northwest United States.

Municipally owned generating and distribution facilities are operated by a number of cities. When cities began to use electric street cars and street lights many cities built their own generators instead of buying electricity. Some cities then began selling electricity to their citizens. Los Angeles, Seattle, and Takoma are examples of cities that supply all of their electric energy with municipally owned utilities.

Electric power in the United States is provided by a variety of types of organizations. Electrical power in some areas is provided by municipal systems, in some by cooperatives, in some by state or federally owned systems, however, the majority of the electrical power used in the United States is generated and distributed by investor owned companies. Although they are commonly called public utilities, they are private, investor owned, state regulated, utilities. In some cases the franchise granting municipality and the state share regulatory authority. These power supplying groups exert a stabilizing influence on each other because their rates must be competitive to attract business to their franchise area and avoid adverse public opinion. The mutually stabilizing influence of the electric power supplying groups and the regulatory authorities normally provide the public with adequate protection from unfair rates and poor service.

In contrast to the United States, in most countries the organization furnishing electric power is the government. The pricing of electrical power in many countries is a political decision dictated by political events. The regulatory and price fixing authority is either elected or appointed. In most of these countries the price of electricity is relatively high because the politicians find it easier to collect electric bills than taxes.

Another factor that is influencing the electrical power industry in the United States is cogeneration. In 1978 Congress passed the Public Utilities Regulatory Policy Act (PURPA), a section of which required power companies to buy cogenerated power (customer generated power in excess of customer needs) at the power companies avoided cost; what the power company would have spent to generate the cogenerated electricity. In the case of large cogenerators a negotiated price is now allowed. PURPA was designed to encourage private power generation in a time of energy scarcity with the expectation that smaller cogenerators using renewable resources such as wind and solar energy would save nonrenewable energy resources such as coal and oil. Within a short time industries that produced a large amount of their own electricity began cogenerating large amounts of electricity for sale to the power companies under PURPA as the market for the cogenerators primary product became depressed. If a company has a use for low temperature, low pressure steam for some process it can produce high temperature, high pressure steam, use the steam to produce electricity and the left over low pressure, low temperature steam for the process more efficiently than it can produce low pressure, low temperature steam. If a company has no use for the low pressure, low temperature steam it cannot usually produce electricity as efficiently as a modern power company unless it uses a combined cycle unit in which gas fire turbine exhaust is used to produce steam, which is far more efficient than a stand-alone steam unit. Large cogenerators usually use nonrenewable energy resources to generate electricity so they provide no savings of these resources unless combined cycle is used.

Cogeneration of large amounts of electricity by large companies caused some problems, not all of which are resolved at this time. One problem is that cogenerators often want to sell electricity to the power companies at night when the power companies' demand is lowest and they have no need for the electricity. Thus the production of electricity by the cogenerator is out of synchronism with power company needs. A second problem is that some large cogenerators are located in areas that do not need the electricity at all, thus some power companies are forced to buy electricity they cannot resell. Often the excess electricity can be wheeled (transmitted using other power companies' transmission facilities) to areas that do need the cogenerated electricity.

Perhaps the most serious problem may occur because state Public Utility Commissions (PUCs) often consider the availability of cogenerated electricity when considering public utility applications for new generating facilities to meet future demands. New generating facilities take a long time to plan and construct (typically 5 years for a coal fired plant, not including the time required for approval by the various government agencies). Cogenerated electricity can disappear at the end of a contract, or at any time if not under contract. If adequate generating facilities are not available to handle the load served by cogenerated electricity at the end of a contract, the cogenerator(s) has a powerful bargaining position. Reliance on cogenerated electricity for a significant fraction of the electric power consumed in an area could produce a rate structure that is not under public control via the state PUC. However, the trend toward deregulation and wheeling rulings by FERC may result in a less territorial view of electricity generation. Many new business entities are being formed as electricity providers, selling their electrical power to the highest bidder, wheeling the power to the buyer.

The Department of Energy, through the FERC, has ruled that the transmission grid will remain a regulated monopoly while both generation and sales of electrical power will be regulated by the market place. Mechanisms that allow electricity to be treated as a commodity are now in place and operating, available for sale on an hour by hour basis, and delivered where it needed. Private power-related industries have formed groups to facilitate the marketing of electricity with the blessing and aid of the Department of Energy. It is now possible for an independent power producer to generate power at one location, feed it into the electric grid, have their power wheeled to a remote location, possibly across several states, and sell the power to any reasonable customer, a municipal system, an REA, an industry, or even to a shopping center. These federal initiatives have resulted in a variety of classes of owners of electrical generating facilities having access to the electrical grids and competing with utilities in electrical power generation.

Most people in the electrical power generation and distribution industry still believe in the principle Samuel Insull stated in a 1921 speech.

In performing this task (of supplying utilities), our first duty and our first interest, an interest beyond everything else, is to square the rights of the public with those of the stockholder. Unless we can bring home to the public that our first thought and desire is to see that they get a square deal we might just as well shut up shop and go out of business and sell our plants for junk. Our income, our earning capacity, is dependent, primarily in my judgement, upon public good will, and you cannot get public good will unless you convince the public that you are treating them honestly and giving them a square deal.

In other words, the most important person associated with the power industry is the customer.

# 1.4

## ELECTRICAL POWER SYSTEM CONSIDERATIONS

Societies must use energy resources in the form in which they appear, whether as water, wind, oil, coal, or uranium, to accomplish the tasks the societies consider desirable. The desirable tasks may be heating, cooling, lighting, manufacturing, or transportation of people and materials. Finding and converting the raw energy resources to usable energy is a vital function, as is the design, production, and use of efficient equipment to convert energy to useful work (motors, heaters, air conditioners, etc.). Both topics are beyond the scope of this book. To be useful energy must be available at the place the work is to be done. Recall that electricity is an efficient and convenient transportation system for energy that allows the raw energy resources and the equipment that converts energy to work to be separated by great distances.

Electricity does exist in nature as lightning and static electricity, but it cannot be controlled well enough to be put to practical use. Thus electricity must be generated by converting another raw energy resource. Electricity can be stored in batteries, but only in relatively small quantities. Therefore, at least for the present time, electricity must be produced at the same time it is used.

### 1.4.1 Reserve, diversity, and economic dispatch

*Reserve* is that portion of an electric utility's available generating capacity that is not producing electricity at a given time. *Spinning reserve* is the generating capacity that is being driven at the proper speed to provide proper voltage, but is not producing power. Spinning reserve can provide power to the system almost instantaneously if the system load is increased or a generator must be taken out of service. The FERC established a requirement that each electric power company construct sufficient excess capacity that it can supply its largest normal load with its largest generating plant off line. This rule has been modified for some circumstances, as will be discussed later. Spinning reserve should be sufficient to meet any sudden load changes anticipated by the utility.

*Diversity* is the term used to refer to load changes during a period of time. Load varies during the day because people get up, go to work, and return in the evening using different amounts of electricity to support their various activities. Similarly, industrial and commercial power use will vary during the day. There are also weekly and seasonal variations in electricity usage. In a warm climate such as that along the Gulf Coast air conditioning results in high electricity consumption in the Summers that peaks daily in the late afternoon. Figure 1.2 shows a 24-hour diversity curve for electric power consumption. The result of diversity is that the electric utility must supply varying amounts of power depending on the time of day, day of the week, and season.

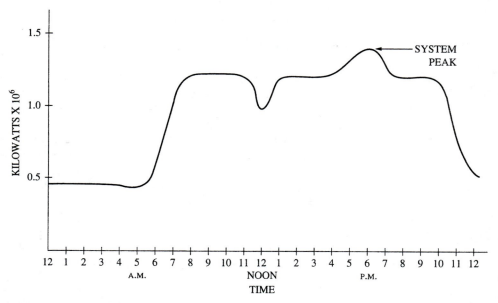

**FIGURE 1.2**  Load diversity

*Economic dispatch* refers to serving the load at all times with as little excess capacity as possible using the most efficient generating units possible. Properly sequencing the timing and size of generating units put into service can result in very large cost savings while adequately serving the load. The more efficient units are used to serve the base load and the less efficient units are used to serve the peak loads. Many factors must be considered other than just generator efficiency for economic dispatch in a large system. Line losses for various lines, fuel cost, availability and cost of interconnected capacity, and many other factors must be considered also, so economic dispatch alternatives are calculated on complicated computer programs. Figure 1.3 illustrates economic dispatch with respect to generator efficiency.

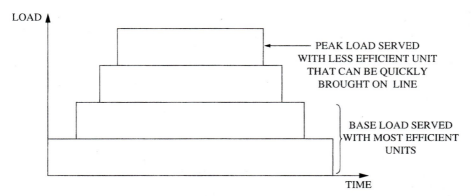

**FIGURE 1.3**   Economic dispatch with load diversity

Short-term power demand peaks require the use of gas turbine driven generators for the speed with which they can be started. Large boilers take a long time to come to heat, and must be brought up in a carefully controlled sequence to avoid severe damage to the boiler plates, tanks, and piping. In like manner the steam turbine and generator must be brought up to speed slowly and carefully to avoid warping the shaft. The weight of large turbines and generators and the speed they must turn (usually 3600 rpm) makes bringing them up to speed a difficult and time-consuming task. Thus the large generators must be started well ahead of the time their power is needed. The shutdown of large units is equally time consuming.

Peak power demand fluctuations cannot be predicted as accurately as normal load fluctuations. The gas turbine, essentially a jet engine adjusted to burn natural gas (jet fuel is used where natural gas is not available) and turn a mechanical load, can be put on line or removed in about 15 minutes. Thus the gas turbine is used in sizes to about 50 MVA to handle short term peak loads at a somewhat lower efficiency than the larger units.

The FERC excess capacity rule mentioned earlier was modified to allow a group of power companies with a strong interconnection system to have jointly a reserve capacity equal to the largest generating plant within the group. This has reduced the necessity for much expensive reserve generating equipment while preserving power system integrity. Transmission equipment is less expensive than generating equipment, so that a strong interconnection system now ties virtually the whole country together. The Eastern and Mid-western system was out of synchronization with the Western system, so a dc tie had to be used to tie the systems together.

## 1.4.2 System components

The electrical power system can be divided into three major parts:

1. Generation, the production of electricity.
2. Transmission, the system of lines that transport the electricity from the generating plants to the area in which it will be used.
3. Distribution, the system of lines that connect the individual customer to the electric power system.

The parts of the electrical power system are illustrated in Figure 1.4. That electricity must be generated and distributed to each customer is obvious, but the necessity of the transmission system is worth a bit of discussion. The trend has been to locate generation plants away from heavily populated areas when possible. Land is less expensive and few people want a generating plant next door. Additionally, remote siting often allows plants to be closer to the source of the fuel they use, providing substantial savings in electricity production. Hydroelectric plants must, of course, be located at an appropriate source of falling water. Transportation of electricity long distances requires higher voltages than are safe to connect to most customers. Electricity is then transported to the general area of use by the transmission lines and distributed to the customers at lower voltages by the distribution lines. The sites at which the power transformers used to step the transmission voltages down to the distribution voltages (which also contain switching and protection equipment) are called *distribution substations*.

## 1.4.3 Voltage levels

Great effort has been expended over the years, as we noted in the history section, to raise the voltage levels for each portion of the electrical power system. The major losses in transporting electricity are proportional to the current squared, thus doubling the current quadruples the loss. Power is the product of current times voltage so as voltage is increased, current is decreased, and losses are decreased for a given amount of power transfer. Thus the highest economically feasible voltage is favored for generating, transmitting, and distributing electrical power.

Rotary machines (generators and motors) have practical voltage limits set by limitations in insulation and cooling technology. Voltages allowable on rotary machines are increasing, but slowly now. Maximum generator voltages are currently about 24 kV and about 12 kV for motors.

The voltage limits for transmission voltages are set by protective devices (mainly circuit breakers) rather than transformers and insulators. Bundled conductors and shielding have removed the past limit of the breakdown of the air around the conductors. Different electric utilities have set different standard voltages, but the following voltages are common:

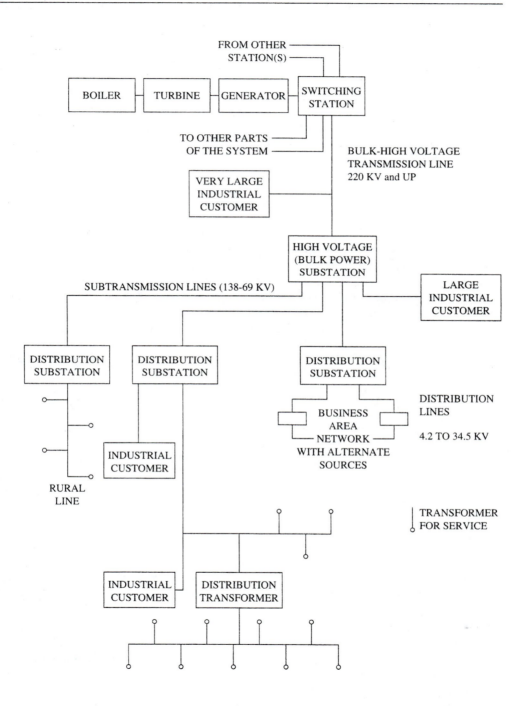

**FIGURE 1.4**   The electrical power system

1. Generators—11 kV to 24 kV.
2. Extra High Voltage Transmission—345 kV, 500 kV, and 765 kV for ac, and around 500 kV ($\pm$ 250 kV), 800 kV ($\pm$ 400 kV), and 1000 kV ($\pm$ 500 kV) for dc.
3. Transmission—138 kV and 230 kV.
4. Sub-transmission—34.5 kV and 69 kV.
5. Distribution—12.5 kV and 34.5 kV (2300 V and 4 kV are still common in small towns).

### 1.4.4 Frequency

Many different frequencies were tried for electrical power systems in the early days of ac power. Some parts of the United States, particularly California, used 25 Hz, with fifty and 60 Hz, and some scattered dc, were used in the Midwest and Eastern parts of the U.S. Direct current was rejected because of the high line loss and large voltage drop along the line at the voltages available then. Twenty-five Hz fell from favor because incandescent lights flickered at such a low frequency. Sixty Hz was chosen in the United States because of the convenient fit with the 60-second minute, and the absence of other problems. The lowest possible frequency is desirable for electricity transmission to lower the reactive voltage drop on the line and reduce radiative loss (like antenna radiation). The lower practical limit on frequency was set by the flicker of the incandescent lamp.

Europe settled on 50 Hz. Most of Africa and some of Asia followed suit because of European colonization. Some of Africa and most of South and Central America use 60 Hz.

## 1.5

### GENERATION

### 1.5.1 Fossil fuel and the steam cycle

Recall that generation involves rotating a magnetic flux (from an electromagnet) inside a set of conductors held in place by a magnetic stator. Many fine works deal with the mechanics and operation of synchronous ac generators, including *Electric Utility Systems and Practices* listed in Appendix E. Modern synchronous generators are extremely efficient. Improvements in the steam boilers and turbines that make up the prime mover system have contributed enormously to the efficiency with which electric power can be generated. Steam turbines operating at 3600 rpm (1800 rpm for nuclear plants) are the most used prime movers for generators. Steam turbines typically have efficiencies around 87%, and can be built as large as 1000 MW. The boiler efficiency contributes much to the overall efficiency of generation. Recall that maximum temperature and pressure differences currently obtainable limit the thermodynamic efficiency to about 40%. Efficiency has improved dramatically over the years. In 1888 15 lb. of coal were required to generate 1 kWh of electricity, in 1903 7 lb. were required, in 1923 only 2.2 lb. were needed, and in 1980 0.3 lb. of coal was required to generate 1 kWh of electricity.

In 1980 the energy sources used for the production of electricity in the United States were: fossil fuel—68.9%, hydroelectric—12.6%, nuclear—9.1%, gas turbine—8.5%, and internal combustion engines—0.9%. About 78% of the electricity in the United States was generated by steam turbine driven generators, and most of the fossil fuel used was coal. Most generating plants in the United States are coal fired. Many countries generate a much higher percentage of their electricity using nuclear fuel. Both nuclear and fossil fuel plants use steam turbines as prime movers.

The process of generation is then the conversion of energy from a primary energy source such as coal by a chemical process such as burning to thermal energy, usually in the form of high pressure steam. The thermal energy is then converted to mechanical energy by a turbine, and from mechanical energy to electricity by a generator. The process is shown in its simplest form in Figure 1.5. Modern plants are a large improvement on this simple scheme, but the basic process is the same.

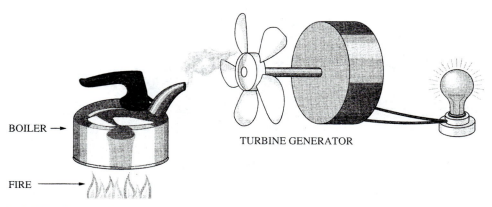

BOILER →

TURBINE GENERATOR

FIRE ———→

**FIGURE 1.5**   Simplified generation system

In modern plants fuel handling is almost fully automatic. For example, in a coal fired plant the coal will be fed automatically from hoppers to conveyers, crushed to a powder, and blown into the burning chamber by automated machinery. Preheated air (heated with exhaust air from the burning process) will be supplied to the burning chamber in the proper amount for efficient burning. The entire burning process will be automatically regulated to optimize combustion.

Many improvements have been made in every area of the electricity generation process through the years. Any small increase in efficiency that can be found in any phase of the generation process, fuel burning, steam production, turbine, or generation is important because of the immense amounts of electricity needed in an industrial society. A small percentage gain in efficiency provides an immense saving of resources. The amount of mechanical power provided by the turbine, thus the electricity provided by the generator, depends on the volume, the temperature, and the pressure drop of the steam as it passes through the turbine. Any step that allows a higher temperature or pressure for input steam, or a lower exhaust temperature or pressure, or uses heat that remains in the steam after it has provided energy to the turbine improves efficiency.

Figure 1.6 is a simplified diagram of a fossil fuel generating plant that shows some of the major improvements that have increased efficiency. Starting at the upper left the furnace input air (forced by a fan) is preheated, using the furnace exhaust to save energy, to improve combustion efficiency. After combustion of the air and fuel mixture an induced draft fan forces the combustion byproducts out the exhaust stacks. This fan is necessary because the air heater and precipitators (for removing pollutants) constrict the exhaust flow. Motors as large as 1800 hp are needed to drive induced draft fans. The steam cycle is a closed loop except for the make up water that replaces the water lost in the inevitable small leaks. The water for the steam cycle must be de-mineralized, to prevent clogging the boiler tubes with mineral build up, and de-aerated, to remove the oxygen from the feedwater (water to be turned into steam), to reduce corrosion.

Starting at the condenser we will trace the steam cycle. The cooling water passing through the condenser causes the spent steam to condense into water. The condensed water occupies less space than steam, so condensation causes a vacuum at the turbine exhaust. The lower exhaust pressure causes the temperature of the steam (the temperature of steam depends on the pressure) to be lower. The temperature difference across the turbine is therefore increased by the condenser and the energy extracted by the turbine is increased.

The condensate from the condenser is pumped to a series of feedwater heaters. The feedwater heaters (up to about seven are used) preheat the water going to the boiler using steam extracted from the turbine near the exhaust. The use of extracted steam is more efficient than heating the feedwater with a separate heater. The preheated water reduces the amount of heat the boiler must add and reduces the mechanical stress on the boiler (cold feedwater would severely stress the boiler). The steam used to preheat the feedwater has already been used to provide energy to the generator turbine or drive turbines that power pumps, so feedwater heating is the second use of the extracted steam. The preheaters will take the steam from about 90°F at about 4 psi to about 500°F at 600 psi. The economizer uses combustion gasses that have already passed through the boiler to further preheat the incoming feedwater at little extra cost. The water then passes through the boiler where it is turned into steam at about 2400 psi at around 650°F.

Superheating steam (increasing the temperature at a fixed pressure) increases the thermal energy that can be extracted from the steam and reduces the possibility of water droplet formation in the steam as it passes through the turbine. Water droplets can corrode turbine blades very rapidly. Most high pressure turbines use input steam at about 1050°F at around 2400 psi. After the steam has passed through the high pressure turbine it is usually reheated to about 1000°F at the

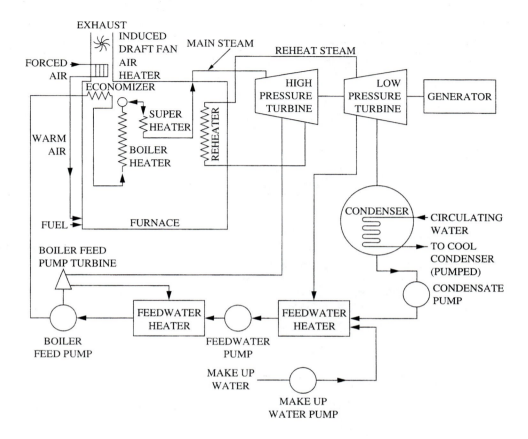

**FIGURE 1.6**   Diagram of the principal components of a fossil-fueled power plant

same pressure as the high pressure turbine exhaust, about 550 psi. After passing through the low pressure turbine the steam goes into the condenser and the process repeats. Most modern large generating plants use three-stage turbines (high, intermediate, and low pressure), and two reheat cycles.

Figure 1.6 does not illustrate the extreme complexity that is characteristic of a modern generating plant. The effort to improve efficiency has resulted in very complex boilers, furnaces, turbines, generators, and control systems. *Electric Utility Systems and Practices* provides a good overview of fossil fuel, hydroelectric, and nuclear generating practices.

## 1.5.2 Hydroelectric power

Falling water was one of the first sources of energy man exploited for work and also one of the first forms of energy used to produce electricity. The work falling water can do depends on the head (how far it falls) and the rate of flow. The energy of the falling water is converted to rotating form to drive generators by specialized water turbines. Water is the most nearly free source of energy available that can provide large amounts of generating power. The costs are capital investment, labor, and maintenance; the fuel costs are zero. Almost all of the major sources of hydro power in the United States are in use, but much potential hydro power remains around the world.

Most of the hydroelectric work has been on high head, large units to provide large blocks of power. Smaller, low head hydro units to exploit smaller sources of hydro power have largely been ignored. Units of this type could be a valuable source of electricity in remote regions of developing nations by providing electricity to a village or a hospital. Such units might also be useful in developed nations to help conserve fossil fuel. The smaller, low head hydro units are a useful power engineering project.

*Pumped storage* refers to one of the few ways massive amounts of energy can be stored. A hydro unit that is used for peak power production can be used to pump water up into a lake or holding reservoir during times that electricity demand is low. A water turbine that can be reversed and used as a pump is required. A synchronous generator can be used as a motor to drive the turbine, but an auxiliary pump and electric motor is more efficient. The pumped water is then used to power the hydroelectric generator during the next demand peak.

### 1.5.3 Nuclear energy

The first commercial nuclear power plant went on line in the United Kingdom in 1956. By 1978 nuclear powered generating plants provided 7.5% of the world's electricity. Nuclear power promised economical electricity from an abundant fuel source. This promise may be fulfilled by nuclear fusion, but controversy has limited the acceptance of nuclear fission for electricity generation in some countries.

Nuclear power plants use controlled nuclear fission to heat water to produce steam. The nuclear decay is simply a water heating method. Nuclear fission can be an economical way to produce steam to generate electricity, but fear of the radioactive fuel and byproducts of nuclear electricity generation has resulted in an immense amount of controversy with respect to the desirability of nuclear power generation. The controversy has resulted in a complex system of regulatory procedures that make the planning, approval, construction, and commissioning of nuclear plants very time consuming and expensive in the United States. Often litigation has added to the expense. Except for the Chernobyl, USSR accident, the safety record of nuclear generating facilities world wide has been very good. The Three Mile Island accident, the worst mishap in the United States, caused no serious injuries. The good safety record has resulted partially from effective regulatory requirements and partially from the concern for safety that has been characteristic of the electrical power industry. However, the long-term storage of spent nuclear fuel and the decommissioning of old nuclear power plants are two major problems that have not yet been solved. Many countries are developing a large nuclear power generating capacity because they lack the fossil fuel reserves of the United States.

Many people in the electric power industry consider nuclear fission to be a temporary energy source for generation. Ultimately nuclear fusion is expected to provide an economical, almost inexhaustible source of energy for generating electricity. Nuclear fusion releases energy when two lighter elements are fused together to produce a heavier one; it is the source of energy for Hydrogen bombs. Controlled nuclear fusion can theoretically be obtained in a very hot plasma (mixture of high energy electrons and ions) confined in a magnetic field, or by the explosion of minute deuterium-tritium pellets in a protected enclosure.

The fusion is started by projecting several laser or particle beams on the pellets from many directions, which causes the pellets to implode. The implosion produces the extreme temperature and pressure needed for nuclear fusion to occur.

Nuclear fusion produces far fewer radioactive byproducts than nuclear fission, and cannot be sustained in the event of an equipment malfunction. It should be much safer than nuclear fission as an energy source. The possibility of commercial fusion-based generating plants in the near future is very small.

### 1.5.4 Alternative energy sources

Nearly any source of energy can be used to generate electricity: wind, water, chemical, thermal, nuclear, or solar. All of these have been used. The use of water, nuclear, and chemical change through burning to produce thermal energy have been briefly discussed. Direct chemical production of electricity from the electrochemical reactions in chemical power cells is now in use. Many smaller cells are connected to produce dc power up to 100 MW. The dc must then be converted, using inverters, to ac for distribution on the power system. Only a few electrochemical generating plants are in use at this time.

In the 1930s wind power was used extensively in rural America. The Delco system consisting of a wind charger and a battery was the most popular. Development is proceeding rapidly on larger, more efficient wind turbine and generator systems with control circuits that can tie them into a standard distribution system. Wind turbines to over 2.5 MW have been in operation for several years.

Geothermal energy refers to energy extracted from steam heated by the Earth's outer core in places near enough to the surface for the steam to be extracted. The geysers of Yellowstone National Park are examples of geothermal energy. The first geothermal plant was established in 1960 by the Pacific Gas and Electric Company in California. The plant capacity was 322 MW in 1972, but the field has a potential of 2000 MW. Geothermal fields such as this one with "dry" steam that can be used after being filtered, with no further processing, are rare. Geothermal steam is normally low pressure, around 100 psi, low temperature, around 350°F, and contains contaminants such as hydrogen sulfide and ammonia. Because of the contaminants the cost of the plants and maintenance is high, although the steam is virtually free. It is estimated that the worldwide potential for geothermal power is very small in comparison to total energy requirements.

Solar energy is produced two ways: by direct conversion to dc with semiconductor solar cells, which can provide as much as several kilowatts of electricity; and by using mirrors to reflect sunlight on to thermal collectors, which heat a working fluid to drive a turbine to generate up to 100 MW of electricity. Solar energy is important for providing power to remote installations, such as space stations, but it is not predicted to provide more than a small fraction of the world's energy needs.

## 1.6

## TRANSMISSION

Large amounts of electric power must be moved from the sites where it is generated to the points where it is distributed for use. As in every other part of the electrical power system this must be done as efficiently as possible. For example, if a transmission line is to move 1000 MW at 95% efficiency and an additional investment can improve the efficiency to 96%, the additional investment must be seriously considered. The 1% saving is 10 MW. At five cents per kWh this represents a saving of 0.05 × 10,000 kW = $500 per hour. If the line has an expected life time of 40 years the total savings will be: $500/h × 24h/day × 365 days/year × 40 years = $175.2 million. Thus a lot of money is spent to obtain as much efficiency as possible in transmission lines.

The sources of loss are the same for both transmission and distribution lines, but because the distances and loads are greater for transmission lines the efficiency must be higher. The losses will be discussed in detail in the transmission section of this text, but are briefly discussed here.

1. *Resistance:* The series resistance of a conductor depends on the resistivity of the conductor material, its length, which is affected by the amount of spiralling of its strands, temperature, and the skin effect. Recall that dc resistance ($R_{dc}$) is directly proportional to the resistivity ($\rho$) and length ($l$) and inversely proportional to the cross-sectional area ($A$) as given by

$$R = \frac{\rho l}{A} \tag{1.1}$$

Stranded conductors are spiraled to improve flexibility and fatigue withstanding ability. Spiralling typically adds about 2% to the length. The area in a stranded conductor is the sum of the areas of each individual conductor. Recall conductor resistance increases almost linearly with temperature.

*Skin effect* refers to the fact that as frequency increases current flow shifts toward the surface of a conductor. Alternating current reverses direction periodically, and each reversal results in the magnetic flux produced by the current also reversing. The magnetic flux density is highest in the center of the conductor, causing the inductive reactance of the conductor to also be highest in the center. The higher inductive reactance at the center forces the current toward the outer surface of the conductor. The skin effect at a fixed frequency is proportional to the conductor diameter. The skin effect can result in the ac resistance of a large conductor being 20% higher than the dc resistance at 60 Hz.

Resistance losses ($I^2R$) are kept low by making transmission voltages as high as practical.

2. *Reactance:* The magnetic flux produced by the ac current produces series inductive reactance ($X_L$) because of both self inductance (which causes skin effect) along a conductor and mutual inductance between conductors. The reactance doesn't dissipate power, but results in a voltage drop ($i \times X_L$) along the line and volt-amperes-reactive (VARs) from $i^2 \times X_L$. Any reactive power on the line must be supplied by the generator in addition to the load power.

3. *Capacitance:* Conductors separated by a distance have capacitance. The capacitance of a transmission line depends on conductor size, spacing, height above the ground, and voltage. Transmission line capacitance must be charged before a line can transfer power, and even though the shunt loading of a line is low, some power is shunted to ground from the line through the capacitive reactance.

4. *Corona:* Corona is caused by the breakdown of the air around a transmission line because of high voltage. The effect is most severe around small conductors and at sharp points and corners. Corona absorbs energy from the line. Bundling of high voltage conductors, separating conductors with spacers placed periodically along the line, has dramatically reduced corona loss.

A typical high voltage transmission line is shown in Figure 1.7. The transmission line towers are a very important part of transmission lines. The tower must support the line under worst case loading with an adequate safety margin, exhibit minimal effect on the transmission line electrical characteristics, and be economical.

A new development that will effect transmission is the development of superconducting magnets for bulk power storage. The scientific investigations into superconducting magnets, which were perfected for guiding particle streams in particle accelerators, such as Super Conducting Super Collider that was planned for Texas but discontinued after construction began, and further developed into superconducting electrical energy storage devices for Star Wars rail guns, have promise in other applications. One application is as Superconducting Magnetic Energy Storage units (SMES) that promise to be useful in the management of the transmission of electricity.

The pilot program for Superconducting Magnetic Energy Storage is currently planned for San Diego Gas and Electric System (SDG&E) at Blythe, California by a consortium that includes SDG&E, Bechtel, The Department of Energy, and others. This unit will demonstrate the capability to store power, use the stored power to stabilize voltage and frequency, prevent subsynchronous reaction (low frequency power perturbations), and provide peaking power. The ability to respond to transient power needs in a few milliseconds by electronic control of electrical power stored in a superconducting magnet allows transmission lines to operate more stably. Thus they will be able to reliably transmit about 10% more power than they can without this ability. This development will allow the more efficient operation of the North American Electrical Grid (including both Canada and Mexico), and improve the transfer of power to meet the load needs of North America following the sun East to West daily, and North to South seasonally.

**FIGURE 1.7**   Transmission line at Fayette Power Project (Courtesy of Lower Colorado River Authority).

## 1.7

## DISTRIBUTION

A distribution system is subject to all of the losses that a transmission system has, but since it carries lower voltages bundling of conductors to prevent corona is not necessary. The shorter distances and lower power per line involved in distribution allows the design emphasis to shift from maximum efficiency (it is still important) to accessibility, safety, and continuity of service. The American system of residential distribution illustrates the safety concern. The three-wire, 240/120 V system has a neutral at ground potential and hot wires at 120 V above and 120 V below neutral as illustrated in Figure 1.8. In this system, 120 V loads are fed from either hot wire to neutral and 240 V loads are connected from one hot wire to the other. This arrangement prevents voltages to ground from exceeding 120 V, provides a lower, safer voltage for low power loads, and provides a higher voltage for higher power loads. The European system is a two-wire, 250 Volt system that works, but lacks the safety and flexibility of the U.S. system.

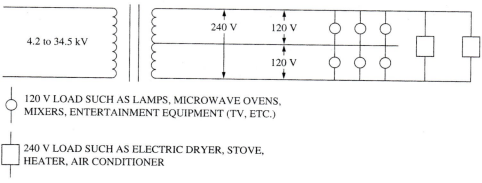

120 V LOAD SUCH AS LAMPS, MICROWAVE OVENS, MIXERS, ENTERTAINMENT EQUIPMENT (TV, ETC.)

240 V LOAD SUCH AS ELECTRIC DRYER, STOVE, HEATER, AIR CONDITIONER

**FIGURE 1.8**   American 240V/120V residential distribution system

Further discussion of distribution will be delayed until Section 2, which is about electrical power distribution. Section 3 is about electrical power transmission.

## 1.8

## SUMMARY

Electricity is a convenient form of energy to transport and the machinery used to generate, transform, and utilize electricity is efficient. This chapter provided the reader with a brief history of the development of electrical power from Thales to the present. Among the more important events were the commercial generation of electricity for street lighting, the Edison dc distribution concept with its central generating station, the invention of ac machinery by Tesla, the generation and transformation of ac by Westinghouse and Stanley, high voltage transmission, and dc transmission. The evolution of the electrical power utilities into state regulated monopolies supplying as much electricity as possible at as cheap a rate as possible (thanks to Insull's vision) was an important force in the United States. The concept of cooperative interconnection between suppliers of electricity led to the ability of electric utilities companies to provide good service at lower prices to consumers because the capital investment needed for reserve power was shared among electrical power companies. Interconnection later made the practice of wheeling cogenerated power possible.

The involvement of the U.S. government in the production and transmission of electrical power in projects such as the TVA, the REA, and BPA helped to provide a check and balance on investor owned electric utilities, electrify rural areas of the U.S. long before they would have otherwise been, and efficiently utilize the hydro-electric potential of the Pacific Northwest respectively. The U.S. government continues to be involved in the electrical power industry via the FERC. The current trend is toward companies formed for generation supplying electricity, and electrical power utility companies managing the transmission and distribution of the electricity.

The perfect source of electricity has yet to be found. Fossil fuel plants deplete nonrenewable resources or pollute, or both (coal and oil), and nuclear fission plants leave hard to manage byproducts. Nuclear fusion, when perfected, appears to be a promising source of economical and plentiful electricity.

## 1.9

### QUESTIONS

1. What is the function of the electrical power system?
2. Who invented the polyphase system of electricity transmission?
3. What were the major contributions of George Westinghouse and William Stanley to the electrical power industry?
4. State the conditions that led to the granting of franchises to a single supplier of electricity to an area instead of open competition.
5. State the problem that single-supplier franchising caused, and the solution to the problem that was ultimately adopted.
6. What major problem did E. M. Hewlett and H. W. Buck solve?
7. What is conductor bundling and what problem did it solve?
8. Name and state the function of the three major parts of the electrical power system.
9. List in order of magnitude the four major sources of energy used to produce electricity.
10. What are the advantages of a strong interconnect system between electric utility companies?

# Basic Concepts, Three-Phase Review, and Per Unit

This chapter reviews some basic concepts from an electrical power point of view. They are the basic circuit elements, voltage, resistance, inductance, capacitance, and their use in representing electrical loads and lines. Phasors and basic three-phase concepts are also reviewed while the per unit concept and impedance diagrams are introduced in this chapter.

## 2.1

### BASIC CONCEPTS

#### 2.1.1 Voltage

Voltage (electromotive force, emf) provides the electric fields that cause electron current in conductors. Electrons drift through the conductor attracted toward the more positive terminal of the voltage source and repelled by the more negative terminal. Voltage can be produced in many ways including chemical reactions in batteries, mechanical flexing of piezoelectric materials such as Barium titanate, separation of electron-hole pairs created by absorption of light in a semiconductor by the static electric field of the semiconductor junction of a photo-electric cell, and moving conductors through a magnetic field. The latter method is the one currently used to generate voltage in the majority of the electric generating plants throughout the world.

The voltage of a generator is determined by the amount of magnetic flux produced by the field windings, the number of armature windings, and the speed at which the rotor is turned. The more of each of these three quantities that is present, the higher the voltage. The equation that describes the relationship between these three quantities, you may recall from your study of electric machines, is

$$E_{GRMS} = 4.44 \, \Phi \, N f \cdot 10^{-8} \, \text{Volts} \tag{2.1}$$

where

$\Phi$ is the number of flux lines per pole for English units, and in Webers for SI

$N$ is the number of turns per phase

$f$ is the frequency in Hertz (1/s where s is the number of revolutions per second)

$E_G$ is the generator phase voltage.

The factor for SI units is 1 instead of $10^{-8}$.

Generation is illustrated in Figure 2.1. If the voltage produced is to be useful the prime mover speed must be very stable to keep the frequency fixed. Recall that after an ac generator is paralleled into an infinite bus that the field excitation controls the reactive current (VARs) and the prime mover torque controls the power that the generator delivers to the bus.

A single generator connected to a load does not have a fixed voltage like a generator connected to an infinite bus. If the prime mover speed is constant, the voltage varies with field excitation, load impedance, and load power factor. Referring to Figure 2.2, as the load current increases, the voltage drop across the internal impedance of the generator causes the generator terminal voltage to drop, if the load power factor is one or less than one and lagging (as most loads are). Recall that armature reaction is the interaction of the field flux and the flux produced in the armature because of load current flow. The armature reaction flux from leading power factor loads aids the field flux, and the armature reaction flux from lagging power factor loads opposes the field flux. The amount of aid or opposition depends on the amount of lead or lag. Since the voltage produced by the armature reaction flux lags the load current by 90 ° it is in phase with the voltage drop across the armature reactance. The measured value of generator synchronous reactance includes the effect of both armature reaction and armature reactance. The generator impedance shown in Figure 2.2 is that of the armature resistance and the synchronous reactance. The terminal voltage ($E_T$) of a generator is

$$E_T = E_G - I_L Z_G \tag{2.2}$$

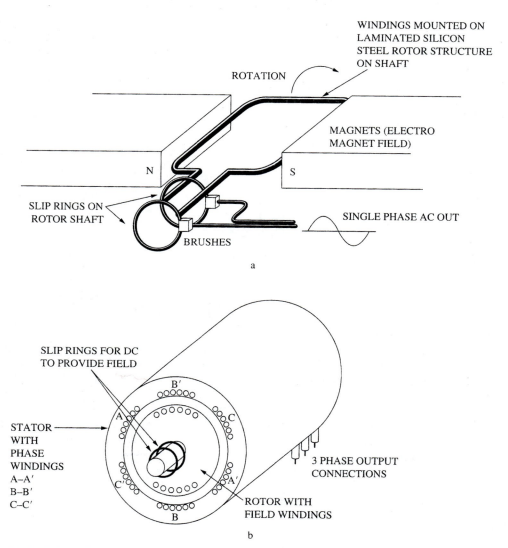

WINDINGS MOUNTED ON
LAMINATED SILICON
STEEL ROTOR STRUCTURE
ON SHAFT

ROTATION

MAGNETS (ELECTRO
MAGNET FIELD)

N                    S

SLIP RINGS ON
ROTOR SHAFT

SINGLE PHASE AC OUT

BRUSHES

a

SLIP RINGS FOR DC
TO PROVIDE FIELD

B′

A            C

STATOR
WITH
PHASE
WINDINGS
A–A′
B–B′
C–C′

C′            A′

3 PHASE OUTPUT
CONNECTIONS

ROTOR WITH
FIELD WINDINGS

B

b

**FIGURE 2.1**   Generation illustration—(a) Single phase and (b) three phase.

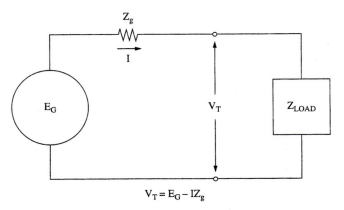

**FIGURE 2.2**   Generator equivalent circuit

## 2.1.2 Current

Current is the flow of charges under the influence of an electric field. The current we most often think of is the flow of electrons in a conductor. The direction of current flow was agreed to be from the positive terminal of a voltage source to the negative before the nature of current in a conductor was understood. After it was discovered that electrons flow from the negative terminal of a voltage source to the positive the older convention was retained as *conventional current* and the actual flow of electrons in the opposite direction was then called *electron current*. The direction of the electron current in most electrical books is opposite to the arrow depicting its direction because most authors use conventional current by tradition. Alternating current, used for most electrical power transfer, changes direction every half cycle, so current direction is for mathematical convenience if it is indicated. The current moving under the influence of the applied voltage provides the electrical power.

## 2.1.3 Impedance

Impedance, as you know, is the total opposition to ac current by the components of a circuit. We are interested in resistance, inductance, and capacitance.

*Resistance* is the property of a component that opposes dc current and the component of ac current in phase with the voltage. Resistance depends on the number of free electrons per unit volume of a material, the probability of collisions of the free electrons in the material with other electrons and atoms, and the physical dimensions of the material. The resistivity of a material ($\rho$) is a combination of the physical characteristics of the material that influence resistance. The resistance in ohms ($\Omega$) is given by the equation

$$R = \rho \frac{1}{A} \text{ Ohms} \tag{2.3}$$

where

$l$ = length and

$A$ = area.

The units of resistance, Ohms, are named after George Ohm, the originator of Ohm's law. The circuit symbol is shown in Figure 2.3. All of the power consumed in a circuit, other than dielectric and hysteresis losses, is consumed by the resistance because of the electrons losing energy during collisions with atoms as they are forced through the circuit by the electric field of the applied voltage.

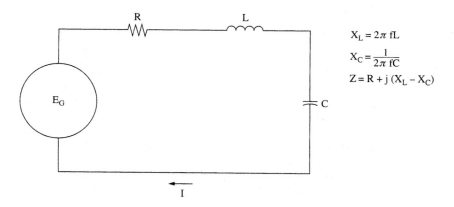

$X_L = 2\pi fL$

$X_C = \dfrac{1}{2\pi fC}$

$Z = R + j (X_L - X_C)$

**FIGURE 2.3**   RLC circuit

*Inductance* is the property of a component that opposes any change in current through that component. Inductance is analogous to inertia in a mechanical system; the tendency of a mass in motion to remain in motion. Inductance (*L*) is, like *R*,

$$E_L = L \frac{di}{dt}$$

or     (2.4)

$$E_L = L \frac{\Delta I}{\Delta t}$$

a constant of proportionality between the voltage across, and the time rate of change of current through, a component.

The amount of inductance is a function of the physical construction of the component. The number of turns of wire, the permeability of the core, and the geometry determine the inductance of a component. No power is consumed by the inductance, but an inductor will consume power in its winding resistance, and its core via hysteresis and eddy currents.

The opposition to current in an inductor results because a change of current in a conductor of the inductor causes a magnetic flux change in the space surrounding the conductor that induces a counter emf in adjacent conductors. The adjacent conductor current is inducing a counter emf in the original conductor at the same time. The counter emf, because the flux is expanding when the current is rising and collapsing when the current is falling, always has a polarity that opposes the external emf, or acts as the circuit emf if the external emf is removed. The counter emf, because it opposes the circuit emf, opposes an increase in current. The emf produced by collapsing magnetic flux when the circuit emf is removed opposes a decrease in current. When the applied voltage is removed from an inductive circuit the emf produced by the collapsing flux will try to keep the current constant by rising to as high a voltage as necessary to do so. This process, often called inductive kick, causes arcing across switch contacts used to remove voltage from circuits with inductance.

*Skin effect* is the name of a phenomenon in which ac current tends to flow at the surface of a conductor. An electron changing velocity in a conductor, because the circuit voltage is changing, produces flux that interacts with the flux of near by electrons opposing their change in velocity. This is the same process that occurs in adjacent conductors on an inductor that results in counter emf. Within the conductor the process occurs on a microscopic scale. The magnetic flux density in the conductor is highest in the center of the conductor because about half the flux from the electrons at or near the surface of the conductor expand (or contract) outside the conductor and do not interact significantly with the magnetic fields of the electrons near the center. The magnetic fields of the electrons nearer the center of the conductor interact with the fields of all the electrons around them. Thus the magnetic flux density is higher at the center of the conductor than at the surface, the opposition to a velocity change of current carriers is higher at the center of the conductor, and the opposition to current flow is higher at the center of the conductor than at the edge. The current density is therefore highest at the surface of a conductor in which ac current is flowing. Effectively, $X_L$ is higher in the center of a conductor carrying ac current so most of the current is forced toward the surface. As the frequency increases, the inductive reactance at the center of the conductor increases, and the current is more strongly confined to the surface (or skin) of the conductor. The current falls off exponentially going from the surface toward the center of a conductor carrying ac current as shown in Figure 2.4.

The skin depth is defined as the distance from the conductor surface at which the current has fallen to $1/e$, or 36.79%, of the value at the surface. The skin depth serves the same mathematical function as the time constant in capacitor and inductor charge and discharge equations. The equation for the skin depth ($\delta$) is

$$\delta = \sqrt{\frac{1}{\pi f \mu_r \mu_o \sigma}}$$

where

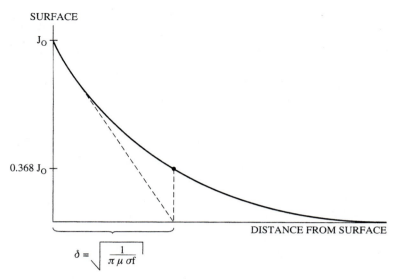

SURFACE

$J_O$

0.368 $J_O$

DISTANCE FROM SURFACE

$$\delta = \sqrt{\frac{1}{\pi \mu \sigma f}}$$

**FIGURE 2.4** Skin depth vs. current density (J)

$\sigma$ = conductivity in Siemens/meter

$\mu_r$ = relative permeability

$\mu_o$ = $4 \pi 10^{-7}$ Henry/meter, the permeability of space

$f$ = frequency in Hertz.

At 60 Hz the skin depth in copper is about 8.53 mm, or about 0.335 in. Thus current capacity cannot be increased by using thicker conductors at power line frequencies. Instead, more conductor area of conductors about 0.5 inch thickness is needed. We will see the skin effect taken into account in many areas of our study of electrical power distribution and transmission.

*Capacitance* results from conductors separated by an insulator, as in a parallel plate capacitor. When the conductors are at a different potential the electric field acting between them concentrates charges at the surface of the conductors that are nearest each other. If a dielectric whose molecules are easily polarized is placed between the conductors it causes the distance between the conductors to appear less than it is physically, because the polarized atoms or molecules within the dielectric transfer the electric field through the dielectric. The ease with which the dielectric can be polarized with respect to a vacuum is the dielectric constant ($\epsilon_r$) of the dielectric, the ratio of the permittivity of the dielectric to that of a vacuum ($\epsilon/\epsilon_o$). Recall that the capacitance of a capacitor is directly proportional to the dielectric constant of the dielectric and the conductor, or plate, area, and inversely proportional to the distance between the conductors. Capacitance, because of its ability to store charge, opposes any change in the voltage across it. Because the voltage across a capacitor opposes circuit voltage at any instant that the circuit voltage is increasing, capacitance opposes ac current flow. The opposition to ac current flow is called capacitive reactance ($X_C$) and is equal to $1/\omega C$. In a manner analogous to inductance, capacitance is the proportionality constant between capacitor current and the rate of change of capacitor voltage.

$$I_c = Q\frac{dV}{dt} \tag{2.6}$$

Other than the energy required to polarize the dielectric and the $I^2R$ loss of the capacitor conductors, capacitors use no energy. They store and return energy to the circuit. Please note that any two conductors separated by a dielectric exhibit capacitance.

*Impedance* is the total opposition to current flow of all the elements of a circuit, as shown in Figure 2.3. Only the resistive elements of a circuit dissipate power.

### 2.1.4 Load

The load is the apparatus or devices to which the electrical system provides power. The electrical system loads may do work, such as electric motors or computers, provide comfort, as air conditioning and electric lighting, or entertainment and information, like television. The terminal characteristics of the load are important for the electrical system. At the rated voltage the load draws a given amount of current at a certain power factor. The current and power factor of the load may not be fixed so the electrical system must provide the maximum amount of apparent power the load will need. For example, an induction motor with a light load will draw mostly magnetizing current so its power factor will be low and lagging. At full load the motor will draw more current and the power factor will be much higher, but still lagging. A heating element or incandescent light will draw current that is in phase with the applied voltage, but draw a much higher current for a few cycles after turn on until they warm up.

Figure 2.5 shows a load that may be inductive, capacitive, or resistive depending on the relative values of reactance of *L* and *C* at the supply frequency. Figure 2.6a shows the impedance diagram and equations for a series RLC circuit, and Figure 2.6b shows the admittance diagram for a parallel RLC circuit. Recall that the j operator represents a 90° phase shift.

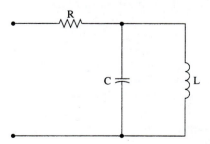

**FIGURE 2.5**   Example of a load equivalent circuit

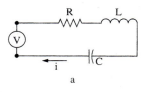

a

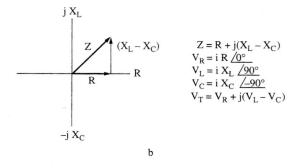

$$Z = R + j(X_L - X_C)$$
$$V_R = i\,R\ \underline{/0°}$$
$$V_L = i\,X_L\ \underline{/90°}$$
$$V_C = i\,X_C\ \underline{/-90°}$$
$$V_T = V_R + j(V_L - V_C)$$

b

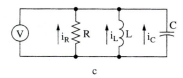

c

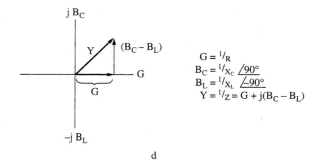

$$G = {}^1/_R$$
$$B_C = {}^1/_{X_C}\ \underline{/90°}$$
$$B_L = {}^1/_{X_L}\ \underline{/-90°}$$
$$Y = {}^1/_Z = G + j(B_C - B_L)$$

d

**FIGURE 2.6** Impedance and admittance diagrams (a) Series RLC circuit (I common to all components) (b) impedance vector diagram (I sets reference direction) (c) parallel RLC circuit (V common to all components) and (d) admittance vector diagram (V sets reference).

The relationship between apparent, reactive, and real power (**S, Q,** and **P,** respectively) for an inductive load is shown in Figure 2.7. Recall that power factor (**PF**) is defined as the ratio of real to apparent power (**P/S**).

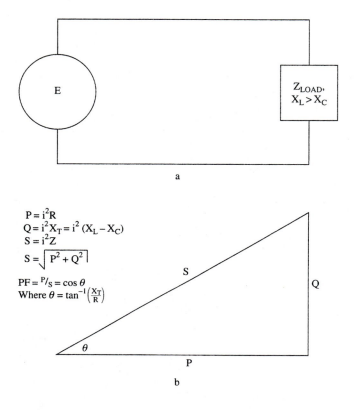

**FIGURE 2.7**   Power triangle (a) The circuit and (b) power triangle

### 2.1.5 The line

The source of electricity and the load must be connected. The transmission and distribution lines provide the connection. These lines will be a major part of our study throughout this book. The line has all of the basic circuit elements: capacitance, inductance, and resistance. The line resistance is increased by the skin effect. The length and load of the line determine the relative impact of each of the basic elements on transfer of power through the line. Figure 2.8 illustrates the equivalent circuit of a generator and a load connected by the equivalent circuit of a medium length line. The equivalent circuit of a short line would not contain the capacitor $C_{LINE}$. Short lines have capacitance but the effect of the capacitance on power transfer is small compared to the other line elements.

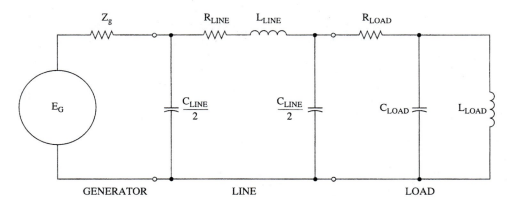

**FIGURE 2.8**  Simple power system equivalent circuit.

## PHASOR REVIEW

Phasors are electrical quantities plotted in polar coordinates (amplitude, angle). We have used the real-imaginary plane illustrated in Figure 2.9 to plot impedance and admittance vectors. The operator j indicates a 90° shift with respect to the real axis. We now wish to use the plane to illustrate ac voltages and currents. Most

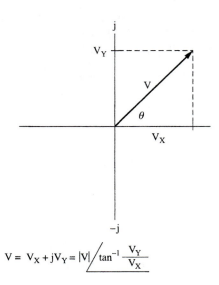

$$V = V_X + jV_Y = |V| \underline{/\tan^{-1} \frac{V_Y}{V_X}}$$

**FIGURE 2.9**  Coordinates

electrical power is supplied as sine waves. A vector one unit in amplitude rotated 360° traces a unit circle as shown in Figure 2.10. The vertical component of the vector on unit in amplitude is proportional to sin θ. If this amplitude is plotted on the vertical axis of a graph and the angle on the horizontal axis a sine wave will result as shown in Figure 2.11a. A sine wave may go through many 360° (2π) rotations per second so the horizontal axis may be in units of time, where time is related to the total angle in radians the vector has traveled ($\theta_T$) by

$$\theta_T = 2\pi f t = \omega t \text{ radians} \tag{2.7}$$

where

$\quad f =$ frequency

$\quad t =$ time

$\quad \omega = 2\pi f$

The sine waves shown in Figure 2.11a are a voltage and a current wave that are 45° out of phase with the current lagging the voltage. Because their peak values are greater than one, the sine function is multiplied by the peak value to obtain the value at a particular time, as shown in Figure 2.11a.

The relationships between voltages, voltage drops, and currents are difficult to illustrate in drawings, such as in Figure 2.11a. The relationship between the current and voltage of Figure 2.11a can be easily shown using a vectorial representation such as the one shown in Figure 2.11b. These rotating vectors are called *phasors* and have an amplitude equal to the rms value of the voltage or current. By mutual agreement, counterclockwise rotation of the phasors has been adopted as the positive rotation direction. The phasor diagram is plotted so that the phase relationships between the phasors is shown. The rotation is not shown but is a way of illustrating which phasors are leading or lagging one another. For

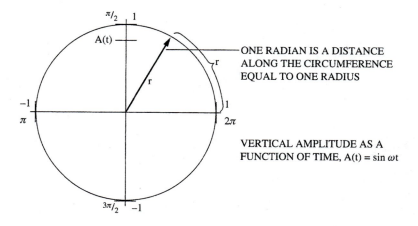

**FIGURE 2.10**   Unit circle

example, the voltage is leading the current in Figure 2.11a so it is drawn 45° to the ccw side of the current in Figure 2.11b. The reference phasor is frequently drawn on the zero degree axis and all other phasors drawn in their proper relationship to the reference phasor, as shown in Figure 2.11c. Phasors are vectors rotating at constant angular velocity. To calculate circuit quantities from phasors we must freeze them in time, and do calculations using normal vector mathematics. Please note that phasors are basically electrical quantities plotted in polar coordinates (amplitude, angle).

Phasors are added and subtracted by resolving the phasors into components with the real component along the horizontal axis and the imaginary component along the vertical axis, as illustrated in Figure 2.11c. The angles depend on the position selected for the reference quantity. Figure 2.11b would be more complicated to resolve into components than Figure 2.11c, but both figures will yield the same solution for problems with rms or average value (**P, Q, S**) solutions. Multiplication and division are most easily performed in polar form.

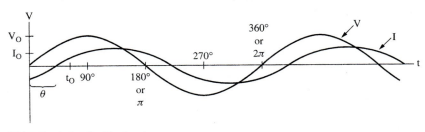

$$V(t) = V_O \sin 2 \pi \, ft = V_O \sin \omega t$$
$$I(t) = I_O \sin \omega t$$

a

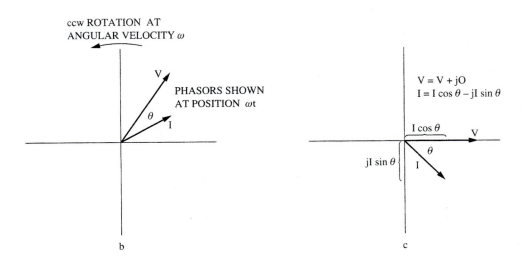

b

c

**FIGURE 2.11** Phasor representation of voltage and current (a) Sinusoidal wave representation (b) phasor representation at $t_0$ and (c) reference chosen for convenience because $\theta$ is constant for the circuit.

## 2.3

### THREE-PHASE REVIEW

Single-phase quantities have been discussed to this point. Most bulk electrical power is transferred as three-phase voltages and currents because the power flow is constant instead of pulsating, as it is in a single-phase or two-phase system. The constant power flow in a three-phase system is shown in Appendix A. Two- and six-phase power are occasionally used for special purposes, and single-phase power is used for the vast majority of single family dwellings, but three-phase power is used in most distribution and transmission systems.

### 2.3.1 Basic Three-phase Concepts

In Figure 2.12a a single-phase circuit with an RL load is shown. If three identical loads are to receive power they can be connected in parallel but the power flow through the loads pulsate and conductors three times larger are needed to carry the current. Three circuits with sources 120° offset from each other, as shown in Figure 2.12b, provide three-phase power and a smooth overall flow of power to the three loads. The maximum current carried by the conductors at a given voltage are reduced by about 42.3% compared to using two lines to feed the three loads in parallel (a capacity only 1.73 times larger than is needed to provide power to one of the loads). The three-phase source allows us to combine the three separate return lines into one, as shown in Figure 2.12c. Section 2.3.3 will show that the return line will carry no current if the loads have the same impedance (called balanced).

### 2.3.2 Three-phase Phasor Notation

Three-phase quantities, like single-phase ac quantities, are handled most easily as vector quantities. The phasors for a three-phase voltage source are shown in Figure 2.13. Notice the operator alpha ($\alpha$) in the figure. This operator indicates a 120° phase shift and follows the algebraic rules of exponents. Thus $\alpha = 120°$ and $\alpha^2 = 240°$. The phasors are considered to rotate counterclockwise as in single-phase phasor diagrams. The phase sequence can be determined by the order in which the phasors will pass the reference axis as they rotate. The phase sequence shown in Figure 2.13 is 1-2-3. Note that

$$E_3 = \alpha E_1 \tag{2.8a}$$

$$E_2 = \alpha^2 E_1 \tag{2.8b}$$

Double subscript notation is often used for three-phase currents to avoid uncertainty as to direction. Wye-connected circuits often have all quantities referenced to neutral so the neutral connection is implied. For example $V_{AN}$ of Figure 2.14a might be written as $V_A$. Double subscript notation is necessary for delta-connected circuit quantities. The first subscript, normally denotes the point of measurement

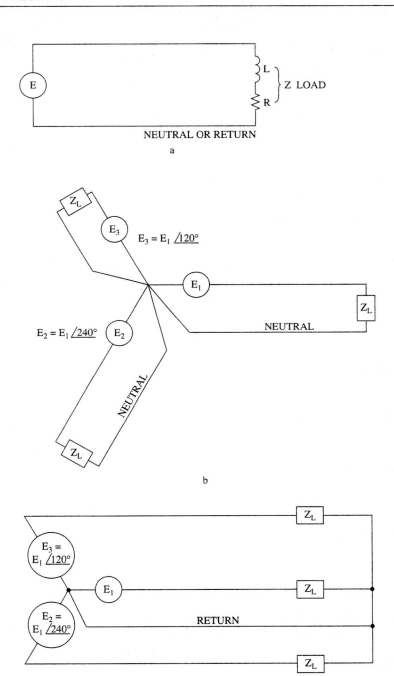

**FIGURE 2.12** Three-phase system (a) Single-phase (b) three phase from single phases at 0°, 120°, 240° and (c) three neutrals replaced with a single neutral

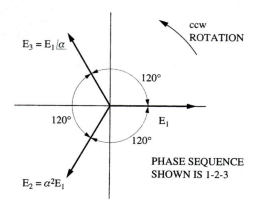

**FIGURE 2.13**   Three-phase phasor convention

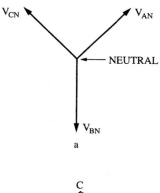

a

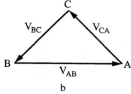

b

**FIGURE 2.14**   Three-phase notation (a) Wye and (b) delta

of a quantity with respect to the second subscript, as shown in Figure 2.14b. $V_{AB}$ indicates the voltage on line A with respect to line B.

To add or subtract phasors they are resolved into components on the real-imaginary plane and the components added or subtracted as shown in Figure 2.15. Figure 2.15 illustrates that the phasor (vector) sum of the voltages in a balanced three-phase system is zero. Multiplication and division of phasors is most easily done with the phasors in polar form. Balanced three-phase systems can be analyzed by calculation of quantities of interest for one phase because the other two phases are behaving identically except that their phases are offset by 120° and 240°.

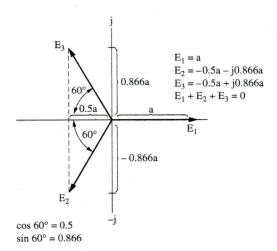

$E_1 = a$
$E_2 = -0.5a - j0.866a$
$E_3 = -0.5a + j0.866a$
$E_1 + E_2 + E_3 = 0$

$\cos 60° = 0.5$
$\sin 60° = 0.866$

**FIGURE 2.15** $V_T = 0$ in a balanced three-phase system

## 2.3.3 Three-phase Line and Phase relationships

Three-phase voltages measured from line to neutral are phase voltages ($V_p$) and those measured from line to line are line voltages ($V_l$). The current that flows in the line is the line current ($I_l$) and that inside the source or load phase is the phase current ($I_p$). The relationship between line and phase quantities are illustrated with two examples, the first for a wye-connected system and the second for a delta-connected system.

**Example 2.1:**

Calculate the line voltage, line current, phase current, and total three-phase current for the circuit of Figure 2.16a. $E_A = 120V \angle 0°$, $E_B = 120V \angle 240°$, and $E_C = 120V \angle 120°$ with ABC phase rotation. $Z_A = Z_B = Z_C = 10 \angle 40°$

**Solution:**

$E_{line} = E_{AB} = E_A - E_B$

Expressing $E_A$ and $E_B$ in rectangular form we have

$E_A = 120\ V + j0\ V$

$E_B = -60\ V - j103.9\ V$

subtracting

$E_{AB} = 120\ V + 60\ V + j103.9\ V$

$\qquad = 207.846V \angle 30°$

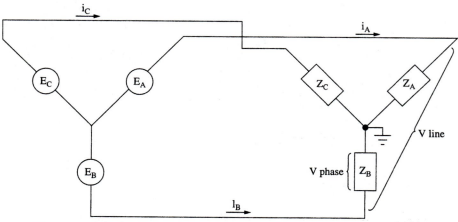

$$V_l = \sqrt{3}\,V_p \text{ offset } 30° \text{ from } V_A$$
$$I_L = I_P$$
$$P = \sqrt{3}\,V_l\,I_l = 3V_p\,I_P$$

a

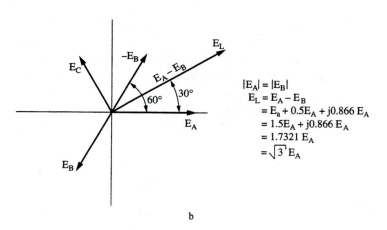

$$|E_A| = |E_B|$$
$$E_L = E_A - E_B$$
$$= E_a + 0.5E_A + j0.866\,E_A$$
$$= 1.5E_A + j0.866\,E_A$$
$$= 1.7321\,E_A$$
$$= \sqrt{3}\,E_A$$

b

**FIGURE 2.16**   Wye (Y) system (a) Circuit and (b) voltage phasor diagram

which is the line voltage. The ratio of the line to phase voltage is of interest.

$$\frac{E_{AB}}{E_A} = \frac{V_l}{V_p} = \frac{207.85V}{120V} = 1.732 = \sqrt{3}$$

Thus for a wye system $V_l = \sqrt{3}\,V_p$. Figure 2.16b illustrates the relationship in vectorial form. Similarly

$$E_{BC} = 207.846V \angle 270° \text{ and } E_{CA} = 207.846V \angle 150°$$

The line and phase current must be equal because the line and phase circuits are in series. Solving for the three-line currents

$$I_A = \frac{V_A}{Z_A} = \frac{120V\angle 0°}{10\Omega\angle 40°} = 12A\angle -40°$$

$$I_B = \frac{V_B}{Z_B} = \frac{120V\angle 240°}{10\Omega\angle 40°} = 12A\angle 200°$$

$$I_C = \frac{V_C}{Z_C} = \frac{120V\angle 120°}{10\Omega\angle 40°} = 12A\angle 80°$$

The total current is found by converting the currents to rectangular form and summing

$$I_{Total} = I_A + I_B + I_C$$

$$= 9.193\ A - j7.713\ A - 11.276\ A - j4.104\ A + 2.084\ A + j11.818\ A$$

$$= (9.193 + 2.084 - 11.276\ )A + j(-7.713 - 4.4104 + 11.818\ )A$$

$$= 0$$

The total current, like the total voltage, in a balanced three-phase system is zero.

To calculate the power (or volt amperes) for a three-phase wye circuit we can multiply three times the phase power or use line voltage and current as follows

$$P = 3\ V_p I_p = 3\ \frac{V_l}{\sqrt{3}}\ I_l = \sqrt{3}\ V_l I_l$$

The line and phase relationships for a wye-connected circuit are summarized in Figure 2.16a.

We will also examine the line and phase relationships of a delta-connected system by using an example.

## Example 2.2:

For the circuit drawn in Figure 2.17a calculate the phase voltage, phase current, line current, and the relationship between the line and phase voltages and currents. The line voltages are $E_{AB} = 120V\angle 0°$, $E_{BC} = 120V\angle 120°$, and $E_{CA} = 120V\angle 240°$ for an ACB phase sequence. All three-phase impedances are the same: $Z_{AB} = Z_{BC} = Z_{CA} = 20\Omega\ \angle 50°$.

## Solution:

The phase voltage is equal to the line voltage because each load is connected directly from one line to the next.

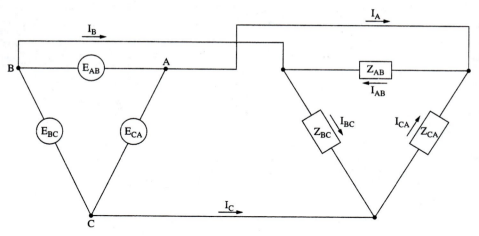

$$V_L = V_P$$
$$I_L = \sqrt{3}\, I_p \text{ offset } 30°$$
$$P = \sqrt{3}\, V_L\, I_L = 3V_P\, I_P$$

a

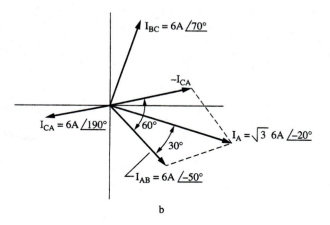

b

**FIGURE 2.17**   Delta ($\Delta$) system (a) Delta connected circuit and (b) current phasor diagram

The phase currents are

$$I_{AB} = \frac{E_{AB}}{Z_{AB}} = \frac{120V\angle 0°}{20\Omega\angle 50°} = 6A\angle -50°$$

$$I_{BC} = \frac{E_{BC}}{Z_{BC}} = \frac{120V\angle 120°}{20\Omega\angle 50°} = 6A\angle 70°$$

$$I_{CA} = \frac{E_{CA}}{Z_{CA}} = \frac{120V \angle 240°}{20\Omega \angle 50°} = 6A \angle 190°$$

The system is balanced so we need calculate only one line current.

$$I_A = I_{AB} - I_{CA}$$

In rectangular form

$$I_{AB} = 3.8\text{ A} - j4.596\text{ A}$$

$$I_{BC} = 2.052\text{ A} + j5.638\text{ A}$$

$$I_{CA} = {}^-5.909\text{ A} - j1.419\text{ A}$$

Thus

$$I_A = (3.8 + 5.909)\text{A} + j({}^-4.596 + 1.0419)\text{A}$$

$$= 9.709\text{A} - j3.554\text{A}$$

$$= 10.339\text{A} \angle {}^-20°$$

The ratio between the line and phase current is

$$\frac{I_l}{I_p} = \frac{I_A}{I_{AB}} = \frac{10.339A \angle {}^-20°}{6.0A \angle {}^-50°} = 1.732 \angle 30°$$

Therefore we see that in a delta system $I_l = \sqrt{3}\, I_p$ offset 30°. The relationships are summarized in Figure 2.17a, and the phasor diagram of the current is shown in Figure 2.17b.

Power is found as follows

$$P = 3V_p I_p \cos\theta = 3V_l(I_l/\sqrt{3})\cos\theta = \sqrt{3}\, V_l I_l \cos\theta$$

Up to this point in Chapter 2 we have been reviewing basic concepts from ac and three-phase circuits. If you are not familiar with an area please refer to an appropriate reference book. A list of references is provided in Appendix E.

## 2.4

## THE PER UNIT CONCEPT

This section is about a convenient method of calculating electrical power quantities.

## 2.4.1 Single-phase Per Unit

A shorthand method of solving power system problems has been developed to eliminate many of the manipulations required in systems with more than one voltage, which is nearly all of them. The method is a modification of working with percentages of a base value called the *per unit system*. The system works as follows.

Suppose we have a system that supplies 50 kVA at 7.2 kV. The current must be

$$I = \frac{kVA}{kV} = \frac{50\ kVA}{7.2\ kV} = 6.94\ A$$

and the system impedance must be

$$Z = \frac{V}{I} = \frac{7.2\ kV}{6.94\ A} = 1.04\ k\Omega$$

We can now define the circuit quantities as our base units so that they are 100%.

100% voltage = 7.2 kV

100% current = 6.94 A

100% impedance = 1.04 kΩ

100% volt amperes = 50 kVA

but

100% kVA = 100% kV × 100% I ≠ 100% kVA

so a percentage system will not work. However, if we define 100% as one per unit (1 pu) in the following manner

1 pu *VA* = 50 kVA

1 pu *V* = 7.2 kV

1 pu *I* = 6.94 A

1 pu *Z* = 1.04 k Ω

then

1 pu kVA = 1 pu kV × 1 pu I

since

50 kVA = 7.2 kVA × 6.94 A

and the problems associated with using percentages are solved.

If one parameter is varied (as often occurs) the change its variance causes in other circuit parameters can be calculated in per unit proportions. If the voltage in the previous paragraph changes to 10 kV the system current does change but the system capacity does not. The new per unit voltage is

$$V\,\text{pu} = \frac{\text{new voltage}}{\text{base voltage}} = \frac{10\text{ kV}}{7.2\text{ kV}} = 1.389\text{ pu V}$$

The new current is

$$I = \frac{\text{base kVA}}{\text{new kV}} = \frac{50\text{ kVA}}{10\text{ kV}} = 5\text{ A}$$

and the new impedance is

$$Z = \frac{E}{I} = \frac{10\text{ kV}}{5\text{ A}} = 2\text{k}\Omega$$

In per unit values these are

$$I\,\text{pu} = \frac{1\text{ pu kVA}}{1.389\text{ pu V}} = 0.7199\text{ pu A}$$

which in actual amperes is

$$I = (\text{base amperes}) \times (\text{pu amperes}) = (6.94\text{ A}) \times (0.7199\text{ pu A}) = 5\text{ A}$$

and

$$Z\,\text{pu} = \frac{\text{new } Z}{\text{base } Z} = \frac{2\text{ k}\Omega}{1.04\text{ k}\Omega} = 1.923\text{ pu }\Omega$$

Please note that the 100% values are called the base. The two parameters of a system that are usually known are its capacity in kVA and its voltage so these are usually designated as base values and all other parameters are calculated from these two values. Thus

$$\text{base } V = \text{base } I \times \text{base } Z \tag{2.9}$$

$$\text{base } VA = \text{base}V \times \text{base } I \tag{2.10}$$

The base impedance is defined by the equation

$$\text{base } Z = \frac{\text{base } V}{\text{base } I} \tag{2.11a}$$

However, by multiplying $Z$ by $V/V$ we get

$$Z = \frac{V}{I} = \frac{V \times V}{VA} = \frac{V^2}{VA} \tag{2.11b}$$

Most of the time it is more convenient to express the base quantities in kV and kVA or MVA. Since

$$kV = \frac{IZ}{1000}$$

$$\mathbf{Z} = \frac{1000 \times kV}{I}$$

Equation 2.11b can be expressed as

$$base\ Z = \frac{(base\ kV)^2 \times 1000}{base\ kVA} \tag{2.11c}$$

and since

$$kVA = \frac{MVA}{1000}$$

$$base\ \mathbf{Z} = \frac{(base\ kV)^2}{base\ MVA} \tag{2.11d}$$

The base impedance must be calculated with the overall base volt amperes and the base voltage of the portion of the circuit in which the impedance of interest is located when the circuit contains transformers and has more than one voltage. It is important to note that the actual impedance of the circuit is set by the electrical properties of the components. The base impedance is not the actual impedance, but the actual system impedances can be expressed as per unit proportions of the base impedance more easily for problem solution. Upon problem solution the per unit quantities are converted back to actual values with the appropriate equation. The conversion equations are

$$Z_{ohms} = Z_{pu}\ Z_{base} \tag{2.12}$$

$$I_{amps} = I_{pu}\ I_{base} \tag{2.13}$$

$$V_{kV} = V_{pu}\ V_{base} \tag{2.14}$$

$$VA_{volt\ amperes} = VA_{pu}\ VA_{base} \tag{2.15}$$

The advantages of the per unit system will become apparent after a few examples.

### Example 2.3:

Refer to Figure 2.18a for the one-line diagram for this problem. We will do the problem single-phase for this example. Figure 2.18b has many of the common one-line diagram symbols. Note that the generator can provide 18 kVA at 10 kV, the load is 10 kVA at 5 kV, transformer 1 (T1) has a capacity of 50 kVA and rated voltages of 10kV/50kV (primary/secondary), and transformer 2 (T2) has a capacity of 15 kVA at 50kV/5kv. We can use any two electrical quantities of this system as the base values. Usually the largest kVA rating and the voltage is chosen as the base (though not always).

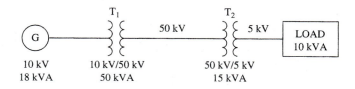

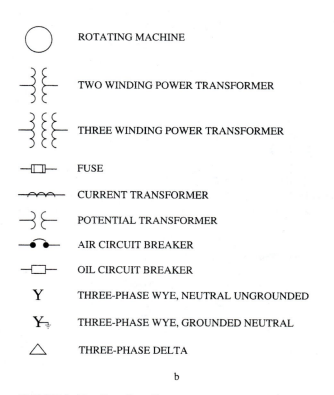

**FIGURE 2.18**   One-line diagram (a) For per unit discussion (b) common one-line symbols

**Solution:**

We will choose the 50 kVA rating of T1 as our base and the highest voltage as our base values.

50 kVA = 1 pu kVA

The remainder of the rated VA capacities are

$$T2 = \frac{15 \text{ kVA}}{50 \text{ kVA}} = 0.3 \text{ pu kVA}$$

$$\text{Load} = \frac{10\text{ kVA}}{50\text{ kVA}} = 0.2\text{ pu kVA}$$

$$\text{Generator} = \frac{18\text{ kVA}}{50\text{ kVA}} = 0.36\text{ pu kVA}$$

Now we must choose the base voltage. The generator voltage or the higher voltage is often the most convenient, so we will choose 50 kV as our base voltage. The transmission line voltage is 50 kV, 1 pu. The other voltages are multiplied by the transformer turns ratio for use in calculations.

To calculate other pu quantities on the generator and load side we must use the change of base proceedures discussed in the next section. The rest of the base parameters must now be calculated from equations 2.9 and 2.10.

$$I_{\text{base}} = \frac{50\text{ kVA}}{50\text{ kV}} = 1\text{ A} = 1\text{ pu A}$$

$$Z_{\text{base}} = \frac{50\text{ kV}}{1\text{ A}} = 50\text{ k}\Omega = 1\text{ pu Z}$$

Remember that the generator impedance, line impedance, and load impedance are expressed in terms of the per unit impedance for calculations. Later in this chapter there are example problems to demonstrate the use of the per unit system.

You may wish to note this page for future reference because the per unit equations are summarized here.

## 2.4.2 Per unit equation summary

$$\text{base } I = \frac{\text{base kVA}}{\text{base kV}} \qquad\qquad \text{(from Eq. 2.10)}$$

$$\text{base } Z = \frac{\text{base } V}{\text{base } I} \qquad\qquad (2.11a)$$

$$\text{base } Z = \frac{\text{base } V^2}{\text{base VA}} \qquad\qquad (2.11b)$$

$$= \frac{(\text{base kV})^2 \times 1000}{\text{base kVA}} \qquad\qquad (2.11c)$$

$$= \frac{(\text{base KV})^2}{\text{base MVA}} \qquad\qquad (2.11d)$$

$$pu\ VA = \frac{VA}{base\ VA} \tag{2.16}$$

$$= \frac{kVA}{base\ kVA}$$

$$pu\ V = \frac{V}{base\ V} \tag{2.17}$$

$$pu\ I = \frac{I}{base\ I} \tag{2.18}$$

$$pu\ Z = \frac{base\ V}{base\ I} \qquad \text{(from Eq. 2.11)}$$

$$= \frac{Z}{base\ Z} \tag{2.19}$$

$$= \frac{Z \times (base\ VA)}{(base\ V)^2} \tag{2.20a}$$

$$= \frac{Z \times (base\ kVA)}{(base\ kV)^2 \times 1000} \tag{2.20b}$$

For the final solution
actual value = (pu value) × (base value)
as shown in Equations 2.12 through 2.15.

### 2.4.3  Change of base

The impedance of transformers, generators, and motors are often given in per unit or percent (pu = % ÷ 100) where the base used is the machine nominal voltage and volt-amperes. The machine base is seldom the base of the system under analysis so a convenient method of changing the base of pu impedance is needed. Consider a system in which base 2 is to be used for analysis, but base 1 is the base of a machine's pu impedance. The actual machine impedance is the same in both system bases, so

$$Z = pu\ Z_1 \frac{V_{B1}^2}{VA_{B1}}$$

and

$$Z = pu\ Z_2 \frac{V_{B2}^2}{VA_{B2}}$$

Since Z is the same in both systems

$$\text{pu } Z_1 \frac{V_{B1}{}^2}{VA_{B1}} = \text{pu } Z_2 \frac{V_{B2}{}^2}{VA_{B2}}$$

Solving for pu $Z_1$ results in

$$Z_{1pu} = Z_{2pu} \frac{V_{B2}{}^2}{V_{B1}{}^2} \frac{VA_{B1}}{VA_{B2}} \qquad (2.21)$$

or in terms of old base and new base

$$\text{new base } Z_{pu} = \text{old base } Z_{pu} \frac{(\text{Old base } V)^2 (\text{new base VA})}{(\text{new base } V)^2 (\text{old base VA})}$$

### 2.2.4 Per unit with balanced three-phase systems

There are two methods of handling per unit calculations for balanced three-phase systems. The first is to use the three-phase line to line voltages and volt-amperes and to calculate the line current in the usual manner from $VA = \sqrt{3}\, I_L V_{LL}$, where subscript LL is line to line. The pu voltage is the same whether line to line or phase to neutral values of voltage are used since $V_{LL} = \sqrt{3}\, V_{LN}$, where subscript LN is line to neutral, as long as consistent values are used in the ratio. The same is true of pu VA since three-phase VA = 3 × single-phase VA. Calculated correctly the base impedance will be the same. The appropriate equations are

$$\text{base } I = \frac{\text{base three-phase kVA}}{\sqrt{3}\ \text{base kV}_{LL}} \qquad (2.22)$$

$$= \frac{\text{base phase kVA}}{\text{base kV}_{LN}}$$

$$\text{base } Z = \frac{(\text{base kV}_{LL}/\sqrt{3})^2 \times 1000}{\text{base three-phase kVA}/3} \qquad (2.23)$$

$$= \frac{(\text{base kV}_{LL})^2 \times 1000}{\text{base three-phase kVA}} \qquad (2.24)$$

$$= \frac{(\text{base kV}_{LL})^2}{\text{three-phase base MVA}} \qquad (2.25)$$

Notice that except for the subscripts these equations (2.23 through 2.25) are the same as equations 2.11 c and d. Actual impedances are given as per phase values. Additionally delta- and wye-connected impedances are not the same. If one has the pu impedance of the load this is not a problem, but if one must calculate the pu impedance of a load from its actual impedance it is easier to calculate with phase to neutral values.

The second method is to convert all circuit parameters to phase to neutral values and convert back to line to line values at the end of the problem. This method presents fewer opportunities for confusion for beginning pu calculations, and is the method used in this chapter except for a few example steps.

Delta-connected loads must be converted to their wye equivalent impedance to obtain their pu impedance. This is simple for balanced three-phase loads. Recall from wye-delta conversion, with ABC = wye values and XYZ = delta values

$$Z_A = \frac{Z_Y Z_Z}{Z_X + Z_Y + Z_Z}$$

In a balanced system $Z_X = Z_Y = Z_Z = Z_D$, where D is for delta

$$Z_A = \frac{Z_D^2}{3 Z_D}$$

$$= \frac{\text{delta phase } Z}{3}$$

and in similar fashion

$$Z_X = \frac{Z_A Z_B + Z_B Z_C + Z_C Z_A}{Z_A}$$

since $Z_A = Z_B = Z_C = Z_{WYE}$

$$Z_D = \frac{3 Z_{WYE}^2}{Z_{WYE}}$$

$$= 3 \times \text{(wye impedance)}$$

The following example may help.

### Example 2.4:

Let the circuit components of Figure 2.18 assume the following values with all values given in three-phase kVA and line to line voltage.

G—23kV at 300 kVA

T1—23kV/69kV at 350 kVA

Line—69 kV at 450 kVA capacity, Z = 10 + j150 ohms

T2—69kV/7.2kV at 120 kVA

Load—75 hp at 7.2 kV full load (assume PF = 1)

calculate the pu values of all voltages, kVAs, and impedances given using phase to neutral values.

### Solution:

The base kVA for our single-phase equivalent one-line drawing is one third of the three-phase kVA

$$\text{base kVA} = \frac{450 \text{ kVA}}{3}$$

$$= 150 \text{ kVA} = 1 \text{ pu}$$

and

$$\text{base kV} = \frac{69 \text{ kV}}{\sqrt{3}}$$

$$= 39.84 \text{ kV} = 1 \text{ pu}$$

so

$$\text{base } I = \frac{\text{base kVA}}{\text{base kV}}$$

$$= \frac{150 \text{ kVA}}{39.84 \text{ kV}}$$

$$= 3.76 \text{ A} = 1 \text{ pu}$$

and the base impedance in the line section is

$$\text{base } Z = \frac{(\text{base kV})^2 \times 1000}{\text{base kVA}}$$

$$= \frac{(39.84 \text{ kV})^2 \times 1000}{150 \text{ kVA}}$$

$$= 10.58 \text{ k}\Omega = 1 \text{ pu}$$

resulting in a pu impedance of

$$\text{pu } Z = \frac{10 + j150 \ \Omega}{10.58 \text{ k}\Omega}$$

$$= 0.000945 + j0.0142 \text{ pu } \Omega$$

The interested student will want to calculate the preceding values using line to line voltages and three-phase kVA.

The line, T1 secondary, and T2 primary voltages are all 1.0 pu because they are at the base voltage. The generator and T1 primary voltage pu voltages are the same and are

$$\text{pu } V_G = \frac{V_G}{\text{base } V}$$

$$= \frac{23 \text{ kV}/\sqrt{3}}{39.84 \text{ kV}}$$

$$= 0.331 \text{ pu } V$$

and the load is

$$\text{pu } V_{\text{L}} = \frac{7.2 \text{ kV}/\sqrt{3}}{39.84 \text{ kV}}$$

$$= 0.104 \text{ pu V}$$

The per unit kVA values are

$$\text{Gen. pu kVA} = \frac{\text{Gen. kVA}}{\text{base kVA}}$$

$$= \frac{100 \text{ kVA}}{150 \text{ kVA}}$$

$$= 0.667 \text{ pu kVA}$$

In similar fashion

$$\text{T1 pu kVA} = \frac{116.67 \text{ kVA}}{150 \text{ kVA}}$$

$$= 0.778 \text{ pu kVA}$$

$$\text{T2 pu kVA} = \frac{40 \text{ kVA}}{150 \text{ kVA}}$$

$$= 0.267 \text{ pu kVA}$$

$$\text{Load pu kVA} = \frac{25 \text{ hp } \times (746 \text{ W/hp})}{150 \text{ kVA}}$$

$$= 0.124 \text{ pu kVA}$$

Note: Motor power factor (one in this problem) must be taken into account in the load pu kVA calculation.

Notice also that to calculate the load pu impedance we must use the over all base kVA and the base voltage of the section of the circuit in which the load is located, in our example the secondary circuit of T2. The impedance of the load, the effective full load impedance of the motor, must be calculated from the nominal machine voltage and kVA.

$$I_{\text{L}} = \frac{\text{three-phase load kVA}}{\sqrt{3} \times \text{load } V_{\text{LL}}}$$

$$= \frac{75 \text{ hp } (746 \text{ W/hp})}{\sqrt{3} \, 7.2 \text{ kV}}$$

$$= 4.48 \text{ A}$$

$$Z_{\text{L}} = \frac{V_L}{I_L}$$

$$= 927 \; \Omega$$

$$\text{base } Z = \frac{\left(\frac{7.2 \; \text{kV}}{\sqrt{3}}\right)^2 \times 1000}{150 \; \text{kVA}}$$

$$= 115 \; \Omega$$

$$\text{load pu } Z = \frac{927 \; \Omega}{345 \; \Omega}$$

$$= 8.06 \; \text{pu ohms}$$

The delta load impedance is 3x927 = 2.78 k $\Omega$.

## 2.5

### GENERATORS, TRANSFORMERS, AND PER UNIT

This section discusses the calculation of generator and transformer pu impedances by way of two examples.

**Example 2.5:**

Referring to the simplified generator equivalent circuit of Figure 2.19 calculate the per unit impedance of the generator.

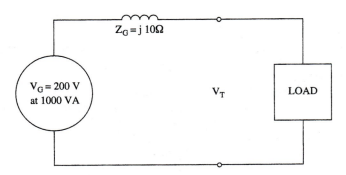

**FIGURE 2.19**   Generator equivalent circuit for Example 2.5

**Solution:**

We know the impedance of the generator so we must find the base impedance of the generator. The generator is rated at 200 V at 1000 VA. These values are our base values for calculating the base impedance.

$$\text{base } I = \text{full load } I = \frac{VA}{V}$$

$$= \frac{1000 \text{ VA}}{200 \text{ V}}$$

$$= 5 \text{ A}$$

$$\text{base } Z = \frac{\text{base V}}{\text{base I}}$$

$$= \frac{200 \text{ V}}{5 \text{ A}}$$

$$= 40 \ \Omega$$

and

$$\text{pu } Z = \frac{Z}{\text{base } Z}$$

$$\text{pu } Z = \frac{j10 \ \Omega}{40 \ \Omega}$$

$$= j0.25 \text{ pu}$$

Notice that 25% of the rated voltage will cause the rated current to flow into a short circuit connected to the generator output. Thus we can find the generator pu impedance by connecting an ammeter across the output terminals and adjusting the voltage until rated current flows. The voltage at the generator terminals with the short circuit removed divided by the rated voltage is the pu generator impedance. Note that

$$5 \text{ A} \times j10 = 50 \text{ V}$$

$$\frac{50 \text{ V}}{200 \text{ V}} = 0.25$$

**Example 2.6:**

Calculate PRI pu $Z = \dfrac{4}{20} = 2$ pu of the simplified transformer equivalent circuit shown in Figure 2.20a and b with the primary side as a base and then with the secondary side as a base.

**Solution:**

Using equation 2.13b we find the base impedance is

$$\text{Pri. base } Z = \frac{(\text{base } V)^2}{\text{base VA}}$$

$$= \frac{(200 \text{ V})^2}{2000 \text{ VA}}$$

$$= 20 \ \Omega$$

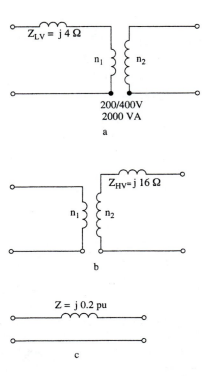

**FIGURE 2.20**  Simplified transformer equivalent circuit for Example 2.6 (a) Primary reference (b) secondary reference and (c) per unit equivalent circuit for impedance diagram

The base impedance on the secondary side of the transformer is

$$\text{sec. base } Z = \frac{400 \text{ V}^2}{2000 \text{ VA}}$$

$$= 80 \text{ } \Omega$$

$$\text{sec. pu } Z = \frac{16 \text{ } \Omega}{80 \text{ } \Omega}$$

$$= 0.2 \text{ pu ohms}$$

Notice that the pu impedance of the transformer is the same on both the primary and secondary side. This fact is one of the major advantages of the pu system. Properly calculated pu impedances are independent of voltage transformation so the transformer impedance can be represented in a drawing of pu impedances as a simple impedance as shown in Figure 2.20c. It is important to remember that the voltage on the primary side of the transformer must be used to calculate the primary side base impedance, and the secondary voltage must be used to calculate the secondary base impedance.

## 2.6

## IMPEDANCE AND REACTANCE DIAGRAMS

A large system drawn with all of the three-phase and the neutral conductors shown would be very difficult to follow. The reader of the diagram would lose sight of the overall system in the detail. The one-line diagram allows one to grasp the overall system without excessive detail. The one-line drawing of Figure 2.21a is used to supply a lot of information about the system in as concise a form as possible. Recall that several common one-line diagram symbols are shown in Figure 2.18b. The one-line diagram is not useful in the calculation of system parameters under load. Because pu impedances are independent of voltage transformation the one-line circuit of Figure 2.21a can be drawn as a pu equivalent circuit, called an *impedance diagram,* as shown in Figure 2.21b. The impedance diagram is useful for such calculations. All of the circuit parameters are given in pu values. The standard network solution methods are easily applied to the impedance diagram since voltage level changes are taken care of by correctly calculating the pu quantities.

The resistance of a generator, transformer, or line is usually approximately 0.1 times the reactance, consequently the resistance can often be ignored in fault calculations for power system protection. The omission of the resistance from the impedance diagram results in the *reactance diagram* of Figure 2.21c. A computer of sufficient speed and calculating power makes such a simplification unnecessary for calculating system quantities.

## 2.7

## PER UNIT EXAMPLES

### Example 2.7:

Draw an impedance diagram of the one-line drawing of Figure 2.22a and show all of the impedances in per unit.

### Solution:

First select the base VA for the entire system. We will choose the highest VA rating in the system, T1. A base VA that is not a rating of any machine or line in the system can be chosen as the base. For our problem

1 pu kVA = 6 kVA/3

= 2 kVA

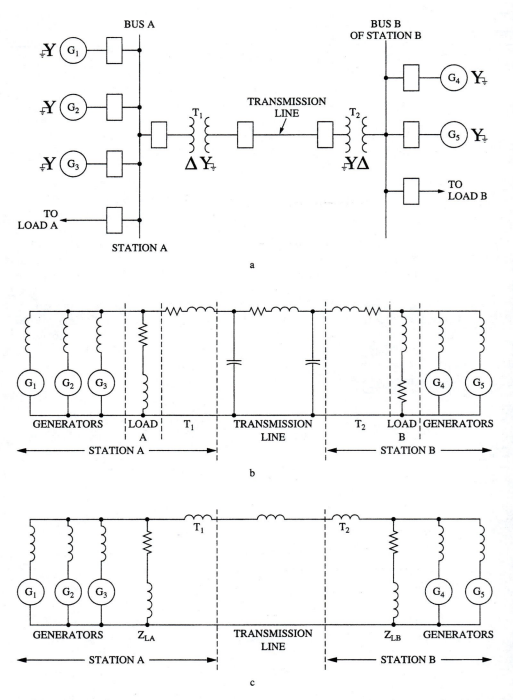

**FIGURE 2.21**   Reactance diagram development (a) One-line diagram (b) equivalent circuit (impedance diagram) and (c) reactance diagram

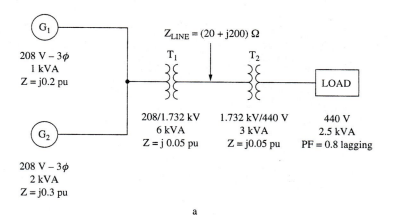

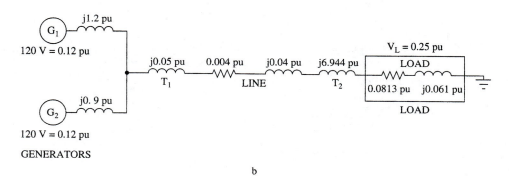

**FIGURE 2.22** Example 2.7 system (a) One-line diagram and (b) impedance diagram

Since transformers are present the base voltage of each section of the system, for base impedance calculation, is set by the transformer ratios after the base voltage for the diagram is selected. It is frequently convenient to use the generator voltage as a base and we will do so for this example.

Gen. base $V = 208/\sqrt{3}$

$$= 120 \text{ V} = 1 \text{ pu}$$

Line base $V = 1732 \text{ V}/\sqrt{3}$

$$= 1 \text{ kV}$$

Load base $V = 440 \text{ V}/\sqrt{3}$

$$= 254 \text{ V}$$

Equation 2.21 will be used to transform the generator impedances to the base kVA. Recall

$$\text{new base pu } Z = (\text{old base pu } Z)\ \frac{(\text{old base } V)^2\,(\text{new base VA})}{(\text{new base } V)^2\,(\text{old base VA})}$$

so that

$$\text{pu } Z_{G1} = j0.2\ \frac{(120\ V)^2\,(2\ \text{kVA})}{(120\ V)^2\,(0.333\ \text{kVA})}$$

$$= j1.2\ \text{pu}$$

$$\text{pu } Z_{G2} = j0.3\ \frac{(120\ V)^2\,(2\ \text{kVA})}{(120\ V)^2\,(0.667\ \text{kVA})}$$

$$= 0.9\ \text{pu}$$

Transformer 1 is at the base voltage and kVA so

$$\text{pu } Z_{T1} = j0.5\ \text{pu}$$

The line per unit impedance is found by dividing the actual impedance by the base impedance. The short form equation for the base impedance, Equation 2.22b, is used.

$$\text{line pu } Z = Z_{\text{LINE}}\ \frac{\text{base } VA}{\text{base } V^2}$$

$$= (2 + j20)\ \Omega\ \frac{2\ \text{kVA}}{1\ \text{kV}^2 \times 1000}$$

$$= 0.004 + j0.04\ \text{pu}$$

Transformer 2 per unit impedance must be found with the base transformation equation.

$$\text{pu } Z_{T2} = j0.05\ \Omega\ \frac{(1\ \text{kV})^2\,(2\ \text{kVA})}{(0.12\ \text{kV})^2\,(1\ \text{kVA})}$$

$$= 6.994\ \text{pu}$$

The wye equivalent load impedance is

$$Z_{\text{LOAD}} = \frac{\left(\dfrac{2.5\ \text{kVA}}{3}\right)}{440/\sqrt{3}}$$

$$= 3.28\ \Omega$$

The base impedance at the load is

$$\text{base load } Z = \frac{\text{base } V\text{p}^2}{\text{base VA}}$$

$$= \frac{254\ V^2}{2000\ \text{VA}}$$

$$= 32.26 \ \Omega$$

$$\text{pu } Z_{LOAD} = \frac{Z}{\text{base } Z}$$

$$= \frac{3.28 \ \Omega}{32.26 \ \Omega}$$

$$= 0.102 \text{ pu } \angle \text{arc cos } 0.8$$

$$= .0813 + j0.061 \text{ pu}$$

The completed impedance diagram is shown in Figure 2.22b. We can now use network theory to solve for the circuit currents. A simpler example circuit for which to solve for current follows.

**Example 2.8:**

Solve for the current in each section of the circuit in Figure 2.23a using per unit values.

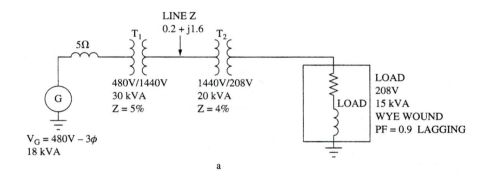

a

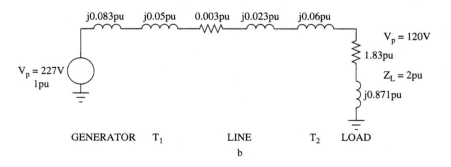

b

**FIGURE 2.23** One-line diagram for Example 2.8 (a) Circuit and (b) impedance diagram

**Solution:**

The first task is to select the base kVA and voltage. The phase to neutral generator voltage is convenient for the base voltage, and the highest kVA is convenient for the base kVA.

$$\text{base voltage} = 480 \text{ V}/\sqrt{3}$$

$$= 277 \text{ V} = 1 \text{ pu}$$

$$\text{line base voltage} = 1440 \text{ V}/\sqrt{3}$$

$$= 831.4 \text{ V}$$

$$\text{load base voltage} = 208 \text{ V}/\sqrt{3}$$

$$= 120 \text{ V}$$

$$\text{base current} = \frac{\text{base VA}}{\text{base V}}$$

$$= \frac{10 \text{ kVA}}{277 \text{ V}}$$

$$= 36.1 \text{ A} = \text{pu}$$

Impedances

The generator impedance from Equation 2.21 is

$$\text{pu } Z_{\text{GEN}} = j0.05 \text{ }\Omega \text{ } \frac{(277 \text{ V})^2 (10 \text{ kVA})}{(277 \text{ V})^2 (6 \text{ kVA})}$$

$$= j0.083 \text{ pu}$$

Because it is at the base kVA and voltage

$$\text{pu } Z_{\text{T1}} = j0.05 \text{ pu}$$

$$\text{pu } Z_{\text{LINE}} = (0.2 + j1.6) \text{ }\Omega \text{ } \frac{(10 \text{ kVA})}{(831.4 \text{ V})^2}$$

$$= 0.003 + j0.023 \text{ pu }\Omega$$

$$\text{pu } Z_{\text{T2}} = j0.04 \text{ }\Omega \text{ } \frac{(10 \text{ kVA})}{(6.67 \text{ kVA})}$$

$$= j0.06 \text{ pu }\Omega$$

Now the load

$$Z_{\text{LOAD}} = \frac{(120 \text{ V})^2}{5000 \text{ VA}}$$

$$= 2.88 \ \Omega \ \angle \text{arc cos } 0.9$$

$$\text{base } Z_{\text{LOAD}} = \frac{(120 \text{ V})^2}{10,000 \text{ VA}}$$

$$= 1.44$$

$$\text{pu } Z_{\text{LOAD}} = \frac{2.88}{1.44}$$

$$= 2.0 \text{ pu } \angle \text{ arc cos } 0.9$$

$$= 1.83 + j0.871 \text{ pu}$$

The completed impedance diagram is shown in Figure 2.23b.

At this point it is instructive to calculate the pu load impedance after it has been referred to the generator circuit through the turns ratios of the transformers ($Z_{\text{RG}}$).

$$Z_{\text{RG}} = 2.88 \ \frac{(831.4 \text{ V})^2 \ (277 \text{ V})^2}{(120 \text{ V})^2 \ (831.4 \text{ V})^2}$$

$$= 15.346$$

$$\text{base } Z \text{ at Gen.} = \frac{(277 \text{ V})^2}{10,000 \text{ V}}$$

$$= 7.6729$$

$$\text{Load pu } Z \text{ at Gen} = \frac{15.346}{7.6729}$$

$$= 2.0 \text{ pu}$$

The same value as calculated earlier.

We now need the total pu impedance to calculate the currents.

$$Z_{\text{total}} = (0.003 + 1.8) + j(0.083 + \ 0.05 + 0.023 + 0.06 + 0.871) \text{ pu}$$

$$= 1.83 + j1.087 \text{ pu}$$

$$= 2.131 \text{ pu}$$

$$\text{pu } I = \frac{\text{pu } V}{\text{pu } Z}$$

$$= \frac{1.0 \text{ pu}}{2.131 \text{ pu}}$$

$$= 0.469 \text{ pu}$$

$$I = I_{\text{pu}} \times I_{\text{base}}$$

$$I_{\text{GEN}} = (0.469) \ (36.1 \text{ A})$$

$$= 16.9\,\text{A}$$

$$I_{\text{LINE}} = (0.469)\,\frac{10\,\text{kVA}}{831.4\,\text{V}}$$

$$= 5.647\,\text{A}$$

$$I_{\text{LOAD}} = (0.4685)\,\frac{10\,\text{kVA}}{120\,\text{V}}$$

$$= 39.1\,\text{A}$$

The currents are line currents.

Again the major advantage of the per unit system is the ease with which it can be used to solve problems such as these examples because the transformer pu impedance is the same on either side of the transformer. The pu system is used throughout this book.

## 2.8

### SUMMARY

Loads, lines, and generators have impedance, and all contribute to the relative phase of the current and voltage in the system. The reference for power factor, the ratio of real to apparent power, is the load for the system or system segment. The power consumed is dissipated as work by electrical machines or heat lost in resistance. The inductive reactance is higher in the center of a conductor because the magnetic flux density is higher. For this reason alternating current is forced toward the outer portion or skin of a conductor. The effective decrease in conductor area causes the ac resistance of a conductor to be higher than the dc resistance. This phenomena is called the *skin effect*.

The selection of three-phase ac for electrical power distribution and transmission resulted because three-phase ac with 120° phase separation results in power that is constant over time. DC and six-phase electrical power is constant over time too, and both are also use for electrical power transmission. Three-phase electrical equipment can be connected in either delta or wye configurations. The line and phase voltage and current relationships of delta and wye connections are discussed in Section 2.3.3.

To ease the calculation of three-phase quantities in balanced three-phase systems, in which each phase has the same load, the per unit (pu) method of representing system quantities was devised. In the pu method one phase of the system is represented with the impedances, voltages, currents, power, VA, and VAR levels represented as a fraction of the system base. The base quantities are normally chosen to be the most significant voltage level and the VA capability of that section of the system, perhaps the generating voltage and capacity. Once the base quantities are chosen all other base quantities, such as current, are referenced to them. The system quantities represented as pu quantities are the ratio of that quantity to the base quantity, such as current to base current. By representing each segment of the system in the base of the most significant segment complex calculation, while still complex, are far simpler than they would otherwise be. The mechanics of pu calculations are shown in Section 2.4.

## 2.9
----

### QUESTIONS

1. Calculate the number of turns a generator must have to provide an rms voltage of 23 kV at a flux of 420 Mlines per pole.
2. The conductivity of copper is $5.8 \times 10^7$ Siemens/meter and the relative permeability is one. Calculate the skin depth at 400 Hz, 100 MHz, and 12 GHz.
3. Repeat problem 2 for aluminum, the conductivity of which is $3.53 \times 10^7$ S/m and $u_r = 1$.
4. If the generator of Figure 2.2 provides a full load output voltage of 16 kV at 500 kVA calculate the terminal voltage with the load removed. The generator impedance is j6 $\Omega$.
5. Calculate the current of the circuit of Figure 2.3 at 60 Hz if $E_G = 480$ V at 60 Hz, $R = 10\ \Omega$, $L = 0.1$ H, and $C = 100\ \mu$F.
6. The load of Figure 2.5 has $R = 10\ \Omega$, $L = 0.1$ H, and $C = 100\ \mu$F. Calculate Z at 60 Hz.
7. The components of Figure 2.6a and b are $L = 0.2$ H, $C = 120\ \mu$F, and $R = 20\ \Omega$. If the voltage for both circuits is 110 V at 60 Hz calculate the total current for each circuit.
8. The load of Figure 2.7 is rated at 500 kVA with a power factor of 0.8. Calculate the load $P$ and $Q$.
9. Draw the phasor representation of the voltage and current of a circuit with an rms voltage of 120 V and an rms current of 12 A and a power factor of 0.7.
10. A three-phase circuit has a 60 Hz voltage of 208 V line to line and a wye connected load of 28 $\Omega\ \angle 60°$ per phase. If the phase rotation is ABC and phase A is at zero degrees, calculate the line current and draw the phasor diagram.
11. Repeat problem 11 for a delta-connected load of 20 $\Omega\ \angle 40°$.
12. The generator of Figure 2.19 has $V_G = 440$ V and $Z_G = $ j12 $\Omega$. Calculate the per unit impedance.
13. The single-phase transformer of Figure 2.20b has a rating of 500 kVA, voltages of 34.5kV/480V, and a high side impedance of j100 $\Omega$. Calculate the high and low side per unit impedance.
14. A 400 V, 5 kVA load has a per unit impedance of 0.2 with its own rating as base. It is to be placed in a system with a base of 10 kVA and 800 V. Calculate the load per unit impedance in the new system.
15. The circuit components of Figure 2.18a have the following specifications: Generator—23 kV at 600 kVA

T1—23kV/138kV at 1 MVA
Line—$Z = (100 + j1500)\ \Omega$
T2—138kV/12.47kV at 300 kVA
Load—200 kVA
Calculate the per unit values of the voltages, kVAs, and line impedance using 1 MVA and 138 kV as base.

16. Repeat problem 15 with 23 kV and 600 kVA as base values.

17. The components in the one-line drawing of Figure 2.23a have the following single-phase ratings:
Generator—12.47 kV at 90 kVA, Z = 10%
T1—12.47kV/34.5kV at 120 kVA, Z = 4%
Line—$Z = (3 + j40)\ \Omega$
T2—34.5kV/480V at 75 kVA, Z = 3%
Load—60 kVA at 480 V, PF = 0.9
Use the generator kVA and voltage as the system base and calculate the per unit impedances and the current flowing through the load, line, and generator.

18. The components of the one-line drawing of Figure 2.23a have the following three-phase ratings:
Generator—23 kV at 240 kVA, Z = 8%
T1—23kV/138kV at 360 kVA, Z = 5%
Line—$Z = (50 + j600)\ \Omega$
T2—138kV/12.47kV at 180 kVA, Z = 4
Load—150 kVA at 12.47 kV, PF = 0.95
Use the generator voltage as the system base voltage and T1 kVA rating as the system base kVA. Calculate the per unit impedances and the line current flowing in the load, line, and generator.

# SECTION II

# Electrical Power Distribution

## CHAPTER 3

# Basic Considerations and Distribution System Layout

### 3.1

#### UTILITY LOAD CLASSIFICATIONS

The electrical power distribution system is that portion of the electrical system that connects the individual customer to the source of bulk power, as shown in Figure 1.4.

Loads are the reason for the electrical power system. The types of loads are: resistive, such as lighting and heating; inductive, such as motor loads; and capacitive, such as rectifier bridges with capacitor filters. Most electrical system loads are predominantly inductive. Utility loads are usually classified by the occupancy of the structures that use the power instead of the type of electrical load.

*Residential loads* are the combined loads of single family dwellings and apartment complexes. Most dwelling units are single-phase although most apartment complexes have three-phase service and the load to each phase is balanced among the apartment units. The types of loads are predominantly lighting, heating (water, space, and cooking), and motors for appliances such as air conditioners, washing machines, and refrigerators. In numbers, residential loads are the largest group of electric utility customers, comprising up to 85% (about 80% typically), and require the largest investment in distribution equipment.

Commercial loads typically make up about 15% of an electric utility's customers. They consist of office buildings and complexes, schools, shopping malls, and stores. The types of loads are primarily lighting, heating, air conditioning, and office equipment such as computers, copying machines, word processors, and coffee makers.

Industrial loads seldom make up more than about 5% of an electric utility's customers but they may use up to 25 to 30% of the kilowatt hours supplied. The loads consist of the same types of loads as commercial businesses plus large motors, control panels, and production equipment.

## 3.2

## POWER FACTOR CORRECTION

Because most commercial and industrial loads are inductive in nature the kVAs drawn from the utility are larger than the kWs, and the current lags the voltage, as shown in Figure 3.1. Only the component of the current that is in phase with the voltage provides useful work. The out of phase component increases the total current that the utility must supply so a premium may be charged to a customer with a power factor that is below a value fixed by the electric utility (usually 0.7). Utilities actually charge their customers for the kVA hours, not kW hours, that they use, although the latter term is more commonly used. Thus customers have an economic interest in improving the power factor of their facilities. The trend has been to improve power factors to between 0.9 to 0.95 or even better as the cost of electricity has risen in recent years. Figures 3.2, 3.3, and 3.4 show the impedance, current, and power diagrams, respectively, for an inductive and capacitive load. The overall power factor of a customer's internal electrical system is brought closer to one by adding capacitance across the line at various points in the customer's internal electrical distribution system. If the system is compact the capacitance may be near the service. Larger systems may require power factor correction for each major bus. In the event the system was capacitive (a rare occurrence) inductance would be added across the line for power factor correction. Automatic equipment

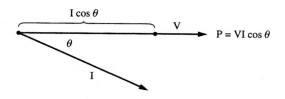

**FIGURE 3.1**　Power from in phase component of current

is available to sense the power factor of a system and switch more or less capacitance across the line as the load changes in nature. The automatic systems must be chosen with care because many will try to over correct the power factor when a significant part of the load is solid state motor drives, possibly causing the system voltage to rise to excessive levels (we will discuss why later). The next two examples review the calculation of reactance for power factor correction.

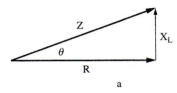

a

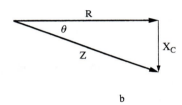

b

$Z = R + j(X_L - X_C)$

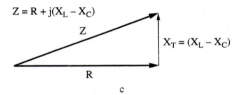

c

**FIGURE 3.2** Impedance diagram (a) Inductive (b) capacitive and (c) general, $Z = R + j(X_L - X_C)$

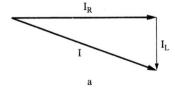

a

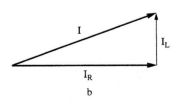

b

**FIGURE 3.3** Current (a) Inductive—lagging and (b) capacitive—leading

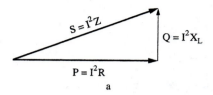

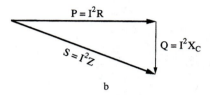

**FIGURE 3.4**  Power triangle (a) with inductive load and (b) with capacitive load

### Example 3.1:

Calculate the value of capacitor needed to correct the power factor of the circuit of Figure 3.5 to unity.

#### Solution:

We will solve the problem by first calculating the total current, which will have the same angle as the total impedance. We will then calculate $S$, the power factor, then $Q$. We will set $Q_C$ equal to the total circuit $Q$, calculate $X_C$, and then $C$.

$$I_1 = \frac{V}{R}$$

$$= \frac{120\ V}{100\ \Omega}$$

$$= 1.2\ A$$

$$Z_2 = (10 + j100)\ \Omega$$

$$= 100.5\ \Omega\ \angle 84.29°$$

$$I_2 = \frac{120\ V}{100.5\ \Omega\ \angle 84.29°}$$

$$= 1.194\ A\ \angle^-84.29°$$

$$I_T = I_1 + I_2$$

$$= 1.2\ A + 0.118\ A - j1.18\ A$$

$$= 1.775\ A\ \angle^-42°$$

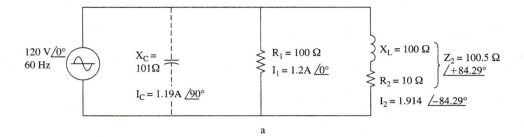

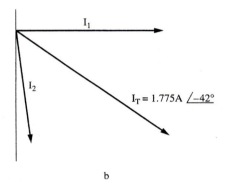

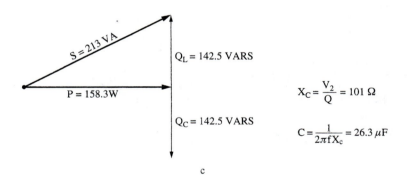

**FIGURE 3.5** Circuit for Example 3.1 (a) Circuit diagram (b) phasor diagram before power factor correction and (c) power diagram

now

$$S = VI$$

$$= 120\ V \times 1.775\ A$$

$$= 213\ VA$$

and the power factor (PF) is

$$PF = \cos 42°$$

$$= 0.743$$

The circuit power is

$$P = S \times PF$$

$$= 158.3 \text{ W}$$

and Q is

$$Q = S \sin 42°$$

$$= 142.5 \text{ VAR}$$

The capacitive Q must cancel the inductive Q so we will set the $Q_C = 142.5$ VAR. Now we calculate $X_C$ from

$$X_C = \frac{V^2}{Q}$$

$$= 101 \ \Omega$$

and

$$C = \frac{1}{2 \pi f X_c}$$

$$= 26.3 \ \mu F$$

### Example 3.2:

Refer to the system of Figure 3.6. A 480 V three-phase system feeds the following loads:

M1—fully loaded induction motor, 80 hp, efficiency ($\eta$) = 91%, $PF$ = 0.86 lagging.

M2—Induction motor rated at 40 hp, loaded to 3/4 rated output, with $\eta$ = 89% and PF = 0.83 lagging at actual load.

R— 50 kW of lighting and heating load.

M3—a synchronous motor added to provide power factor correction and drive a 70 hp load. $\eta$ = 93% at full load.

Calculate the leading kVAR that the synchronous motor must provide to correct the overall power factor to 98% lagging.

### Solution:

Our solution method will be to calculate first the lagging kVAR of the system without the synchronous motor. We will then calculate the lagging kVAR allowed with the synchronous motor and the system operating at PF = 0.98. The solution is the difference in the two values.

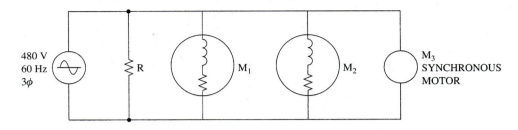

M₁ = FULLY LOADED INDUCTION MOTOR
80 hp, PF = 0.86, η = 91%
M₂ = INDUCTION MOTOR, ³/₄ LOADED
40 hp, PF = 0.83, η = 89%
R = LIGHTING AND HEATING LOADS, 50 kW
M₃ = SYNCHRONOUS MOTOR, η = 93%, LOAD = 70 hp

a

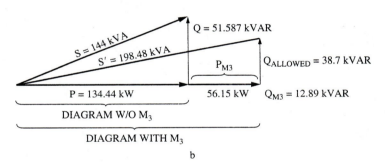

b

**FIGURE 3.6** Circuit for Example 3.2 (a) Impedance diagram and (b) power diagram of system

Without the synchronous motor.

$$P_R = 50 \text{ kW}$$

$$P_{M1} = \frac{P_{OUT}}{\eta}$$

$$= \frac{80 \text{ hp} \times 746 \text{ W/hp}}{0.91}$$

$$= 65.582 \text{ kW}$$

$$Q_{M1} = S \sin\theta$$

$$= (P/PF)\sin(\arccos PF)$$

$$= \frac{644.582 \text{ kW}}{0.86} \sin(\arccos 0.86)$$

$$= 76.529 \text{ kVA} \sin 30.68°$$

$$= 38.914 \text{ kVAR}$$

Similarly for M2

$$P_{M2} = \frac{0.75 \, (40 \text{ hp}) \, (746 \text{ W/hp})}{0.89}$$

$$= 25.15 \text{ kW}$$

$$S_{M2} = \frac{P}{PF}$$

$$= \frac{18.86 \text{ kW}}{0.83}$$

$$= 30.3 \text{ kVA}$$

and

$$Q_{M2} = S \sin(\arccos PF)$$

$$= 30.3 \text{ kVA} \sin 33.9°$$

$$= 16.9 \text{ kVAR}$$

The total power and reactive kVA are

$$P_T = 50 \text{ kW} + 65.582 \text{ kW} + 25.15 \text{ kW}$$

$$= 140.5 \text{ kW}$$

$$Q_T = 38.914 \text{ kVAR} + 16.9 \text{ kVAR}$$

$$= 55.8 \text{ kVAR}$$

and if needed

$$S_T = \sqrt{P^2 + Q^2}$$

$$= 151.4 \text{ kVA}$$

With the synchronous motor (primed values)

$$P_{M3} = \frac{70 \text{ hp} \times 746 \text{ hp/W}}{0.93}$$

$$= 56.15 \text{ kW}$$

$$P'_T = 140.5 \text{ kW} + 56.151 \text{ kW}$$

$$= 196.9 \text{ kW}$$

$$S'_T = \frac{P'_T}{PF}$$

$$= \frac{196.9 \text{ kW}}{0.98}$$

$$= 200.9 \text{ kVA}$$

$$Q' = S' \sin(\text{arccos } PF)$$

$$= 200.9 \text{ kVA} \sin 11.48°$$

$$= 39.98 \text{ kVAR}$$

synchronous motor $Q = Q_T - Q'$

$$= 55.80 \text{ kVAR} - 39.98 \text{ kVAR}$$

$$= 15.82 \text{ kVAR}$$

We may not always choose to correct the power factor to one for one of two main reasons. The cost of correcting the power factor to one as opposed to a more modest correction between 0.95 and 0.99 may be much higher, and the return on investment for the optimal correction may be too small to justify. The second reason is that a correction to a power factor of unity under one set of operating conditions may result in a leading power factor under differing load conditions. A leading power factor could result in an excessive line voltage increase in many systems. Such systems would need automated equipment to change the power factor correction as the load changes to operate well at unity power factor.

The reason that line voltage rises with a leading power factor is illustrated in Figure 3.7. The secondary circuit of any transformer contains the transformer secondary leakage reactance and resistance as well as the reactance and resistance of the line feeding the load, as shown in Figure 3.7a. Current drawn from the source causes an in phase voltage drop across the resistances and a voltage drop across the reactances that leads the current 90°, as shown in Figure 3.7b. If the source voltage is held constant, then as the power factor becomes less lagging the voltage drops across the reactance and resistance of the transformer secondary and line will rotate in the direction of the arrows in Figure 3.7b and c. The algebraic sum of the voltages around a closed loop must equal zero so the load voltage will rise, as shown in Figure 3.7d.

## 3.3

## UTILITY FACTOR

Utility factor is a measure of how much of the total capacity of a utility is in use. If all of the power a utility can generate is being used the utility factor is 100%. The requirement for reserve power, and the load diversity (illustrated in Figures 1.2

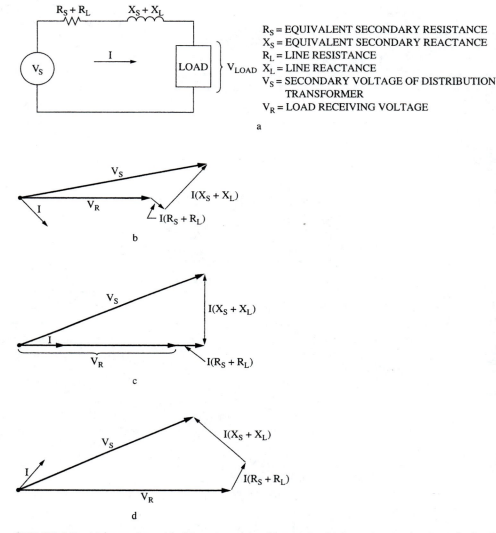

$R_S$ = EQUIVALENT SECONDARY RESISTANCE
$X_S$ = EQUIVALENT SECONDARY REACTANCE
$R_L$ = LINE RESISTANCE
$X_L$ = LINE REACTANCE
$V_S$ = SECONDARY VOLTAGE OF DISTRIBUTION TRANSFORMER
$V_R$ = LOAD RECEIVING VOLTAGE

**FIGURE 3.7** Voltage rise with PF < 1 and leading (a) Equivalent circuit of source (b) load PF < 1 and lagging (c) load PF = 1 and (d) load PF < 1 and leading

and 3.8) keep the utility factor less than one. Both the desire to provide reliable service and the Federal Energy Regulatory Commission (FERC) requirements dictate that reserve capacity be available in case of failure by some generating equipment.

The reserve requirements are for both spinning reserve and stand-by reserve. Spinning reserve refers to reserve power available from generators that are spinning but are not producing full rated output power. Spinning reserve is available almost immediately. Spinning reserve must be available to meet historically expected immediate load demand changes whether planned or unexpected. The

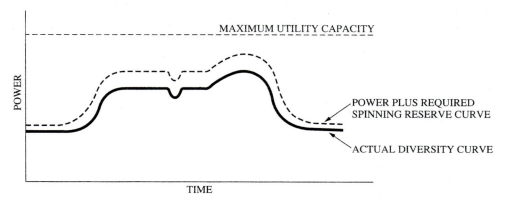

**FIGURE 3.8** Diversity diagram

spinning reserve requirement is typically 10% of the anticipated or actual load, whichever is greater, at a given time. The increased use of large natural gas fueled turbines that start rapidly has resulted in the spinning reserve being reduced to 70% in many areas. Records of electricity usage are kept so that the anticipated power requirement for a given time is fairly accurately known. A stand-by reserve equal to the largest single unit in the system is required. The stand-by reserve requirement does not include the spinning reserve. The system is defined for standby reserve as all solidly interconnected utilities so interconnected area utility companies can can share stand-by reserve. The interconnection of electric utility systems provides more reliable service to customers, and saves the duplication of investment on stand-by reserve capability by interconnected utilities.

Service factor refers to the amount of time a load can be fed by the system. A service factor of 100% means the system feeding a particular load (or set of loads) was never out of service. Lightning storms, other weather events, and scheduled maintenance cause some portion of the system to be out of service for some period of time, thus, some portion of the system will experience a service factor of less than 100%.

## 3.4

## DISTRIBUTION SYSTEM LAYOUT

Recall that the transmission line system carries bulk power at high voltages to the general areas of usage. Subtransmission lines carry large amounts of power from the bulk power substations to the immediate area of use at intermediate voltages, typically 138 or 69 kV. The distribution system carries electrical power from the distribution substation to the individual customer at voltages that range between 34.5 kV and about 4.2 kV. The distribution layout refers to the physical arrangement of the distribution lines.

### 3.4.1 Major Distribution Layout Classifications

A *radial* subtransmission and distribution layout is shown in Figure 3.9. The distribution lines extend from the substation to the last load with service drops to customers along the way. Often the system is shown with distribution lines extending from a distribution substation like spokes on a wheel, thus the name. The major advantages of the radial layout are that it is simpler and more economical to install than other types of layouts. The major disadvantage is that any problem usually leaves a number of customers out of service until the problem is solved. In fact, radial layouts are never used in subtransmission systems. A modified radial subtransmission layout is used in which two parallel radial subtransmission lines have a provision for switching the load to the good line in the event of a failure of one of the lines.

The *loop* arrangement is shown in Figure 3.10. The loop connection is more expensive than the radial because it requires more equipment, but any point on the line has service from two directions. If one is out, the customer can be fed from the other direction. Switches must be placed periodically around the loop so that a malfunctioning section can be repaired without removing much of the line from service. The loop layout is very reliable and expensive.

A *combination* of loop and radial, shown in Figure 3.11, is often used to provide the most reliable service to critical customers, such as business and industrial, by a loop, and reasonably economical service to residential neighborhoods. The radial part of the system is arranged so that only a few residential customers will be out of service at one time for any foreseeable fault condition.

*Networks* are designed to provide very reliable service to areas with dense loads such as downtown and suburban business districts containing many multi-story buildings. The network consists of underground secondary lines connected (phase A to phase A, B to B, etc.) at corners with transformers feeding the network every one to two blocks. The network equipment is contained in underground vaults with access through man holes in streets and alleys. Figure 3.12a shows a segment, two blocks on a side, of a larger network. Although networks can be very large, *network sections* are seldom larger than four square blocks. The network is fed from two or more feeders and no two adjacent network transformers are fed by the same feeder. The idea is that if one feeder is down the network can be fed by the other feeders. The primary feeders are typically between 12.5 kV and 34.5 kV. The network secondary lines are at 208/120 V wye for older networks, and 480/277 V wye for newer networks.

The network vaults contain a network unit that consists of a primary switch, a transformer, and a network protector. A network unit one-line diagram is shown in Figure 3.12c. The primary switch is a three-position primary switch with positions to connect the transformer primary to the feeder, open the line between the primary and the feeder, or ground the feeder line for testing. The transformer is usually rated between 1000 and 3000 KVA. The network protector is usually an air circuit breaker with a fuse back up. The circuit breaker is operated by relays, which are electro-mechanical or electronic sensing systems that detect, via current and potential transformers, unacceptable electrical conditions on the lines they

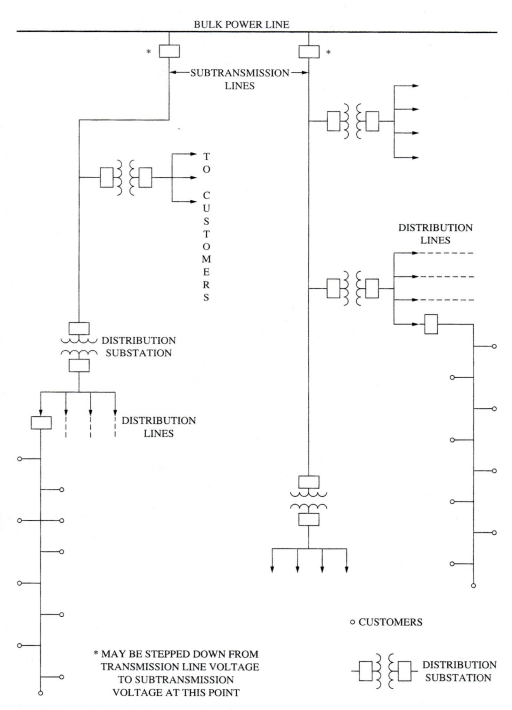

**FIGURE 3.9** Radial distribution and subtransmission

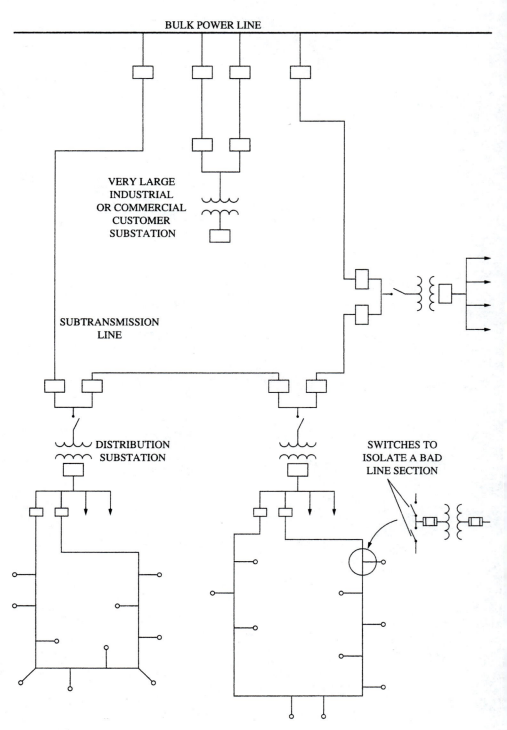

**FIGURE 3.10**   Loop distribution and subtransmission

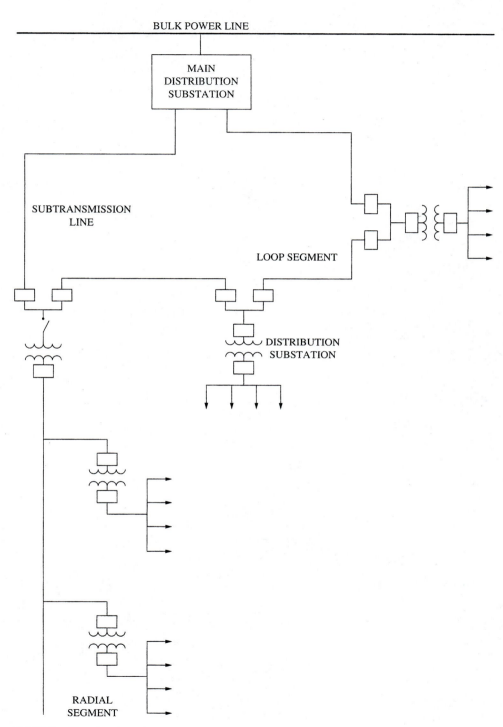

**FIGURE 3.11**  Combination

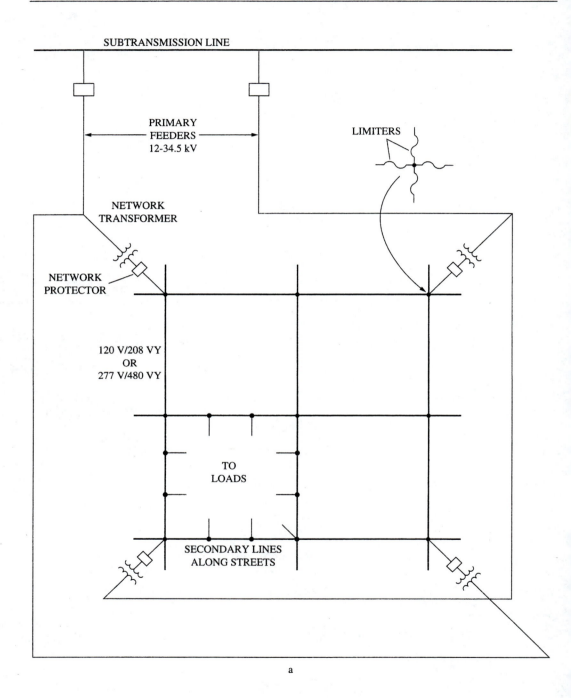

**FIGURE 3.12** Network for high density loads (a) Single network section for four square blocks (b) Network grid, single block connections (note feeder alternation in each direction) and (c) Network unit

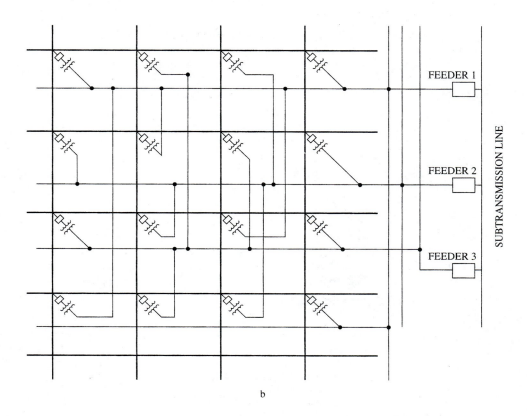

b

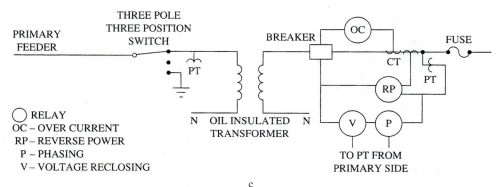

RELAY
OC – OVER CURRENT
RP – REVERSE POWER
P – PHASING
V – VOLTAGE RECLOSING

c

monitor. Relays provide trip signals to the circuit breakers when they detect fault conditions. Some relays can provide reclose signals to breakers when the fault conditions are removed. Relays are discussed in more detail in later chapters. The network relays monitor for secondary overcurrent, reverse power flow through the transformer, and primary voltage loss. They will open the breaker under any of these conditions. The relays also monitor the primary voltage for restoration at the proper phase after they open, and reclose when those conditions are met unless the circuit breaker has been prevented from reclosing manually.

A network with three feeders and a network unit on each corner is shown in Figure 3.12b. Note the alternation of feeders to the network units.

The limiters indicated in Figure 3.12a are short sections of copper connected in series with each phase at each corner with a transformer. The copper sections contain a constricted section designed to burn open when supplying substantial fault current so that the insulation on the secondary conductor will not be damaged between the network unit and the fault.

*Spot Networks* are designed to provide 480/277 V service to larger buildings within 208/120 V networks. The network units are similar but are located in one place. Two or more feeders are used, as in networks, but the loads are served by a bus (a set of high capacity lines) instead of a grid.

*Individual feeders* are often used for critical loads. Many very large buildings in high density load areas, hospitals, as well as large industrial customers will have their own high voltage service. These large customers are usually connected at around 12.5 kV. Figure 3.10 illustrates loop service to a very large industrial or commercial customer.

## 3.4.2   Overhead and Underground

Overhead distribution, such as the line shown in Figure 3.13, is used because of the low initial cost and good reliability. The equipment for overhead distribution costs less than the equipment required for underground distribution, and it is less costly to maintain. Overhead distribution lines are more subject to storm, lightning, and wind damage than underground lines, but they are more easily and cheaply repaired when damage occurs.

Underground lines have been readily accepted because they are not as vulnerable to weather conditions, and they are out of sight so the neighborhood sky line is less cluttered. Underground distribution lines require waterproof insulation of high quality and cost. Additionally, underground lines are higher in electrical loss than overhead lines. The price of equipment for underground distribution is decreasing as more underground service is installed, but it is unlikely that it will become as inexpensive as equipment for overhead distribution. The distribution lines to a neighborhood are overhead, and the laterals (the lines that feed the padmount transformers for the houses) are underground, as are the lines that provide service to the individual houses. The utility company will charge the developer the difference in cost between overhead and underground service, which the developer then includes in the price of the property. A padmount transformer for underground distribution is shown in Figure 3.14.

**FIGURE 3.13** Overhead distribution line (Courtesy of Gulf States Utilities Co.).

**FIGURE 3.14**   Pad mount transformer for underground distribution Primary connections (left) and secondary connections (right) shown (Courtesy of Houston Lighting & Power).

### 3.4.3 Line Losses

Line losses are discussed in greater detail in the transmission section of this book. At this point please note that the primary losses on the relatively short distribution lines are the resistive $I^2R$ losses, and the voltage drop across the line inductive reactance. Capacitive effects are usually negligible on distribution lines because they are not very long. The resistance increases with length, as does inductive reactance. The magnetic flux cancellation between lines increases as their spacing decreases, so close spacing of lines decreases the voltage drop across the line reactance. (We will calculate the line inductance and capacitance in the transmission section.) Underground lines are spaced more closely than overhead lines, and their insulation has a higher dielectric constant and loss than air. Underground distribution has higher losses than overhead distribution, because of increased dielectric losses.

## 3.5

### VOLTAGE LEVELS

Distribution voltage levels have standardized at a nominal 12 kV with variations from 11.6 to 13.2 kV depending on the utility's preference and history, with the wye phase voltages ranging from 6.7 to 7.6 kV. The most common distribution voltage is 12.5/7.2 kV. Early in the development of electrical distribution 2300 V delta systems were common, but it was quickly found that a 4160/2300 V wye system provided a higher distribution voltage (with lower $I^2R$ losses) with little change in equipment.

Electric utilities are now in the process of increasing the distribution voltages to 34.5/19.9kV. The change will provide more efficient distribution of electricity with lower losses. Better, less expensive insulating devices and materials have made possible the switch to higher voltages for distribution.

## 3.6

### SUMMARY

The loads that electrical power utilities service are classified by customer type. Residential customers comprise the most numerous group, about 80%, commercial customers the next most numerous, about 15%, and industrial customers comprise about 5% of the total number of customers. The industrial customers typically use about 30% of the electrical power supplied. Often commercial and industrial customers correct power factor within their facilities to reduce the demand penalty that many electrical power providers charge for providing VAR capacity in excess of the power used.

Electrical power must be distributed to customers that a utility serves. The major distribution substations layouts are:

1. Radial, in which all users are fed by lines connected directly to a central supply. It is an inexpensive layout to build.
2. Loop, in which any user within the system is fed from two directions or sources. Loop systems are very reliable but more expensive.
3. Combination, in which critical users are supplied by a loop and noncritical users by radial feeds. It is a compromise between reliability and economy.

4. Networks, in which high density users such as business districts and down-town areas are fed by a grid of interconnected lines with well protected multiple service points. Networks are reliable but expensive.
5. Spot networks, in which a small high density load areas are fed by a small networks, with similar protection capabilities as larger networks, which have two or more feeds.
6. Individual feeders are used for large customers who then distribute the power to their loads from the feeder. The distribution layouts used by an electrical power utility are similar to those used by large industrial and multi-building commercial customers to distribute electricity from their service to the various sections of their businesses.

Overhead distribution is less inexpensive and lower in loss than underground distribution. However, esthetic and reliability considerations make underground distribution attractive in many areas.

## 3.7

## QUESTIONS

1. How are loads classified by electric utilities?
2. Define power factor correction.
3. What situation has made the improvement of power factor to 0.95 or better a trend?
4. Why must the power factor of a system not become leading?
5. A circuit with a 120 V, 60 Hz, single-phase source has a 7.5 hp capacitor start motor running at full load. The motor full load efficiency is 78%, and the power factor is 0.75. Calculate the value of capacitor required to correct the full load motor power factor to unity.
6. A 277/480 V three-phase wye system has the following load:
   M1—Induction motor, 50 hp, $\eta$ = 0.91, PF = 0.89 at full load.
   M2—Induction motor, 25 hp, $\eta$ = 0.9, PF = 0.9 at full load.
   R—Heater load, PF = 1.
   M3—A synchronous motor to drive a second 25 hp load.
   Calculate the leading kVARs the synchronous motor must provide to correct the system power factor to unity. The synchronous motor will run at full load, 25 hp, with $\eta$ = 0.94.
7. Calculate the kVA and kVAR of the system of problem 6 if M3 is an induction motor of the same type as M2 instead of a synchronous motor.
8. Calculate the difference in electricity cost between the systems of problems 6 and 7 if they run 16 hours a day for 10 years and electricity costs eight cents per kVA hour (normally called kW hour).
9. If the synchronous motor of problem 6 costs $500 more than the induction motor of problem 7, how long will it take to pay for the difference in cost if the system of problem 6 is used instead of the system of problem 7?

10. List the major advantage and disadvantage of each of the following distribution layouts: radial, loop, and combination.
11. State the primary purpose of the network distribution layout.
12. State the function of spot networks.
13. Why must network protectors sense reverse power flow through the transformer?
14. List two advantages and two disadvantages for each: overhead and underground distribution.
15. State the development that is chiefly responsible for higher distribution voltages.

# CHAPTER 4

# Distribution Transformers

## 4.1

### NEED FOR TRANSFORMERS

Transformers, as you know, are electrical machines that use magnetic induction to change alternating current (ac) voltage levels. Without high transmission and distribution voltages the power losses and voltage drops associated with line resistance would make electrical power transfer very inefficient. Currently the highest practical generating voltage is around 25 kV, so transformers are needed to step up voltage for economical transmission, and step down voltage to levels that are safe for the customer to use. Transformers are also a key element in the equipment used for high voltage direct current (dc) transmission. The transformer is the machine that made ac transmission and distribution the standard early in the history of the electrical power industry.

Although the student is already familiar with basic transformer theory, this chapter briefly reviews some basics, and then discusses topics of interest relating to transformers for distribution.

### 4.1.1 Description

Transformers, whose schematic and one line symbols are shown in Figures 4.1a and b, are essentially two sets of windings of insulated wire called magnet wire (transformer wire in larger sizes) wound around a high permeability silicon steel

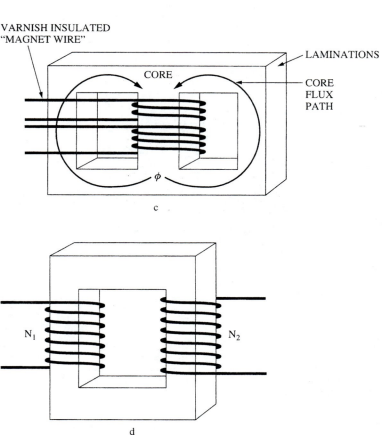

**FIGURE 4.1** Transformer (a) Schematic symbol (b) one-line symbol (c) shell constructed single-phase transformer and (d) core constructed single-phase transformer

core, assembled from thin laminations, as shown in Figure 4.1c. The wire is insulated with a thin coat of high quality plastic varnish, and the core laminations are made of silicon steel with good magnetic properties (low loss, high $\mu$) that are between 0.009 and 0.014 inches thick for 60 Hz. The transformers are usually constructed in shell or core form. Figures 4.1c and d show shell and core constructed single-phase transformers respectively. The shell type has less magnetic flux leakage than the core type but requires more core material. Shell constructed three-phase transformers are less frequently used because of their added cost, and the good performance of core constructed transformers. Figure 4.2 shows three-phase core and shell constructed transformers. The core laminations are supplied in two forms, flat and ribbon. The flat laminations are stamped out of thin sheets, as shown in Figure 4.3a, that are stacked like sheets of paper to form the core with the laminations cut to overlap at joints. Laminations are also supplied in ribbons that are rolled out and are put together like the layers of an onion skin as shown in Figure 4.3b (again the ends overlap). All laminations have a very thin layer of insulating material on their surface so that they are not in electrical contact with other laminations.

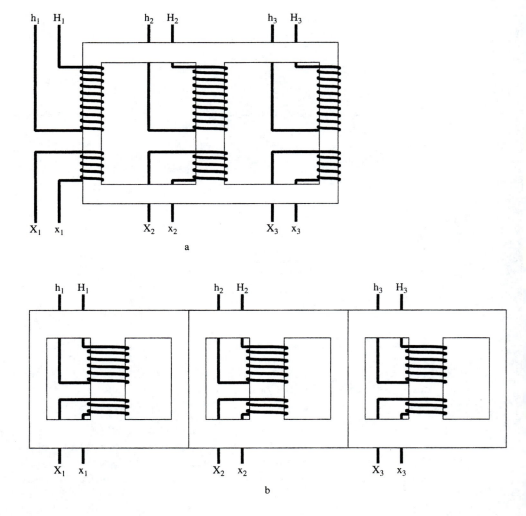

**FIGURE 4.2** Three-phase transformer (a) core-constructed transformer and (b) shell-constructed transformer

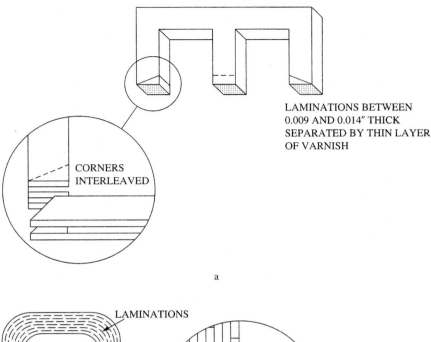

LAMINATIONS BETWEEN
0.009 AND 0.014″ THICK
SEPARATED BY THIN LAYER
OF VARNISH

CORNERS
INTERLEAVED

a

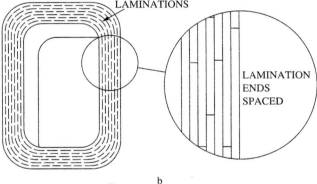

LAMINATIONS

LAMINATION
ENDS
SPACED

b

**FIGURE 4.3** Core laminations (a) Stacked laminations and (b) ribbon, also called form core and wrap core laminations

The windings are helical, as in Figure 4.1c and d, with insulation between the core and windings and each layer of windings, for low voltage transformers. Higher voltage transformers are wound with either layer or disk type windings for greater separation (Figure 4.4), with disk favored for very high voltages. The entire core and winding is placed in a support structure and the winding leads brought out to terminations. A sketch of an oil immersed transformer is shown in Figure 4.20 later in the chapter.

Transformers are classified by their function such as unit, transformers used in generating stations to step up the generated voltage; power transformers, used in

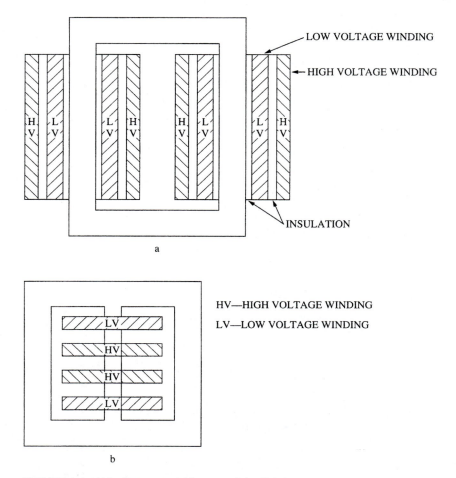

**FIGURE 4.4**   Winding types (a) layer and (b) disk

substations and large customer service; and distribution, generally the last transformer between the customer and the power system. The lines of classification by
use are often fuzzy since the same transformer may be suitable for more than one
job. For example, the distribution transformer for a very large industrial customer
may be identical to a substation power transformer. Transformers are also classified by cooling. Dry types are air cooled, primarily by convection. Oil immersed
transformers, in which the windings and core are immersed in oil, are both cooled
and helped in insulation by the oil. Almost all utility power and distribution transformers are oil. Dry types are used primarily where minimum cost is a factor and
the transformer is supplied by the customer such as in apartment house and
building distribution systems. Because the cooling of dry transformers is by convection they are very intolerant of overloads. Dry types must be in an enclosure
for safety. Figure 4.5 is a photograph of an oil immersed transformer.

**FIGURE 4.5**  Photograph of a power transformer (Courtesy of Houston Lighting & Power).

Oil immersed transformers are inside tanks of steel. Great effort is expended to assure the oil in the tanks remains free of water and impurities to ensure the insulating property of the oil, and if a transformer is opened for maintenance the oil is tested for impurities, and cleaned or replaced if necessary. The oil is in intimate contact with the transformer core and coil and removes heat from them. The oil circulates by convection and loses its heat to the environment through the steel transformer tank. External fins through which the oil can flow improves the cooling, as does forcing air across the cooling fins with fans (called forced air, FA). Pumping the oil through the fins (called forced oil, FO) improves it more and doing both even more (FOA). Each cooling improvement increases the capacity of the transformer. Power transformer ratings often follow the following pattern: MVA-25/33/42, which means the capacity is 25 MVA with convection circulated oil (OA), 33 MVA with forced air added (FA), and 42 MVA with both forced oil and air (FOA). The rating ratio is 1/1.33/1.67 for OA/FA/FOA. Many transformers are rated for only one type of cooling, such as FOA and all systems are expected to be used while it operates. Very high power transformers may have heat exchangers in which heat from the oil is transferred to water and the water is then cooled. Oil transformers are much more tolerant of overloads than dry types. Overloads that result in excessive heating of either type shortens the transformer life by degrading the insulation (and oil in oil types).

Dry type transformers are available in voltages to 15 kV and power ratings to around 150 kVA, and oil immersed transformers are available in sizes over 1000 MVA and 1000 kV. Both are available in single and three-phase systems.

### 4.1.2 Ideal Relationships and Dot Convention

The high voltage connections of a transformer are labeled H, and the low voltage connections are labeled X, as shown in Figure 4.6a and b. Referring to Figure 4.6c the primary side turns are $N_1$ and the secondary side turns are $N_2$. The turns ratio is defined as

$$a = N_1/N_2 \qquad\qquad (4.1)$$

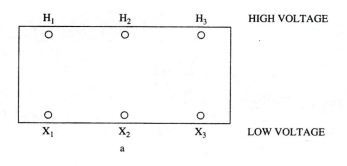

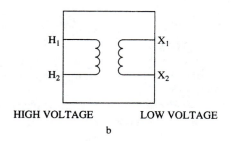

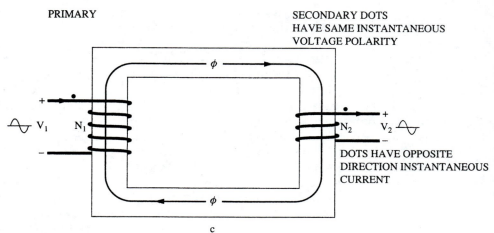

**FIGURE 4.6** Nomenclature conventions (a) single phase (b) three phase and (c) dot convention

From Faraday's law we know that the voltage induced on a conductor in a magnetic field is

$$e = \frac{\Delta \phi}{\Delta t} = \frac{d\phi}{dt} \tag{4.2}$$

where

$e$ = instantaneous conductor voltage

$\phi$ = magnetic flux lines linking the conductor

$\Delta$ = change in time, labeled $dt$ as $\Delta t$ becomes small

$d\phi/dt$ = time rate of change of flux linking the conductor

If the conductor is a loop then it is called a turn, and the conductor voltage is the number of turns, N, times $(d\phi/dt)$. If the same $d\phi/dt$ links two sets of turns then the ratio of the voltage induced on the two sets of turns is

$$\frac{V_1}{V_2} = \frac{N_1 \, (d\phi/dt)}{N_2 \, (d\phi/dt)} = \frac{N_1}{N_2}$$

The flux linking the two sets of turns is very nearly equal when the two windings are on the same core because the high permeability of the core confines the magnetic flux to the vicinity of the windings. By the conservation of energy principle the VA out of the transformer cannot exceed the VA in. If the transformer losses are assumed to be zero,

$$VA_{out} = VA_{in}$$

$$I_2 V_2 = I_1 V_1$$

$$\frac{I_2}{I_1} = \frac{V_1}{V_2} = a \tag{4.3}$$

Which is the inverse of the voltage ratio.

The agreed upon convention is for the dots on a transformer to indicate terminals with the same instantaneous polarity. If in Figure 4.6 the dot on the primary is at the peak of the positive half cycle, the secondary terminal is at the same voltage. The current is flowing into the primary terminal because it is a voltage drop in the primary circuit. The secondary of the transformer is a source in the secondary circuit so it must supply current, thus current is flowing out of the dotted secondary terminal, as shown in Figure 4.6.

### 4.1.3 Losses in Transformers

There are three sources of loss in a transformer: copper, hysteresis, and eddy current. They use real power and reduce transformer efficiency.

Copper losses are the power losses caused by current flowing through the winding resistance, $I^2R$ losses. Skin effect causes the ac resistance of the transformer windings to increase with frequency, so copper losses increase with frequency if the winding current remains constant. Proper sizing of the winding wire and proper cooling for the transformer minimize copper losses.

Hysteresis losses are the result of the energy needed to magnetize the core first in one direction and then the other as the applied ac voltage reverses in polarity. The magnetic domains must form in one direction, and then in the other. Hysteresis losses increase with frequency because the domains must be reversed more frequently. These losses are minimized by the proper choice of magnetic material. A new magnetic material has been developed that is amorphous in structure instead of having the small crystalline grain structure of most current magnetic materials. This means the magnetic domains are almost atomic in size and require very little energy to reverse. The amorphous magnetic materials have very low hysteresis losses, but currently can be manufactured only in ribbon form. For this reason, and their relatively high cost, they are not currently in wide use.

Eddy current is current induced in metal near a magnetic field. If the field surrounds the metal object, the induced current will be a loop at a right angle to the flux. The core of a transformer is ideally positioned with respect to the windings for eddy current to be induced in it. The laminations break the core into many small thin pieces of metal, thereby reducing the voltage that is induced to drive eddy currents in the core (the conductor length is reduced within the flux linkages), and increasing the resistance through which eddy currents must flow because the cross-sectional area of each lamination is small. Additionally the silicon steel is higher in resistance than non-silicon steel, which also helps to reduce eddy current. Any reduction in eddy current reduces the losses by the square of the reduction. These losses can be less than 0.5% of the total power transferred by a transformer.

## 4.1.4 Transformer Efficiency

Transformers are among the most efficient machines ever invented. A transformer that is properly designed, built, and cooled can be more than 99% efficient. The following efficiency values are average measured values for three-phase power transformers operating at full load with a unity load power factor: 2000 kVA—15 kV, 98.97%; 34.5 kV, 98.89%; 69 kV, 98.83%; 138 kV, 98.57%; 10,000 kVA—15 kV, 99.23%; 34.5 kV, 99.27%; 69 kV, 99.17%; and 138 kV, 99.12%. Maximum efficiency occurs when the copper loss is equal to the core loss. If amorphous magnetic metals can be made suitable for large power transformers efficiencies may someday routinely be greater than 99.5%.

## 4.1.5 BIL (Basic Insulation Level)

A transformer BIL is the peak transient voltage level that the transformer can withstand for a specified time. The insulation class of a transformer is the maximum rms working voltage of the transformer. The BIL is between 5 and 30 times the insulation class. The lower ratio is for higher voltage transformers, with only

1.2 kV and 5 kV insulation class transformers being tested to a BIL of more than 8.6 times their insulation class. The high voltage withstanding capability is necessary because of lightning and switching transients. Lightning strikes impress very high voltage transients of short duration on transmission and distribution lines that propagate down the lines to the transformers. If the BIL levels were lower the reliability of the electrical power system would be lower.

The transient test waveforms for BIL tests are shown in Figure 4.7a. The full wave, rising to its crest voltage in 1.5 $\mu$s and decaying in 40 $\mu$s, is modeled after a "typical" lightning strike. The waveform of Figure 4.7a is called the full wave, and the waveform of Figure 4.7b is called the chopped wave. It has a crest about 15% higher than the full wave and is interrupted by a spark-gap breakdown between one and three us after the wave begins. The BIL transient test consists of one reduced full wave (full wave of 50 to 70% full wave crest voltage), two chopped waves, and one full wave. The full wave crest voltage is specified as 95 kV for a

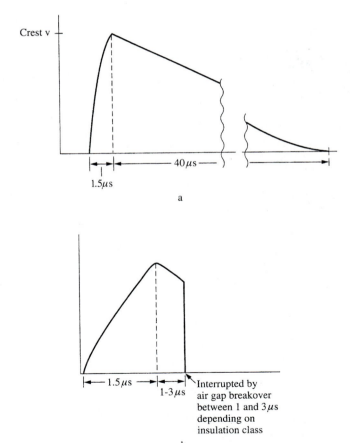

**FIGURE 4.7** Basic insulation level (BIL) test waveforms (a) full wave and (b) chopped wave

15 kV oil transformer under 500 kVA, and 110 kV for one over 500 kVA. A 69 kV oil transformer would be tested to a full wave crest voltage of 350 kV regardless of its kVA rating. A 1.2 kV transformer is tested to a crest voltage of 30 kV. Test specifications are established for every standard insulation class.

Other insulation tests are performed. A low frequency *applied potential* test is run that consists of applying a low frequency voltage between each winding and ground for 1 minute. Example values used for this test are (insulation class, test voltage): 1.2 kV, 10 kVrms; 15 kV, 34 kVrms; 34.5 kV, 70 kVrms; and 69 kV, 140 kVrms. An *induced potential* test is one in which twice the rated voltage is applied at 120 Hz (to reduce core excitation) to one of the windings for one minute.

Dry type transformers are tested to lower BIL crest voltages and applied potential voltages than equivalently rated oil transformers. The oil improves the ability of the transformer to withstand high voltage. For example, a 1.2 kV dry type is tested to a low frequency applied potential of 4 kV.

## 4.2

## TRANSFORMER EQUIVALENT CIRCUIT

In this section we will develop an equivalent circuit for transformers. We will also consider the excitation current and leakage flux effect on the ideal transformer in the equivalent circuit.

### 4.2.1 Excitation Current

Let $\mathcal{F}$ be the magnetomotive force (MMF) in ampere turns

$$\mathcal{F} = NI \tag{4.4}$$

The mutual flux is the result of both the primary and secondary MMFs, so

$$\mathcal{F}_T = \mathcal{F}_P + \mathcal{F}_S$$
$$= N_P I_P + N_S(-I_S)$$
$$= N_P I_P - N_S I_S$$

solving for $I_P$ we find

$$I_P = \frac{\mathcal{F}}{N_P} + \frac{N_S}{N_P} I_S$$

That portion of the primary current represented by $\mathcal{F}/N_p$ is called the excitation current, that portion of the primary current needed to establish the magnetic flux and supply power for the core losses. Thus we can write the equation for primary current as

$$I_P = I_{ex} + I_S/a \tag{4.5}$$

At no load the primary current is all excitation current, except for the $I_{ex}^2 R_p$ loss (resistive power dissipated in the transformer windings is called copper loss) in the primary winding. At loads over $^1/_2$ the rated load $I_{ex} << I_p$ Because the excitation current supplies both magnetization current and core loss current we can express the excitation current as

$$I_{ex} = I_m + I_{h+e} \quad (4.6)$$

where

$I_m$ = magnetization current component of the excitation current which lags the primary voltage by 90°.

$I_{h+e}$ = current to supply core losses that is in phase with the primary voltage.

The core losses consist of hysteresis and eddy current loss.

## 4.2.2 Leakage Inductance

Not all the flux generated by the primary winding links the secondary winding via the core. Some flux leaks out of the core coupling only the primary or only the secondary winding. The actual pattern of the leakage flux is quite complex and depends on the physical configuration of the core. The leakage flux is shown as $\Phi_{1P}$ and $\Phi_{1S}$ in Figure 4.8. If $\Phi$ is the mutual flux linking both the primary and secondary windings

$$\Phi_{1P} = \Phi_P - \Phi \quad\quad\quad (4.7)$$

$$\Phi_{1S} = \Phi_S - \Phi \quad\quad\quad (4.8)$$

In general $\Phi_1 << \Phi$, consisting of only 1 to 7% of the total flux. High voltage transformers have more leakage flux than lower voltage transformers because their windings must be further apart to provide proper insulation for the high voltage.

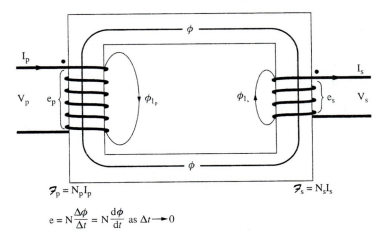

**FIGURE 4.8**   Transformer flux paths

The medium of the leakage flux, which has leaked out of the core, is air so the leakage flux paths do not saturate. Leakage flux is linear with current. The leakage flux changes at the same rate as the mutual flux so it induces a voltage in the winding with which it is associated. Let e be the induced voltage, then

$$e_{1P} = N_P \frac{d\Phi_{1P}}{dt}$$

$$e_{1S} = N_S \frac{d\Phi_{1S}}{dt}$$

These induced voltages are shown in Figure 4.9a. Recall that flux expressed in terms of permeance ($\mathcal{P}$) is

$$\Phi = \mathcal{P} NI$$

We now substitute this expression into the preceding instantaneous voltage equations and move the constants out of the portion that changes with time to obtain

$$e = N^2 \mathcal{P}(dI/dt)$$

but $L = N^2\mathcal{P}$, so the expression becomes

$$e_{1P} = L_{1P} \frac{dI_P}{dt} \tag{4.9}$$

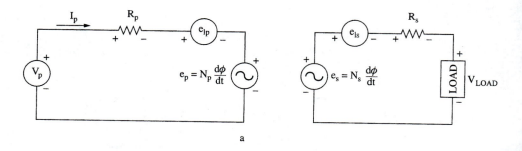

a

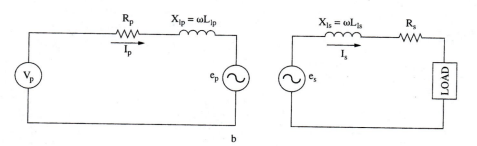

b

**FIGURE 4.9**   Transformer equivalent circuit with leakage flux (a) as voltage drop and (b) as inductive reactance

$$e_{1S} = L_{1S} \frac{dI_S}{dt} \quad (4.10)$$

This means that the leakage flux cutting the windings of the primary and secondary must generate a voltage across them, and that the voltage is proportional to the change of current per unit time through the windings, which acts as leakage inductances because of the leakage flux. The reactance of the leakage inductance is included in the transformer equivalent circuit, as shown in Figure 4.9b.

The primary voltage induced on the secondary is reduced by two factors, the primary resistance and the primary leakage reactance. Similarly the voltage across the secondary terminals is reduced from the induced secondary voltage by the secondary winding resistance and leakage reactance. The following equations summarize this.

$$\begin{aligned} e_p &= V_p - I_p R_p - e_{1P} \\ &= V_p - I_p R_p - L_{1P}(dI_p/dt) \end{aligned} \quad (4.11)$$

and similarly

$$V_S = e_S - I_S R_S - L_{1S}(dI_S/dt) \quad (4.12)$$

## 4.2.3 Equivalent Circuit

Only two more items remain to complete the equivalent circuit. The magnetizing current can be represented by an inductive reactance ($X_m$) across the primary because the magnetizing current lags the applied voltage by 90° and is drawn from the primary. The current drawn from the primary to supply the hysteresis and eddy current losses is in phase with the primary voltage (these are real power losses) and can be modeled with a resistance ($R_c$) across the primary. These are shown in Figure 4.10a.

In use in a reactance diagram all of the elements will be referred to either the primary or the secondary side of the transformer to simplify calculations. Recall that any secondary impedance referred to the primary side $Z_{rp}$ is

$$Z_{rP} = \frac{V_P}{I_P} = \frac{V_S(N_P/N_S)}{I_S(N_S/N_P)} = Z_S(N_P/N_S)^2 = Z_S a^2$$

and conversely a primary impedance referred to the secondary is

$$Z_{rS} = Z_P/a^2$$

The equivalent circuit with all elements referred to the primary is shown in Figure 4.10b and with all elements referred to the secondary is shown in Figure 4.10c.

In most calculations such a complete equivalent circuit is not necessary so the circuit is simplified by omitting $R_c$ and $X_m$, and combining all of the series resistors into one equivalent and all of the series inductive reactances into one, as shown in Figures 4.10d and e. In Figure 4.10d, referred to the primary

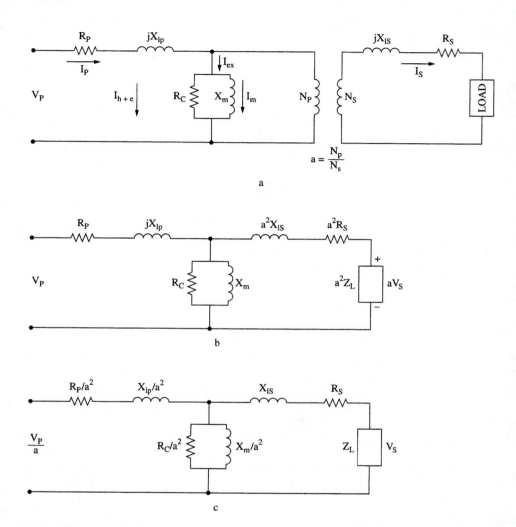

**FIGURE 4.10** Transformer equivalent circuits (a) Complete equivalent circuit (b) all elements referred to the primary (c) all elements referred to the secondary (d) approximate equivalent circuit referred to the primary and (e) approximate equivalent circuit referred to the secondary

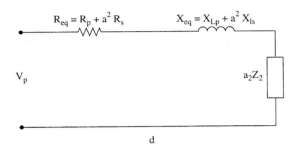

d

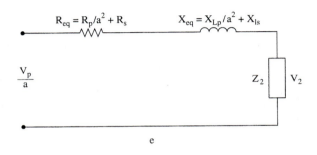

e

$$R_{eq} = R_P + a^2 R_S \qquad (4.13)$$

$$X_{eq} = X_P + a^2 X_S \qquad (4.14)$$

therefore

$$Z_{eq\,rP} = R_{eq\,rP} + X_{eq\,rP} \qquad (4.15)$$

In Figure 4.10e, referred to the secondary

$$R_{eq} = R_P/a^2 + R_S \qquad (4.16)$$

$$X_{eq} = X_P/a^2 + X_S \qquad (4.17)$$

$$Z_{eq\,rS} = R_{eq\,rS} + X_{eq\,rS} \qquad (4.18)$$

Typical values of $I_m$ range from 0.01 (per unit) pu to 0.05 pu, but can be as high as 0.1 pu in extremely high voltage transformers. $I_{h+e}$ is normally about 0.03 pu in a good transformer. Thus $X_m$ is around 20 to 100 pu, and $R_C$ around 33 pu. Transformer impedance runs between 0.01 pu to 0.05 pu in distribution transformers, but it can be as high as 0.15 pu in high voltage power transformers. These values are only approximate to give the reader a feel for the values expected. A good transformer may vary markedly from these values depending on the transformer design.

### 4.2.4 Examples

**Example 4.1:**

An ideal transformer is used to feed a single-phase 240 V load of 5 kVA with a PF = 0.8 lagging. The primary voltage is 4160 V. Calculate a, $I_S$, and $I_P$.

**Solution:**

$$a = \frac{V_P}{V_S}$$

$$= \frac{4160 \text{ V}}{240 \text{ V}}$$

$$= 17.333$$

$$I_S = \frac{\text{kVA}}{\text{V}}$$

$$= \frac{5 \text{ kVA}}{240 \text{ V}}$$

$$= 20.83 \text{ A}$$

$$I_P = \frac{I_S}{a}$$

$$= \frac{20.83 \text{ A}}{17.333}$$

$$= 1.2 \text{ A}$$

**Example 4.2:**

The transformer of Example 4.1 has: $X_P = j46 \ \Omega$, $R_P = 3.7 \ \Omega$, $X_S = j0.15 \ \Omega$, and $R_S = 0.012 \ \Omega$. Reference the equivalent circuit of Figure 4.10e to the secondary and use it to calculate the actual voltage across the load.

**Solution:**

First calculate $Z_L$

$$Z_L = \frac{\text{rated V}}{I \text{ at rated V}}$$

$$= \frac{240 \text{ V}}{20.83 \text{ A}}$$

$$= 11.52 \ \Omega$$

In polar form it is

$Z_L = 11.52 \ \Omega \ \angle \arccos 0.8$

$\quad = 11.52 \ \Omega \ \angle 36.87°$

$\quad = 9.216 \ \Omega + j6.912 \ \Omega$

We now calculate all primary values referred to the secondary. These values are primed.

$X_P' = \dfrac{X_P}{a^2}$

$\quad = \dfrac{j46 \ \Omega}{(17.333)^2}$

$\quad = j0.153 \ \Omega$

$R_P' = \dfrac{R_P}{a^2}$

$\quad = \dfrac{3.7 \ \Omega}{(17.3333)^2}$

$\quad = 0.012 \ \Omega$

$V_S' = \dfrac{4160 \ V}{17.333}$

$\quad = 240 \ V$

The equivalent values are now

$X_{eq} = X_P' + X_S$

$\quad = j0.153 \ \Omega + j0.15 \ \Omega$

$\quad = j0.303 \ \Omega$

$R_{eq} = R_P' + R_S$

$\quad = 0.012 \ \Omega + 0.012 \ \Omega$

$\quad = 0.024 \ \Omega$

$Z_{eq} = 0.024 \ \Omega + j0.303 \ \Omega$

Now we calculate the total impedance

$Z_T = Z_{eq} + Z_L$

$\quad = 9.24 \ \Omega + j7.215 \ \Omega$

$\quad = 11.72 \ \Omega \ \angle 37.98°$

$$I = \frac{V_S'}{Z_T}$$

$$= \frac{240 \text{ V}}{11.72 \ \Omega}$$

$$= 20.47 \text{ A}$$

Finally

$$V_L = IZ_L$$

$$= 20.47 \text{ A} \times 11.52 \ \Omega$$

$$= 235.84 \text{ V}$$

In this case the more accurate equivalent circuit made little difference with respect to the ideal transformer, about 1.8%.

Actually $X_m$ and $R_C$ would have caused the primary current to be higher than $I_S/a$. Assuming $I_m = 0.05$ pu and $I_{h+e} = 0.01$ pu we will calculate the approximate difference.

$$\text{primary base } I = \frac{\text{rated kVA}}{\text{rated V}}$$

$$= \frac{5 \text{ kVA}}{4160 \text{ V}}$$

$$= 1.2 \text{ A}$$

$$I_{h+e} = 0.01 \ (1.2 \text{ A})$$

$$= 12 \text{ mA}$$

$$I_m = 0.05 \ (1.2 \text{ A})$$

$$= 60 \text{ mA}$$

$$I_{ex} = \sqrt{I_{h+e} + I_m}$$

$$= \sqrt{(12 \text{ma})^2 + (60 \text{ma})^2}$$

$$= 61 \text{ mA} \ \angle 78.7°$$

Therefore the primary current would have been 1.2 A $\angle 39.6°$ + 0.061 A $\angle 78.7° = 1.248$ A $\angle 41.37°$, an error of 4% and 1.77°.

## 4.3

## HARMONICS AND INRUSH CURRENT

Transformer core material is not linear and requires initial magnetization upon energization. Therefore transformers generate harmonics in operation, and draw more current when energized than in normal operation. These characteristics are discussed in this section.

## 4.3.1 Harmonics

The nonlinear nature of transformer core material is evident in the B-H curve in Figure 4.11, in which magnetic flux density ($B$), in flux lines per unit area, is plotted against magnetic field intensity ($H$), in ampere turns per unit length. One might ask why the transformer cannot be operated around the origin of the curve where the curve is almost linear, and far away from saturation where the curve is very nonlinear. The reason is that the amount of core material needed for a distribution transformer to provide a reasonable power transfer would be so much as to make the transformers prohibitively expensive. To make power transformers reasonable in cost the operation along the $B$-$H$ curve has to be close enough to saturation that nonlinearity is inevitable. Not only is the $B$-$H$ curve nonlinear, but it also exhibits hysteresis; that is, the position of a point on the curve depends on the direction of approach. The area inside the curve is a measure of the losses to hysteresis and eddy current in the core. Recall that the excitation current of the core provides the magnetizing current and the core loss current.

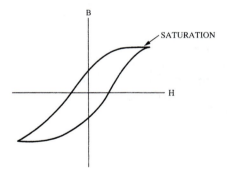

**FIGURE 4.11**   B-H curve of core material

### Voltage and Flux Relationship

Referring to the primary induced voltage recall that

$e_p = N_p(d\Phi/dt)$

Solving for $\Phi$ we find

$$d\Phi = (1/N_p)e_p dt$$

The flux can be found by summing $e_p$ over all of the small dts. The process of summing the dts, which are infinitesimally small $\Delta ts$, is called integration. An integral can be thought of as the area under a curve between two points. The symbol for integration is $\int$. Thus the total flux is the sum, or integral, of the area of each

$dt \times e_p$ within the $dt$. The turns are fixed so they can be multiplied by the integral after it is found. The flux is then

$$\Phi = \frac{1}{N} \int e_p dt$$

Let $e_p = (\sqrt{2})\, E_p \sin(\omega t + \alpha)$, then

$$\Phi = \frac{(\sqrt{2})E_p}{N_P} \int \sin(\omega t + \alpha)dt$$

From memory or from a table of integral solutions we find that

$$\sin(\omega t + \alpha) = -(1/\omega)\cos(\omega t + \alpha) + \text{constant}$$

therefore

$$\Phi = -\frac{(\sqrt{2})E_p}{\omega N_P}\cos(t + \alpha) + \Phi_c \tag{4.19}$$

where $\Phi_c$ is a constant that depends on the conditions at $t = 0$, a transient flux that dies away quickly. The maximum flux will occur when $\cos(\omega t + \alpha) = 1$, or at 90° from the voltage maximum

$$\Phi_{MAX} = \frac{(\sqrt{2})E_p}{\omega N_P} = \frac{V_P}{(\sqrt{2})\,\pi\, fN_P} \tag{4.20}$$

From equation 4.20 we see that the maximum flux looks like the primary voltage except that it lags the voltage by 90°, that it is directly proportional to the volts per turn on the primary, and that it reduces with frequency. The relationship between the voltage and the flux is shown in Figure 4.12. The secondary voltage will look like the flux from which it is induced.

A mathematician named Fourier showed that any nonsinusoidal waveform was made of harmonics (multiples of the fundamental frequency) of the waveform with varying amplitude, the fundamental, and a dc component if the waveform is not symmetrical around the time axis. His mathematical equations were subsequently

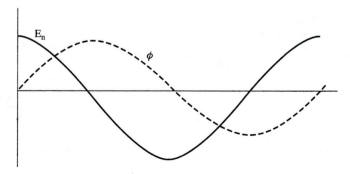

**FIGURE 4.12**   Magnetic flux and primary winding voltlage

proven correct by experimentation. The fundamental frequency amplitude, the harmonic amplitudes, and the dc component if there is one, is called the *Fourier series*. By calculating the Fourier series one can discover all of the harmonics that make up a waveform and their amplitude. The series is of the form, where f(t) is the equation of the waveform with time

$$f(t) = a_0 + \Sigma_{n=1}^{\infty}[a_n\cos(n\omega t) + b_n\sin(n\omega t)]$$

where

$a_0$ = dc amplitude of the waveform

$a_n$ = amplitude of the sine terms

$b_n$ = amplitude of the cosine terms

Many excellent books go into the details of how to calculate the amplitude of the terms in the Fourier series. It turns out that certain waveforms do not contain all possible harmonics. Square waves, for example, contain only odd harmonics of the fundamental. Sharply peaked, triangular shaped waves also contain only odd harmonics.

Consider Figure 4.13. In this figure the primary voltage and the resulting flux are plotted against the *B-H* curve. The magnetizing force is shown in ampere turns instead of ampere turns per unit length to make the drawing clearer (in other words the unit length is unity). The magnetizing current that is drawn by the primary to produce the flux is drawn below the $B-N_pI_p$ curve. Several points are

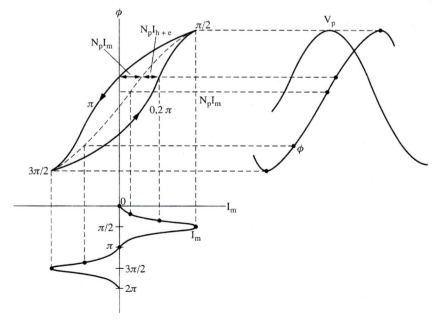

**FIGURE 4.13**  Magnetizing current

plotted on the curve. Notice that the resulting magnetizing current is sharply peaked. A Fourier analysis of the waveform reveals that it contains, in addition to the fundamental (about 90% of the energy), harmonic energy divided as follows: third harmonic (about 45%), fifth harmonic (about 15%), seventh (about 3%), and lesser amounts of higher harmonics. These values are typical at normal transformer load ranges. If $I_{h+e}$ were plotted from the outer perimeter of the $B-N_pI_p$ curve a flat topped sinusoidal wave as shown in Figure 4.14 would result. Little error results from considering $I_{h+e}$ sinusoidal and this is normally done. Figure 4.14 illustrates the excitation current waveform that results from its two components, magnetization and core loss currents.

In a three-phase system the fundamental waves are 120° out of phase so the current sums to zero at the neutral and the three-phase voltages sums to zero, as we noted in Chapter 2. Let the three-phase voltages of a system be

$E_1 = E \sin(\omega t)$

$E_2 = E \sin(\omega t + 120°)$

$E_3 = E \sin(\omega t - 120°)$

then the third harmonic voltages (primed values) will be

$E_1' = E \sin(3\omega t)$

$E_2' = E \sin(3\omega t + 360°)$

$E_3' = E \sin(3\omega t - 360°)$

which are all in phase, and do not cancel. However, as long as the third harmonic currents can flow the third harmonic voltages will not be a problem because in a wye system with the neutral solidly connected to the source system neutral the line to line third harmonic voltages will sum to zero, and in a delta (in which the third harmonic current can flow easily) the internal third harmonic voltage and voltage drop across $Z_{eq}$ will sum to zero in each phase. If the third harmonic current cannot flow the excitation current is forced to be sinusoidal and the core

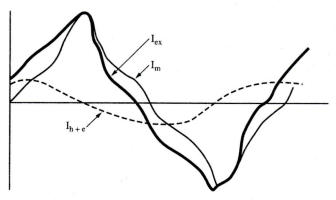

**FIGURE 4.14**   Excitation current, $I_{ex} = I_m + I_{h+e}$

nonlinear characteristics will be exhibited in the secondary voltage; it will become very nonsinusoidal. The secondary voltage will then have a waveshape similar to the normal excitation current waveshape, and will have a peak voltage nearly twice the normal fundamental voltage. A wye-wye transformer without a grounded neutral has no path for third harmonic current, thus it has a large third harmonic voltage on the secondary.

A transformer with a delta winding has a third harmonic current path and no third harmonic voltage on the line terminals. Because of third harmonic flux cancellation in the core, three leg core constructed three-phase transformers have only a little third harmonic (about 5%) even when connected wye-wye. Wye-wye transformers to be used in applications where third harmonic content is not desirable (this includes most) have a delta connected tertiary sized about one third of the transformer VA rating. The tertiary allows the third harmonic current to flow, and is often used to provide auxiliary power for other substation equipment.

### 4.3.2 Inrush Current

When a transformer is first energized the initial current is usually much higher (up to 16 times) than the full load current. The initial current surge, which lasts only a few cycles, is the initial magnetization current of the transformer, and is called the inrush current. The inrush current is normally high enough to saturate the core, which is the reason the inrush current is so high. Shell constructed transformers normally have lower inrush current than core constructed transformers because they have more core material, and are harder to saturate. The inrush current of shell constructed transformers is approximately the same whether energized from the high or the low voltage side, but may be twice as high in a core constructed transformer when it is energized from the low side. The inrush current waveform is illustrated in Figure 4.15.

The constant of equation 4.19 is the mathematical expression for the initial magnetic flux required by the transformer. Before energization the only flux is the small amount of residual flux left from previous operation. Evaluating equation 4.19 at $t = 0$ we obtain

$$\Phi_R = -\frac{(\sqrt{2})E_p}{N_p \omega} \cos \alpha + \Phi_c$$

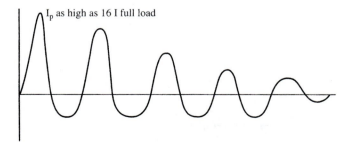

**FIGURE 4.15**   Inrush current waveform

where

$\Phi_R$ is the residual flux

$\Phi_C$ is the initial condition flux

$\alpha$ is the angle of $e_p$ at $t = 0$

We may now write

$$\Phi_c = \Phi_{MAX} \cos \alpha + \Phi_R$$

If at $t = 0$, $e_p = 0$, and $\alpha = 0$, then

$$\Phi_c = \Phi_{MAX}(1) + \Phi_R$$

causing the initial total flux to be from equation 4.19

$$\Phi = \Phi_{MAX} \cos(\omega t) + \Phi_{MAX} + \Phi_R$$
$$= 2\Phi_{MAX}$$

Thus the flux required at $t = 0$ is about twice the flux required for normal operation. This much flux saturates the core, causing the reactance of the transformer to be lower than normal. The abnormally low reactance causes the inrush current to be high. The initial flux level is higher than required for normal operation and decays in a few cycles. The abnormally high primary current causes the internal voltage drops across the primary winding reactance and resistance, the first holding the $e_p$ induced on the secondary down and the second dissipating the transient energy. The highest inrush current results when the transformer is energized as the line voltage passes through zero.

If the transformer is magnetized when the line voltage is maximum, $\alpha = 90°$, $\cos \alpha = 0$, then we find

$$\Phi_c = \Phi_R$$

which is very small and from equation 4.19 $\Phi = -\Phi_{MAX} + \Phi_R$, which does not saturate the transformer core, so the inrush current is almost zero, allowing $I_p = aI_s$.

It is difficult to time transformer energization to coincide with the peak of the energizing voltage, and the three phases peak 120° apart. Transformer protective devices are designed to recognize and ignore the inrush current since it is inevitable.

## 4.4

## THREE-PHASE TRANSFORMER CONNECTIONS

There are four major three-phase transformer connections for two winding transformers. They are wye-wye, delta-delta, wye-delta, and delta-wye. Additionally each configuration can be obtained from a three-phase transformer or three single-phase transformers. Figure 4.16 illustrates these four configurations, their line to phase relationships, and their turns ratio.

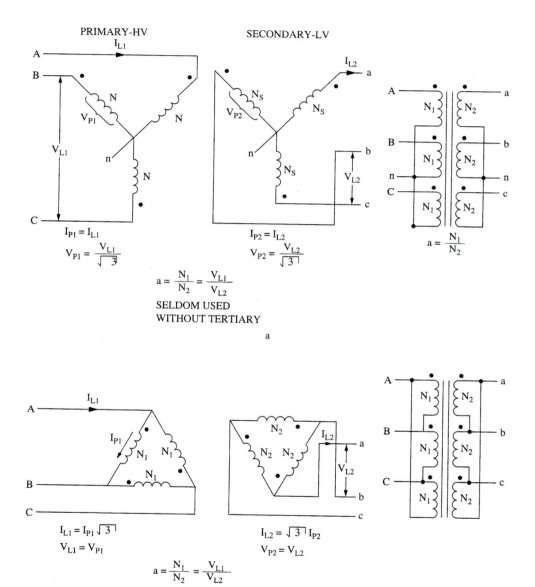

**FIGURE 4.16** Three-phase transformer connections include: (a) Y-Y, (b) Δ-Δ

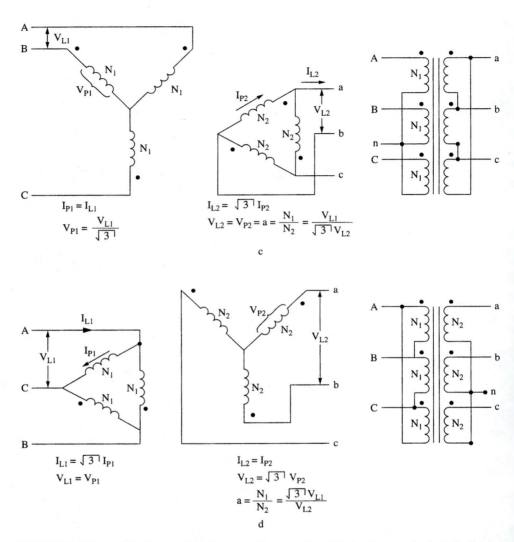

**FIGURE 4.16** (continued)  (c) Y-Δ, shown with secondary 30° lagging, and (d) Δ-Y, shown with secondary 30° lagging.

Three-phase transformers are usually less expensive than three single-phase transformers because less total core material is needed for the three-phase transformer and the packaging costs are similarly reduced. Additionally they take up less space, are lighter, require less on site external wiring for installation, and are a bit more efficient than three single-phase transformers. When the initial kVA required by an installation is substantially less than the subsequent kVA requirement the initial capital investment can be lowered substantially by using two single-phase transformers connected in an open delta. Recall that an open delta can supply only 57% of the kVA that three transformers of the size used for the open delta can supply without line voltage imbalance.

Figure 4.16a shows the wye-wye (Y-Y) connection. Recall that this connection has significant third harmonic content on the secondary lines (almost twice the normal peak line voltage) unless the neutral is solidly grounded, and the source neutral is solidly grounded, or a delta connected tertiary is built into the transformer. Even with solid neutral grounding the third harmonic current in the neutral can interfere with signal and telephone lines that run parallel to the neutral. The Y-Y configuration is seldom used without a delta connected tertiary with a capacity of about one third that of the transformer. Recall the tertiary is often used for auxiliary power at the substation. There is no phase shift between the primary and secondary of a Y-Y connected transformer.

The Δ-Δ connection has no harmonic problem and no phase shift from primary to secondary. The only disadvantage with respect to a wye connection is that the delta insulation class must be for the line to line instead of line to neutral voltage. When closing a delta secondary care must be taken to assure proper connection because a wrong connection causes a large fault current to flow (Figure 4.17). Before making the last delta connection measure the voltage across the open connection. If the voltage is less than the line voltage (third harmonics prevent the voltage being zero if the primary is wye connected) the connection is correct and the connection can be made after the power is removed. If the connection is incorrect the measured voltage will be larger than the line voltage and the winding must be reversed before it is connected.

Figures 4.16b and c show the Y-Δ and Δ-Y connections respectively. There is a 30° phase shift in both connections. United States industry convention is to connect the secondary so it lags the high voltage primary by 30°. Figure 4.16 shows the $V_2$ side lagging the $V_1$ side 30°. When possible the Y is connected to the high voltage side because the insulation requirements are lower (recall that Y phase voltage is $1/\sqrt{3}$ that of the line voltage). The Y may be necessary on the low voltage side because of the distribution system requirements as in 480/277 V and 208/120 V installations.

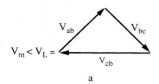

a

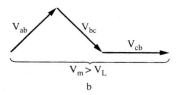

b

**FIGURE 4.17** Delta secondary connection (a) properly connected and (b) incorrectly connected

## 4.5

### NEED FOR TRANSFORMER PROTECTION

Transformers, especially power transformers, are expensive. The investment in a transformer needs to be protected. The transformer must be protected from over-current due to faults on its secondary circuit, and from over voltage, which is usu-ally caused by lightning. The protective devices are discussed in detail in a later chapter, but they are briefly introduced here.

Lightning protection is provided by lightning arresters. At one time these were spark gaps of controlled gap length, the voltage being set by the gap length. Bulk metal oxide semiconductors are now popular because they can absorb large amounts of power with a modest terminal voltage rise. Zinc oxide lightning arresters are the most popular now. The voltage at which the lightning arrester begins conducting, absorbing power, and preventing further voltage rise on the line is set at a voltage below the maximum insulating voltage of the transformer and above the maximum operating voltage of the transformer. Notice in Figure 4.18 that lightning arresters are used on both the primary and secondary side of larger power transformers because lightning can strike on either side. Only the primary side of pole top transformers for buildings will have lightning arresters.

Overcurrent protection is provided by circuit breakers and their associated pro-tective relays, and fuses. A fuse opens on overcurrent. Higher voltage fuses are expulsion type, which means that they contain a means of blowing out the arc caused by their operation to interrupt a fault current. Circuit breakers open elec-trical contacts when they receive a trip signal from one of their associated relays. The opening is done by driving the contacts apart with powerful springs (which are compressed with an air driven piston in large power breakers) or, on smaller circuit breakers, with springs compressed with a motor or a manual ratchet. The arc across the contacts is extinguished by oil in oil circuit breakers, pressurized air blown across the contacts in air breakers, and sulfur hexaflouride in $SF_6$ breakers. Circuit breakers are used on the secondary side of the transformer with a fuse back up on the primary side in case the breaker fails to operate (a secondary over-current will cause a primary overcurrent), or on both the primary and secondary sides of power transformers. The primary protection also protects the primary feeder if something should fail inside the transformer. The transformer in Figure 4.18 uses a secondary circuit breaker with a primary fuse for back up. Residential distribution transformers usually have a primary fuse for protection.

Current transformers and potential transformers, to be discussed more later, pro-vide line current and voltage information respectively for the relays. The relays are electromechanical or microprocessor controlled electronic systems that detect over-current, over voltage, and other fault conditions on the line. If fault conditions are detected the relays send a trip signal to the circuit breaker to open it. Some relays can send a reclose signal to properly equipped relays after sensing that a fault con-dition has been removed. Protective relaying equipment is very expensive and is usually used only on larger, more expensive substation and network transformers.

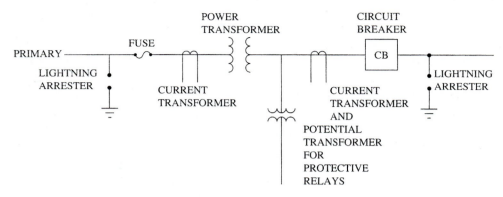

**FIGURE 4.18**   Power transformer one-line drawing with protection

4.6

## TYPES OF DISTRIBUTION TRANSFORMERS

Transformers used in distribution include: power transformers; autotransformers, which may also be power transformers; distribution transformers; and instrument transformers. We will now examine each type.

### 4.6.1 Power Transformers

Power transformers are normally oil immersed transformers used for substations and connection to large commercial and industrial customers. Smaller power transformers may be dry types used primarily for connection to commercial customers who purchase their own transformers. Figure 4.19 is a photograph of a power transformer for a distribution substation. Figure 4.20 is a diagram of the same general type of transformer. Note the high and low voltage bushings for connection, the cooling fins, and the location of the current transformers in the bushings (usually two per bushing).

Power transformer capacities vary over a wide range. A smaller distribution substation may have a power transformer rated at 1000 kVA, and a large one may have a power transformer rated at 30,000 kVA. Subtransmission substations will have even larger power transformers. Power transformers are made with capacities to over 1000 MVA. Impedances will range from 1 to 3% at around 2.5 kV at 3 kVA, around 4% at 15 kV and intermediate kVA ratings, to as high as 20% at 230 kV and high kVA ratings. Lower voltage transformers typically have lower impedances. Power transformers are designed for very high efficiency, and are routinely maintained. As noted in the last section, power transformers are well protected both for safety and economy. Almost all larger power transformers are three-phase as opposed to three single-phase units.

**FIGURE 4.19**   Photograph of power transformer (Courtesy of Houston Lighting and Power)

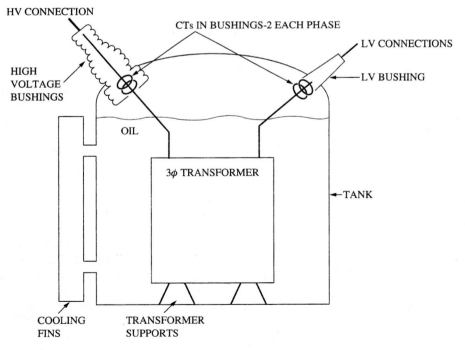

**FIGURE 4.20**   Power transformer diagram

## 4.6.2 Autotransformers

Autotransformers are one winding transformers that are often used in transmission and subtransmission substations. Figure 4.21a shows an autotransformer schematic symbol. Notice that the primary and secondary sides are not isolated from each other. This means that if the secondary neutral, which is common to the primary neutral, is opened for any reason the full primary voltage could appear on the secondary side, as shown in Figure 4.21b, with disastrous results. Unsafe conditions would result for people as well as damage to the low voltage side protective devices and equipment. For this reason autotransformers are never used as the final transformer in distribution substations.

From Figure 4.21a we see that the primary and secondary kVA must be equal in a lossless transformer, so if the turns ratio is

$$a = \frac{N_1 + N_2}{N_1}$$

and the same $d\Phi/dt$ is cutting both windings

$$V_2 = V_1/a$$

$$I_2 = I_1 a$$

In other words the ideal transformer relationships are approximately true for autotransformers as well as isolation (conventional) transformers.

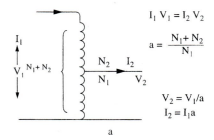

$$I_1 V_1 = I_2 V_2$$

$$a = \frac{N_1 + N_2}{N_1}$$

$$V_2 = V_1/a$$

$$I_2 = I_1 a$$

a

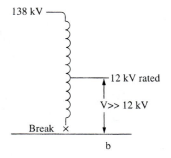

b

**FIGURE 4.21** Autotransformer (a) Schematic diagram, step down and (b) Safety consideration

Autotransformers differ from isolation transformers in that the lack of isolation between the primary and secondary sides allows some of the transferred power to be conducted to the secondary side instead of being transferred by magnetic induction. The relative amounts of power transferred by conduction and induction vary, as we shall see, with the turns ratio. Because some of the transferred power is conducted autotransformers need less core material for each kVA transferred than do isolation transformers. Thus autotransformers are smaller, lighter, more efficient, and less costly for a given kVA rating than isolation transformers. For this reason they are used in transmission and subtransmission substations where their lack of isolation poses no safety hazard to the public.

The *Autotransformer Rating Advantage* is because of the reduced core loss. Refer to Figure 4.22a. Let $N_1$ be the common turns and $N_2$ be the series turns. Then

$$\frac{V_2}{V_1} = \frac{V_1 + (N_2/N_1)\,V_1}{V_1}$$

$$= 1 + (N_2/N_1)$$

$$= N'$$

giving

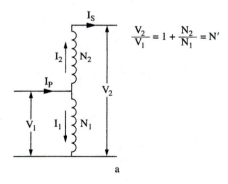

$$\frac{V_2}{V_1} = 1 + \frac{N_2}{N_1} = N'$$

a

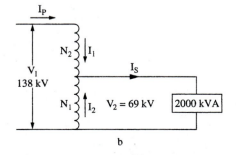

b

**FIGURE 4.22**  Autotransformer diagram for discussion of (a) ratio and (b) rating

$$\frac{N_2}{N_1} = N' - 1$$

The ampere turns must be equal in each winding, so

$$I_1 N_1 = I_2 N_2$$

therefore

$$I_2 = I_1 \frac{N_1}{N_2}$$

$$I_P = I_1 + I_2 = I_1 + I_1(N_1/N_2)$$

$$= I_1[1 + (N_1/N_2)]$$

$$= I_1 \left(1 + \frac{1}{N' - 1}\right)$$

$$= I_1 \left(\frac{N' - 1}{N' - 1} + \frac{1}{N' - 1}\right)$$

$$= I_1 \frac{N'}{N' - 1}$$

Let

$S_I$ = VA rating as an isolation transformer = $I_P V_1$
$S_A$ = VA rating as an autotransformer = $I_1 V_1$

$$\frac{S_A}{S_I} = \frac{V_1 I_1}{V_1 I_P}$$

$$= \frac{V_1 I_1}{V_1 I_1 [N'/(N' - 1)]}$$

$$= \frac{N' - 1}{N'}$$

Therefore

$$S_A = S_I \frac{N' - 1}{N'} \qquad (4.21)$$

Referring to Figure 4.22b we find the induced kVA of the autotransformer for a load of 2000 kVA at 69 kV is

$$N' = N_2/N_1 + 1$$

$$= (69 \text{ kV}/69 \text{ kV}) + 1$$

$$= 2$$

and

$$S_A = 2000 \text{ kVA} \frac{2-1}{2}$$

$$= 2000 \text{ kVA (0.5)}$$

$$= 1000 \text{ kVA}$$

Thus the autotransformer needs only the core to induce 1000 kVA to supply a load of 2000 kVA. As N′ gets larger (larger step-up or step-down ratio) the advantage becomes less favorable. For example if $N' = 5$, $S_A = 0.8S_I$. Notice that $N_1$ is the winding shared by the primary and secondary circuits. Figure 4.23 shows the kVA transferred by conduction and by induction in both the step-up and step-down connections.

The impedance of an autotransformer is handled like that of an isolation transformer if it is in percent or per unit. The impedance of autotransformers is lower than that of isolation transformers, often as low as 4% for high voltage transformers. Therefore autotransformers tend to have good voltage regulation but allow high fault current.

Autotransformers are usually wye connected with a built-in tertiary with a capacity of about 30% of the autotransformer kVA rating.

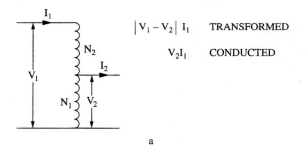

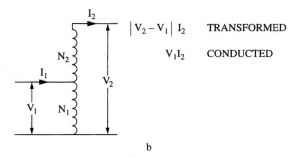

**FIGURE 4.23**   Transformed and conducted VA (a) step down and (b) step up

**Example 4.3:**

What is the required rating of an autotransformer to be used to step 900 kVA down from 69 kV to 46 kV?

**Solution:**

Using equation 4.20

$$S_A = S_l[(N' - 1)/N']$$

where

$$N' = V_1/V_2$$

$$= \frac{69 \text{ kV}}{46 \text{ kV}}$$

$$= 1.5$$

$$S_A = 900 \text{ kVA} \frac{1.5 - 1}{1.5}$$

$$= 300 \text{ kVA}$$

## 4.6.3  Distribution Transformers

Distribution transformers are those that are used to provide the final link with the customer. The distribution voltage is brought down to a level that is safe to use on the customer's premises. The primary voltage is between 34.5 kV and 2.3 kV, single or three phase depending on the customer size, and the secondary is normally either 480Y/277 V, 208Y/120 V three phase or 240/120 V single phase. A second transformer or a tertiary winding always provides 120 V for a building. Very large commercial and industrial customers may be fed at 12.5 kV or even higher voltages, and have a distribution substation of their own on their property. Transformers for services of this size are referred to, and are, power transformers.

Distribution transformers may be dry type, in which the windings are *not* immersed in oil. These are usually used in moderate to small commercial installations where cost is the primary consideration.

The most familiar distribution transformer is the pole top, shown in Figure 4.24. These are normally in the 15 to 100 kVA range. Three-phase service with pole top transformers is usually provided by a bank of three transformers, although three-phase pole top transformers are also used. Pole types are oil immersed, sealed, and seldom have any auxiliary cooling other than raised fins on the case. They are built with enough excess cooling capacity to sustain 200% overloads for hours. A typical high quality pole top transformer has a life expectancy of up to 40 years. They are protected with a minimum of a primary fuse, and a lightning arrester at or near the transformer primary.

**FIGURE 4.24**  Photograph of pole top distribution transformer. Three-phase cluster on single pole (Courtesy of Gulf States Utilities Corp.)

The growth of underground service has made the pad-mount transformer almost as familiar a sight as the pole top. Pad-mount transformers are oil immersed, mounted in a steel case with a front that closes to prevent injury, and installed on a concrete pad. A photograph of a pad-mount transformer with the front open is shown in Figure 4.25. Residential pad-mount transformers are usually single phase, primary with two or more secondaries, each providing power for three or four residences (each pad-mount transformer feeds eight to 12 houses). Commercial pad-mount transformers have a three-phase primary. The minimum protection is primary fusing. Elbow connections have been very important in making underground service economical. A cable with a conductor-insulator-concentric neutral-insulated sheath construction can be directly buried (no conduit), brought up from underground, and connected to the mating concentric connection on the transformer in a fairly small space with an elbow connector. The primary connections are completely insulated outside of the transformer (called dead front primary) with the access door open. Elbows rated at below 200 A are usually load break.

**FIGURE 4.25** Photograph of pad-mount transformer with the front open (Courtesy of Houston Lighting and Power).

## 4.6.4 Instrument Transformers

Instrument transformers provide line current and voltage information to protective relays and control systems at low power levels. Two types of transformers are in this class: current transformers and potential transformers.

Figure 4.26 is a diagram of a *current transformer* (*CT*), it is a secondary winding of many turns wound on a torroid core. The primary is the conductor on which current is to be measured. The cores need not be torroids, but torroids provide the

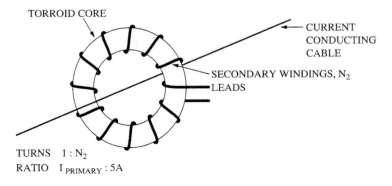

**FIGURE 4.26** Current transformer

lowest leakage for better mutual flux coupling. Current transformers are wound for a secondary current of 5 A at the maximum current they are to measure because the instruments to which they provide information are 5 A full scale (or range). Thus the CT ratios are expressed as *primary amperes:5 A* or sometimes primary current as a ratio:1, where the 1 is taken as 5 A. Therefore, a 100:1 CT is read as 500 A:5 A. CTs must be very well insulated between the primary and secondary because the primary line may be at an extremely high voltage. For this reason high voltage CTs are quite large. Current transformers are designed for use with low impedance current sensors and coils. When they are not attached to the instrument their secondary leads are shorted together to prevent induced secondary turn-to-turn voltage from exceeding the insulation rating of the secondary wire.

*Burden* refers to the load imposed on an instrument transformer, usually in VA. The accuracy is specified with a specific range of burden. The voltage drop across the internal impedance of a CT causes the accuracy to decrease with increased burden. The CT secondary current is offset from the primary current in phase by a small amount because of the excitation current drawn by the CT. The angular error introduced by the CT magnetizing current is usually less than 1°. Figure 4.27 is a photograph of a current transformer.

**FIGURE 4.27**   Photograph of current transformer (Courtesy of Houston Lighting and Power).

*Potential transformers* step line voltage down to the 0-150 V range for which most instruments are designed. Like CTs, potential transformers (PTs) are very large for their VA rating because of the amount of insulation required for the high voltage side. A photograph of a PT is shown in Figure 4.28.

**FIGURE 4.28** Photograph of potential transformers (Courtesy of Gulf States Utilities)

The major errors incurred in PTs are ratio error and phase error. PTs are designed with compensation for their series impedance by setting $N_1/N_2$ to provide the correct voltage ratio at a specific burden. The voltage ratio varies slightly with burden change. Phase error results from the PT magnetizing current. Phase error is seldom more than 0.5° and usually closer to 0.1° for a PT with a ratio accuracy of $\pm 0.3\%$ (said to be 0.3 accuracy class).

Figure 4.29 shows a less expensive and less accurate method of providing line potential information. A capacitive voltage divider is used to provide the measurement voltage. Often the capacitance between the high voltage bushing and a metal sleeve around a portion of the bushing serves as $C_1$, and the capacitance between the sleeve and the case as $C_2$.

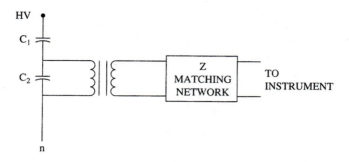

**FIGURE 4.29**   Capacitive potential divider

## 4.7

### TRANSFORMER VOLTAGE REGULATION

Voltage regulation is a measure of how close the secondary voltage remains to the nominal value as the load varies in amount and power factor. The voltage drops across the internal impedance of the transformer cause the voltage variation at the secondary terminal. Let S be secondary, FL be full load, and NL be no load, then voltage regulation (VR) at full load is

$$\text{Voltage Regulation (VR)} = \frac{V_{SNL} - V_{SFL}}{V_{SFL}} \text{ pu} \tag{4.22}$$

and percent voltage regulation = pu voltage regulation × 100 Voltage regulation can be calculated for any load but full load regulation is of the most interest.

### 4.7.1  Voltage Regulation Calculation

Figure 3.7 showed why secondary voltage varies with load and power factor changes. We now wish to find a reasonably easy and accurate method of calculating the secondary voltage as the load changes. The voltage drop across the transformer internal impedance will be, to a very close approximation, the same whether the primary or secondary voltage is held constant. We will hold the secondary voltage constant for our calculations, in other words $V_S$ is the reference phasor. The induced secondary voltage, $E_S = V_P/a$ is the variable voltage. The equivalent circuit referred to the secondary is shown in Figure 4.30a.

Figure 4.30b shows the phasor diagram of the secondary circuit with a lagging power factor load. The $IR_{eq}$ voltage drop is in phase with the load current, and the $IX_{eq}$ voltage drop is 90° ahead of the load current. The dotted lines dropping to the dotted line extending at the same angle as the load current make the phasor diagram into a right triangle. From the right triangle we can write

$$E_S = (V_S\cos\theta + IR_{eq}) + j(V_S\sin\theta + IX_{eq})$$

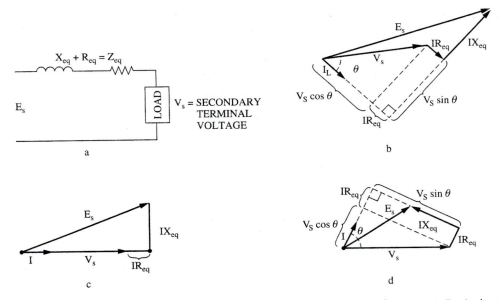

**FIGURE 4.30** Diagram of transformer voltage regulation, $V_s$ as reference (a) Equivalent circuit (b) PF lagging (c) unity power factor and (d) PF leading.

The same equation can be used for the PF = 1 load of Figure 4.30c except that $\cos\theta$ = 1, and $\sin\theta$ = 0. A close inspection of Figure 4.30d reveals a right triangle that yields a similar equation except the opposite side is $V_s\sin\theta - IX_{eq}$. A general equation can now be written as follows

$$E_S = (V_S\cos\theta + IR_{eq}) + j(V_S\sin\theta \pm IX_{eq}) \tag{4.23}$$

where last term is $+$ for a lagging power factor, and $-$ for a leading power factor.

An additional advantage to using the secondary terminal voltage as a reference vector is that $aE_S$ is the primary voltage necessary to provide the given secondary voltage

### Example 4.4:

A 2300V/240V, 75 kVA transformer has $X_{eq}$ = 0.07 $\Omega$ and $R_{eq}$ = 0.02 $\Omega$ referred to the low side. Calculate $E_S$ with a constant $V_S$ = 240 V for load power factor of 0.8 lagging, unity, and 0.7 leading, and the voltage regulation for each.

### Solution:

First we find the rated current and the voltage drops across the transformer internal reactance and resistance.

$$I_S = \frac{kVA}{V}$$

$$= \frac{75 \text{ kVA}}{240 \text{ V}}$$

$$= 312.5 \text{ A}$$

$$I_S X_{eq} = 312.5 \text{ A}(0.07 \ \Omega) = 21.88 \text{ V}$$

$$I_S R_{eq} = 312.5 \text{ A}(0.02 \ \Omega) = 6.25 \text{ V}$$

Now $E_S$ is calculated from equation 4.23 and VR from equation 4.22 for each load power factor.

PF = 0.8 LAGGING

$$E_S = (V_S \cos\theta + I_S R_{eq}) + j(V_S X_{eq} + I_S \pm I_S X_{eq})$$
$$= [240 \text{ V}(0.8) + 6.25 \text{ V}] + j[240 \text{ V}(0.6) + 21.88 \text{ V}]$$
$$= 198.25 \text{ V} + j165.88 \text{ V} = 258.49 \text{ V}$$

$$\text{VR} = \frac{E_S - V_S}{V_S}$$

$$= \frac{258.49 \text{ V} - 240 \text{ V}}{240 \text{ V}}$$

$$= \frac{18.49 \text{ V}}{240 \text{ V}}$$

$$= 0.077 \text{ pu or } 7.7\%$$

Note that since

$$a = \frac{V_P}{V_S}$$

$$= \frac{2300 \text{ V}}{240 \text{ V}}$$

$$= 9.583$$

for a secondary voltage of 240 V with a 75 kVA load at a PF = 0.8 the primary voltage must be

$$V_P = aE_S = 9.583(258.49 \text{ V}) = 2477 \text{ V}$$

PF = 1

From equation 4.23 again

$$E_S = (240 \text{ V} + 6.25 \text{ V}) + j(0 \text{ V} + 21.88 \text{ V})$$
$$= 247.22 \text{ V}$$

$$VR = \frac{247.22 \text{ V} - 240 \text{ V}}{240 \text{ V}}$$

$$= 0.03 \text{ or } 3\%$$

$$PF = 0.7 \text{ LEADING}$$

$$E_S = [240 \text{ V}(0.7) + 6.25 \text{ V}] + J[240 \text{ V}(0.714) - 21.88 \text{ V}]$$

$$= 229.6 \text{ V}$$

Notice the minus sign in the last term because the power factor is leading.

$$VR = \frac{229.6 \text{ V} - 240 \text{ V}}{240 \text{ V}}$$

$$= -0.043 \text{ or } -4.3\%$$

Output voltage higher than the induced secondary voltage is indicated by negative voltage regulation.

The impedance of the transformer in this example was about 7%, which is high for a transformer of this capacity and voltage. The higher than typical impedance made the results easy to see. Normal impedance for a transformer of the voltage and capacity of the one in the example is about 3 to 4%. The curious and interested student may wish to work the example problem with a transformer whose impedance is in this range. The problem is identical for a single-phase transformer or a balanced three-phase transformer as long as per phase impedance and current are used.

## 4.7.2 Voltage Regulation—Per Unit Quantities

Refer to Figure 4.31. Let a be calculated from the transformer name plate voltages (as is the standard practice). Then by Kirchoff's law

$$V_1/a = V_2 + I_2 Z_{eq}$$

Dividing both sides by the base voltage of the secondary side, $V_{2B}$ we obtain

$$\frac{V_1}{a V_{2B}} = \frac{V_2}{V_{2B}} - \frac{I_2 Z_{eq}}{V_{2B}}$$

but a $V_{2B} = V_{1B}$ and $V_{2B} = I_{2B} Z_{2B}$ so we obtain from the preceding equation

$$V_{1pu} = V_{2pu} + I_{2pu} Z_{eqpu}$$

Let $V_{2pu}$ be the reference phasor as before, then

$$I_2 = I_{2B}$$

but

$$I_{2pu} = 1$$

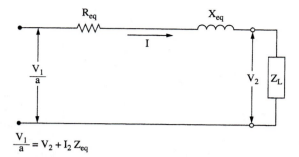

$$\frac{V_1}{a} = V_2 + I_2 Z_{eq}$$

**FIGURE 4.31**    Voltage regulation calculation, per unit values

Now the equation becomes

$$V_{1pu} = 1\angle 0° + 1\angle \pm \theta \, (Z_{eqpu})$$                                                      (4.24)

The voltage regulation equation can now be written

$$VR = \frac{|V_1 / a V_{2B}| - V_{2B}}{V_{2B}}$$

Dividing by results in

$$VR = \frac{|V_1|}{|V_{1B}|} - 1$$

$$= |V_{1pu}| - 1$$

**Example 4.5:**

A transformer has a per unit $R = 0.03$ and $X = 0.07$. The load power factor is 0.85 lagging. Calculate the voltage regulation.

**Solution:**

First calculate $Z_{pu}$

$$Z_{pu} = 0.03 + j0.07 = 0.076\angle 66.8°$$

From equation 4.24

$$V_{1pu} = 1\angle 0° + 1\angle \text{-}31.78°(Z_{pu})$$

$$= 1\angle 0° + 0.076 \angle (66.8° - 31.78°)$$

$$= 1 + 0.062 + j0.044$$

$$= 1.062 + j0.044$$

$$= 1.063\angle 2.35°\text{pu}$$

Finally

$$VR = |V_{1pu}| - 1$$
$$= 1.063 - 1$$
$$= 0.063 \text{ or } 6.3\%$$

## 4.8

## VOLTAGE DROP ALONG THE LINE

Customer taps along the line draw current through the line causing the voltage to drop as one gets further from the substation. This is illustrated in Figure 4.32, where Figure 4.32a is the lateral from a line near a substation to pad-mount transformers, each providing service to four residences. Figure 4.32b and c illustrate the voltage drop along the line with a heavy load and a light load respectively. Note that if the voltage at the substation is set at the nominal voltage the customers at the end of the line have too low a voltage under heavy load. If the voltage is set so that the customers at the end of the line receive the nominal voltage under heavy load the customers near the substation have too high a voltage, and the voltage is too high for all of the customers at light load. A compromise voltage setting must be chosen so that the voltage is at an acceptable level for all of the customers regardless of the load, and the line drop must be acceptably low under all load conditions. A favorable compromise of voltage drop and voltage setting is not always possible for all load conditions, so other means of voltage regulation have been devised for such conditions.

### 4.8.1 Capacitors for Voltage Regulation

Since most loads are lagging and the line reactance is much higher than the line resistance, switching *shunt capacitors* across the line will increase the voltage by reducing the inductive VARs drawn as in power factor correction. One can also think of the capacitors as regulating voltage as in transformer voltage regulation. However, if so much capacitance is switched across the line that the current becomes leading, the method is not cost effective because the VARs begin to rise again. Shunt capacitors are only helpful with lagging load power factors.

The capacitors are installed near load points. To raise the voltage along the lateral of Figure 4.34a the capacitors are installed at points A, B, and C. Capacitor installations are usually in three banks, one for each of the three phases. Switched capacitors can be, and sometimes are, installed on long single-phase distribution lines. A three-phase capacitor bank is shown in Figure 4.33. Fixed value banks must be selected so that the change of voltage between light and heavy loads is not excessive, and is centered around the nominal voltage. Switched capacitor banks, which contain a number of smaller capacitors that are switched in parallel to obtain the needed capacitance, are used when a compromise cannot be found because the

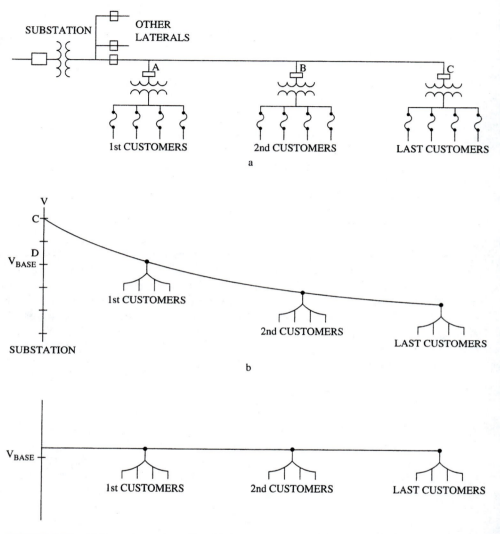

**FIGURE 4.32** Voltage drop along line (a) Representation of distribution circuit (b) line voltage drop with heavy load and (c) line voltage drop with light load.

load variation is too large. Switched capacitor banks are expensive because they must have sensing equipment to monitor the line voltage and control equipment to activate the proper switching. They are installed only when essential.

All capacitor banks require protection. The entire bank will be protected with a fuse or circuit breaker, and each capacitor will be fused, as shown in Figure 4.34. The capacitors can be connected either delta or wye. The ungrounded wye is the preferred configuration because if a capacitor should short out in one leg of the wye the other two legs will limit the fault current.

a

b

**FIGURE 4.33**   Capacitor bank (a) poletop (b) substation (Courtesy of Houston Lighting and Power)

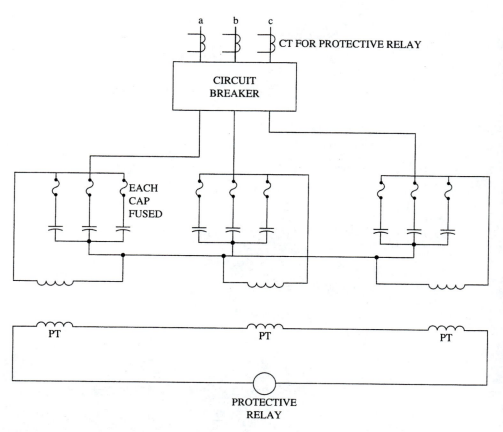

**FIGURE 4.34**    Ungrounded wye capacitor bank.

The charging of the capacitors can cause a large current transient when energized. Small reactors (with $X_L \ll X_C$) in series with the capacitors effectively reduce the inrush current by limiting the rate of rise of the current.

Shunt capacitors are usually installed in radial feeders. Their chief effect is to correct for load power factor, and their only current is from the VARs. *Series capacitors* are connected in series with the line and carry full line current. The capacitor voltage as well as current varies with the load. The capacitive reactance of the series capacitance is used to cancel the inductive reactance of the line to reduce the voltage drop along the line. The principle is shown in Figure 4.35: in a the line impedance results in the voltage drop and receiving voltage shown in b; in c the capacitive reactance cancels a portion of the line inductive reactance causing the receiving voltage to rise, as shown in part d of the figure. The series capacitance is set to undercompensate the reactance a little to prevent excessive line voltage rise during heavy loading transients such as occur with large motor starting. Series capacitors can be switched or fixed. They are protected in much the same way as are shunt capacitors.

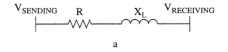

$$a$$

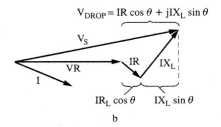

$$V_{DROP} = IR\cos\theta + jIX_L\sin\theta$$

$$b$$

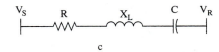

$$c$$

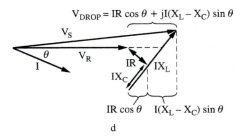

$$V_{DROP} = IR\cos\theta + jI(X_L - X_C)\sin\theta$$

$$d$$

**FIGURE 4.35** Series capacitor. (a) Line equivalent circuit (b) Phasor diagram with lagging power factor load (c) Equivalent circuit with capacitor and (d) Phasor diagram with capacitor (same $V_S$ as in b).

## 4.8.2 Tap Changers

Another method of line voltage regulation is tap changing. Taps are connections on a transformer winding that change the turns ratio slightly. The ratio change is normally $\pm 10\%$, although $\pm 7.5\%$, and $\pm 5\%$ are also available. The ratio change steps vary from 2.5% to 32 steps to cover the now normal range of $\pm 10\%$ (0.625% per step). Normally the taps are located on the primary because less current has to be switched by the tap changing connections than would be necessary on the secondary. Tap changers can be manual or automatic. Some distribution transformers and distribution substation transformers have manual tap changers so that added load can be compensated for, such as a new shop in a shopping center in the first case or a new street of residences in the second. Motor driven automatic tap changers are necessary for voltage regulation with widely fluctuating loads. This is called tap changing under load (TCUL) or load tap changing (LTC). LTC is used in distribution and subtransmission substations to keep the secondary line voltage at the proper level in response to load and primary voltage changes.

Many designs exist for tap changers. An under load tap changer with a reversing switch, type URS, is shown in Figure 4.36a. The reversing switch allows the tapped section to be connected series aiding or series opposing with respect to the un-tapped section of the winding, so that twice the range can be covered with the same number of taps. The transfer switches that switch between the taps are make-before-break type to prevent arcing at the tap changing contacts. For example, to move from tap 2 to tap 3, transfer switch A moves from tap 2 to tap 3 while transfer switch B remains on tap 2. Only after transfer switch A is on tap 3 does transfer switch B move to tap 3. A mechanical arrangement to accomplish the switching is illustrated in Figure 4.36b. The preventive autotransformer connected to line $H_1$ of Figure 4.36a has a high impedance between transfer contacts A and B to prevent high circulating current from damaging windings as the taps are changed, and a low impedance between line $H_1$ and the transfer switch contacts. The reversing switch does not switch any current because the transfer switch contacts are on tap R before it is switched.

Automatic load tap changers add approximately 50% to the price of a transformer, but the voltage regulation ability they provide improves the quality of service substantially.

### 4.8.3 Voltage Regulating Transformers

Regulating transformers are designed to provide a boost in voltage magnitude along a line, or a change in phase. They are used primarily to control the flow of power between two systems with different sources or along a tie feeder between two load centers that are fed by the same bulk substation. Phase regulating transformers are used to control the power flow around loops with two or more sources. They are almost always motor driven and have extensive protection and control systems.

A voltage regulating transformer is shown in Figure 4.37. The autotransformer provides a $\Delta V$ to add to the line via the series transformers.

Figure 4.38a shows a phase regulating transformer. The core arrangement causes the windings that produce $\Delta V$ to be 90° out of phase with the line to neutral voltage. Thus $V_{AN} + \Delta V_{AN}$ produces a phase shift with very little amplitude change, as shown in Figure 4.38b.

**4.9**

## TRANSFORMER SHORT AND OPEN CIRCUIT TESTS

The information obtained from the short circuit and open circuit tests of a transformer allow the calculation of the transformer reactance and efficiency.

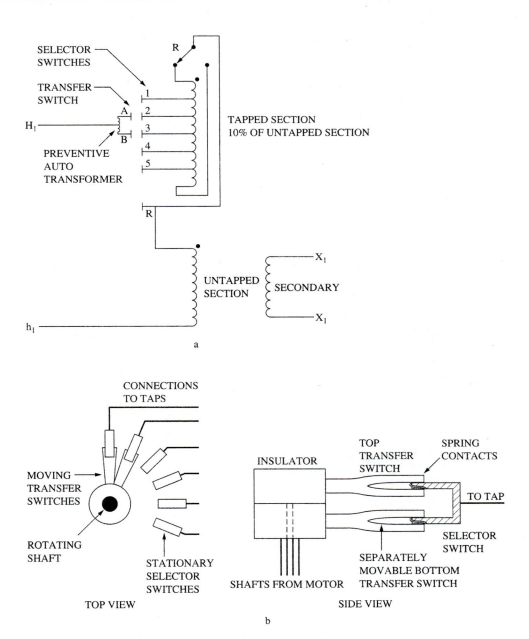

**FIGURE 4.36** URS tap changer (a) Diagram and (b) mechanical switching arrangement

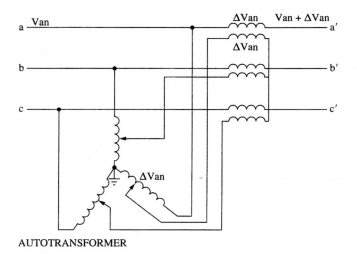

AUTOTRANSFORMER

**FIGURE 4.37**   Voltage regulating transformer

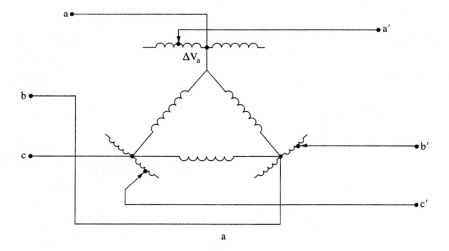

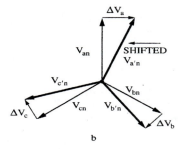

**FIGURE 4.38**   Phase regulating transformer (a) Diagram, parallel drawn windings are on the same core (b) Phasor diagram

## 4.9.1 Short Circuit Test

Refer to Figure 4.39a. The short circuit test is performed by shorting out the low voltage winding and applying voltage to the high voltage winding until rated current flows in the high voltage winding, at which time rated current should flow in the low voltage winding. The voltage and power is recorded. Because core losses are proportional to $V^2$, and the voltage required to cause rated current to flow with a shorted winding is low, the core losses are very low during the short circuit test. The core losses are low enough in the short circuit tests to be ignored for all but the most accurate calculations.

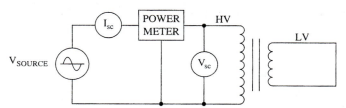

$V_{SOURCE}$ adjusted until rated current flows in windings.
Values referred to measured windings.

$$Z_{eq} = \frac{V_{sc}}{I_{sc}} \qquad R_{eq} = \frac{P}{I_{sc}^2}$$

$$X_{eq} = \sqrt{Z_{eq}^2 - R_{eq}^2}$$

a

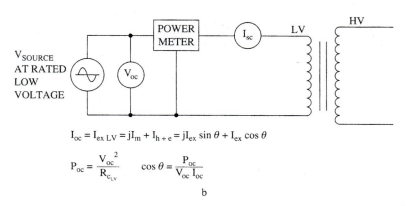

$$I_{oc} = I_{ex\,LV} = jI_m + I_{h+e} = jI_{ex}\sin\theta + I_{ex}\cos\theta$$

$$P_{oc} = \frac{V_{oc}^2}{R_{c_{LV}}} \qquad \cos\theta = \frac{P_{oc}}{V_{oc}\,I_{oc}}$$

b

**FIGURE 4.39** Transformer test connections (a) Short circuit test (b) open circuit test

The per unit impedance of the transformer can be found from

$$\text{pu } Z = \frac{V_{SC}}{\text{rated } V} \qquad (4.26)$$

Where the subscript SC stands for short circuit. For the actual impedance value

$$Z_{eq} = \frac{V_{SC}}{I_{SC}} \tag{4.27}$$

$$P_{SC} = I_{SC}^2 R_{eq} + \text{(full voltage core loss)}(V_{SC}/V_{RATED})^2 \tag{4.28}$$

The last term is frequently ignored so $P_{SC}$ is the full load copper loss, and

$$R_{eq} = P_{SC}/I_{SC}^2 \tag{4.29}$$

$$X_{eq} = (Z_{eq}^2 - R_{eq}^2)^{1/2} \tag{4.30}$$

These values are referred to the side on which they are measured.

Measurements are very similar for three-phase transformers. Three-phase power, line current, and line voltage are usually measured. The calculations are done on a per phase basis whether wye or delta. Recall from Chapter 2 that delta impedance is three times wye impedance. To obtain phase values

$$P_P = P_{SC}/3$$

$$R_{eq} = P_P/I_P^2$$

$$Z_{eq} = V_P/I_P$$

$$X_{eq} = (Z_P^2 - R_P^2)^{1/2}$$

**Example 4.5:**

A Y-Y connected three-phase transformer, 12.47 kV/480 V, 25 kVA, has the following short circuit measurements from the primary side.

Three phase power $P = 400$ W

$V_{SC} = 620$ V

$I_{SC} = 3.47$ A

Calculate the equivalent $X$, $Z$, and $R$ referred to the primary side.

**Solution:**

$$Z_{eq} = \frac{V_P}{I_{SCP}} = \frac{620 \text{ V}/\sqrt{3}}{3.47 \text{ A}} = \mathbf{103.16 \ \Omega}$$

$$\text{pu } Z = \frac{Z}{Z_B} = \frac{Z}{V_B/I_B} = \frac{103.16 \ \Omega}{(12.47 \text{ kV}/\sqrt{3})/3.47 \text{ A}} = 0.05 \text{ pu}$$

Notice that

$$\text{pu } Z = \frac{V_{SC}}{V_B} = \frac{620 \text{ V}}{12.47 \text{ kV}} = 0.05 \text{ pu}$$

$$R_{eq} = \frac{P/3}{I_P^2} = \frac{133.3 \text{ W}}{(3.47 \text{ A})^2} = 11 \ \Omega$$

$$X_{eq} = (103.16 \ \Omega^2 - 11 \ \Omega^2)^{1/2} = 102.57 \ \Omega$$

These values of $R_{eq}$ and $X_{eq}$ are referred to the high voltage side.

## 4.9.2 Open Circuit Test

Figure 4.39b shows the connection for the open circuit test. Normally the low voltage winding is supplied with rated voltage, which should result in rated voltage on the high voltage side. The current drawn should be low enough that the copper losses are very low and the power measured is almost all from core losses. The current drawn during the open circuit test will be the excitation current.

$$I_{OC} = I_{h+e} + I_m$$
$$= I_{OC} \cos\theta + jI_{OC} \sin\theta \qquad (4.31)$$

where

$$\cos\theta = \frac{P_{OC}}{I_{OC}V_{OC}}$$

now

$$R_C = \frac{V_{OC}^2}{P_{OC} - I_{OC}^2 R_{eq}} \qquad (4.32)$$

$$= \frac{V_{OC}}{I_{OC} \cos\theta}$$

$$X_m = \frac{V_{OC}}{I_{OC} \sin\theta} \qquad (4.33)$$

These quantities are referred to the low voltage side and need to be referred to the side of the transformer on which they will be used. Phase current, voltage, and power should be used for three-phase transformers.

### Example 4.6:

For the transformer of example 4.5 the open circuit measurements are

$P_{OC} = 600 \ \text{W}$

$I_{OC} = 5 \ \text{A}$

$V_{OC} = 480 \ \text{V}$

Calculate the transformer $X_m$ and $R_C$ referred to the primary. Also calculate the transformer efficiency at full load. Load PF = 0.8.

**Solution:**

Calculating $\cos\theta$ first

$$\cos\theta = \frac{P_{OC}/3}{V_{POC}I_{OC}}$$

$$= \frac{200\ \text{W}}{(480\ \text{V}/\sqrt{3})\ (5\ \text{A})}$$

$$= 0.144$$

$$a = \frac{V_P}{V_S}$$

$$= \frac{12470\ \text{V}}{480\ \text{V}}$$

$$= 25.98$$

$$a^2 = 674.9$$

$$I_m = I_{OC}\sin(\text{arc}\cos 0.144)$$

$$= 5\ \text{A}\sin 81.7°$$

$$= 4.94\ \text{A}$$

$$X_{mS} = j\frac{V_{OC}}{I_m}$$

$$= j\frac{480\ \text{V}/\sqrt{3}}{4.94\text{A}}$$

$$= j56\ \Omega$$

$$X_m = a^2 X_{mS}$$

$$= 37.8\ \text{k}\Omega$$

Now for the hysteresis losses

$$I_{OC}{}^2 R_{eqS} = (5\ \text{A})^2(11\ \Omega/674.9) = 0.407\ \text{W}$$

This is only 0.079% of the open circuit power and can be neglected without serious inaccuracy in most applications.

$$R_{CS} = \frac{V_{OC}{}^2}{P_{OC}/3}$$

$$= \frac{(480\ \text{V}/\sqrt{3})^2}{200\ \text{W}}$$

$$= 384\ \Omega$$

$$R_C = 674.9 \ (384 \ \Omega)$$

$$= 259.2 \ k\Omega$$

Recall that efficiency $\eta = P_{OUT}/P_{IN}$

$$\eta = \frac{P_{OUT}}{P_{OUT} + P_{cu} + P_{h+e}} \tag{4.34}$$

$$= \frac{25 \ kVA(0.8)}{25 \ kVA(0.8) + 400 \ W + 600 \ W}$$

$$= 0.952 \ or \ 95.2\%$$

## 4.10

# PER UNIT IMPEDANCE OF THREE WINDING TRANSFORMER

Transformers with two secondary windings often have different kVA ratings for all three windings. The percent or pu impedance of each winding is given with its own rating as a base unless otherwise stated. The impedance diagram must have a common base for all the impedances.

Let the three windings of a transformer be the primary (P), secondary (S), and tertiary (T). Additionally let

$Z_{PS}$ = leakage impedance measured in the primary with the secondary shorted and the tertiary open.

$Z_{PT}$ = leakage impedance in the primary with the tertiary shorted and the secondary open.

$Z_{ST}$ = leakage impedance in the secondary with the tertiary shorted and the primary open.

Then Z referred to the voltage of one winding is

$$Z_{PS} = Z_P + Z_S$$

$$Z_{PT} = Z_P + Z_T$$

$$Z_{ST} = Z_S + Z_T$$

The simultaneous solution of these three equations is

$$Z_P = (1/2)(Z_{PS} + Z_{PT} - Z_{ST}) \tag{4.35}$$

$$Z_S = (1/2)(Z_{PS} + Z_{ST} - Z_{PT}) \tag{4.35}$$

$$Z_T = (1/2)(Z_{PT} + Z_{ST} - Z_{PS}) \tag{4.36}$$

Note the order of the terms of the equations. The equivalent circuit is shown in Figure 4.40a.

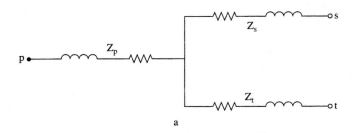

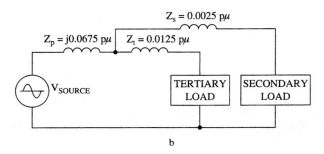

**FIGURE 4.40** Three winding transformer equivalent circuit with all impedances on a common base (a) Equivalent circuit and (b) in an impedance diagram

**Example 4.7:**

A three winding, three-phase transformer has windings rated as follows

Primary: Y, 69 kV, 30 kVA

Secondary: Y, 34.5 kV, 20 kVA

Tertiary: Δ, 2.3 kV, 10 kVA

$Z_{PS} = 7\%$ on 69 kV, 30 kVA base

$Z_{PT} = 8\%$ on 69 kV, 30 kVA base

$Z_{ST} = 4\%$ on 34.5 kV, 20 kVA base

Assume that $Z_{eq} = X_{eq}$ and solve for the equivalent circuit components of Figure 4.40a, and connect them properly in an impedance diagram showing the loads and source.

**Solution:**

First all Zs must be on a common base for the equations to be used. Using the change of base equation from chapter two we obtain the tertiary impedance on the 69 kV, 30 kVA base.

$$Z_{B2} = Z_{B1} \frac{\text{base 1 kV}^2}{\text{base 2 kV}^2} \frac{\text{base 2 kVA}}{\text{base 1 kVA}}$$

$$= 0.04 \left( \frac{34.5 \text{ kV}}{69 \text{ kV}} \right)^2 \frac{30 \text{ kVA}}{20 \text{ kVA}}$$

$$= 0.04(0.25)(1.5)$$

$$= 0.015 \text{ pu}$$

Substituting into equations 4.35, 4.36, and 4.37 we obtain

$$Z_p = 1/2 \,(j0.07 + j0.08 - j0.015) = j0.0675 \text{ pu}$$

$$Z_s = 1/2 \,(j0.07 + j0.015 - j0.08) = j0.0025 \text{ pu}$$

$$Z_T = 1/2 \,(j0.08 + j0.015 - j0.07) = j0.0125 \text{ pu}$$

The impedance diagram is shown in Figure 4.40b.

## 4.11

## SUMMARY

Transformers are electrical machines that work by electromagnetic induction. The two major types of transformers are dry and oil immersed. The additional cooling and insulation properties of the oil result in greater power handling capabilities for the oil immersed. The power handling capability of oil transformers increases as the cooling means for the transformer oil go from natural convection of the oil through the cooling fins (OA) to adding fans to force air through the cooling fins (FA) to adding pump circulation of oil through the cooling fins (FOA). The ratio of OA/FA/FOA with respect to power handling capability is normally 1/1.33/1.67. Dry type transformers can be obtained in sealed housings filled with sulfur hexa-fluoride, an effective insulating gas. The insulating gas and housing, while costly, allow the transformer to be mounted in less space, which may be essential when space is at a premium.

The magnetic flux expanding and collapsing in the primary coil couples to the secondary coil(s) inducing voltage in the secondary in direct proportion to the ratio of the secondary turns to the primary turns. Energy must be preserved in the transformation so secondary current is transformed in inverse proportion to the ratio of secondary to primary turns. Transformer relationships are covered in sections 4.1.2 and 4.2.

The losses in a transformer include copper losses, power dissipated in winding resistance, hysteresis losses, power dissipated in work done reversing the magnetic field in the core once every half cycle, and eddy current losses, from power dissipated by current induced in the magnetic core laminations. These losses can be as low as 0.5% of the total transformed power, so transformers are very efficient machines.

Transfomers must be insulated to voltages well above the voltage of the system in which they are used because electrical systems are subject to transient voltages caused by switching, lightning, or faults on the lines. The insulation for transients is called the basic insulation level (BIL) and tests for BIL are performed with a waveform specified to resemble a lightning induced transient.

When power is first applied to a transformer the core must be magnetized before transformation can occur. Thus, energizing a transformer results in a high initial current, called the inrush current, to establish the initial maginetic field.

Transformer secondary voltages change both with the load power drawn and with the power factor of the load. The change in secondary voltage from no load to full load referenced to full load voltage is called voltage regulation. Voltage regulation calculation is discussed in section 4.7.

Autotransformers are constructed with a tapped single winding for each phase, similar to a taped inductor. The secondary power is partially induced and partially conducted, the fraction of each determined by the autotransformer winding ratio. The conduction of power to the secondary means that the core of an autotransformer can be smaller than that of a two winding transformer for the same power. This results in a more efficient, less expensive transformer. The loss of an autotransformer ground can result in the full primary voltge on the secondary line. Thus, autotransformers are never used when the secondary is connected to a utility customer facility.

Transformer reactance and efficiency can be calculated using the results of the short and open circuit test. These tests are explained in section 4.9.

## 4.12

---

### QUESTIONS

1. State the reason that oil immersed transformers can handle higher kVA loads and higher transient voltage levels than equivalent sized dry type transformers.
2. A three-phase power transformer has the following name-plate rating; MVA 100/133/167. What does it mean?
3. State the meaning of the dots on the primary and secondary windings of a transformer schematic diagram.
4. A 69 kV/12.5 kV, 15 MVA, three-phase, Y-Y transformer is operated at full load. Calculate a, $I_S$, and $I_P$. Assume a well grounded neutral.
5. State the major sources of loss in a transformer.
6. Define BIL.
7. The transformer of problem 4 has $I_m = 0.05$ pu, and $I_{h+e} = 0.02$ pu. Calculate the actual value of the excitation, magnetization, and hysteresis (plus eddy current loss) currents.
8. Calculate the pu $X_m$ and $R_c$ for the transformer of problem 4.
9. If the pu $Z = 0.06$, and the pu $R = 0.01$ for the transformer of problem 4, calculate $R_{eq}$ and $X_{eq}$ referred to the primary and the secondary.

10. For the transformer of problem 4, use the information of problem 9 to calculate the required primary voltage for rated secondary voltage and the voltage regulation with the following loads: PF = 0.8 lagging, unity power factor, PF = 0.75 leading. All three are 15 MVA loads.

11. A 4200V/208V, 30kVA, Δ-Y connected three-phase transformer has secondary per phase $X_{eq} = 0.072\ \Omega$ and $R_{eq} = 0.0144\ \Omega$. It is loaded to 30 kVA with loads of PF = 0.76 lagging, PF = 1, and PF = 0.7 leading, respectively. For each load calculate the primary voltage necessary to maintain rated secondary voltage, and the voltage regulation.

12. Repeat the voltage regulation calculations of problem 11 with per unit quantities.

13. State the function of a delta connected tertiary winding on a Y-Y transformer.

14. State the cause of transformer inrush current and the conditions under which it is highest and lowest.

15. State the purpose of transformer protection, and the major types of equipment used for the protection.

16. An autotransformer, such as the one in Figure 4.21a, has a primary voltage of 34.5 kV and a secondary voltage of 12.5 kV. It is to supply a 60 kVA load. What equivalent two winding kVA rating must the autotransformer have? Also what portion of the load kVA is transformed and what portion is conducted?

17. Calculate the turns ratio needed by a CT and a PT to monitor a 69 kV line carrying 2000 A. Let the PT secondary be 120 V.

18. A three-phase 4200 V/480 V, Δ-Y connected, 15 kVA transformer has the following short circuit and open circuit measurements. All values are line or three phase.
$V_{SC} = 210$ V
$I_{SC} = 2.06$ A
$P_{SC} = 200$ W
all taken from the high voltage side
$V_{OC} = 480$ V
$I_{OC} = 0.9$ A
$P_{OC} = 300$ W
all taken from the low voltage side
Calculate: $Z_{eq}$, $X_{eq}$, $R_{eq}$, $I_m$, $I_{h+e}$, and the efficiency of the transformer.

19. A three winding transformer, Y-Y-Δ connected, has the following ratings: primary, 138 kV at 12 MVA; secondary, 34.5 kV at 8 MVA; tertiary, 4.2 kV at 4 MVA. It has the following measured impedances
$Z_{PS} = 8\%$ on 138 kV, 12 MVA base
$Z_{PT} = 6\%$ on 138 kV, 12 MVA base
$Z_{ST} = 4\%$ on 34.5 kV, 8 MVA base
Calculate $Z_P$, $Z_S$, and $Z_T$

# CHAPTER 5

# Distribution Equipment

A large variety of equipment is necessary to economically and efficiently distribute electrical power. The equipment includes conductors, insulators, conductor support towers and poles, transformers, and protection devices. We cannot cover all of the distribution equipment in this chapter because the chapter would be too long. Some equipment, such as transformers, has already been covered, and some equipment will be covered in later chapters. This chapter concentrates on protective devices.

Practical, reliable, and safe distribution depends on protective devices to sense fault conditions and disconnect malfunctioning equipment. Protective devices must protect people and equipment from malfunctions, whether the malfunction is caused by equipment failure, weather (lightning and wind), or carelessness such as cutting an underground cable. Quickly disconnecting malfunctioning equipment minimizes the damage to the equipment and thereby shortens the time that the equipment is out of service. The development of protective devices that can operate at higher voltages, and interrupt higher fault currents (called interrupting capacity), has been crucial to the implementation of higher voltage, more efficient transmission and distribution systems.

The most important protective devices for distribution are circuit breakers, reclosers, sectionalizers, fuses, relays to sense fault conditions, and lightning arresters. Each of these, as well as switches and metering equipment, are discussed in this chapter.

## 5.1

### CIRCUIT BREAKERS

Circuit breakers for electrical power distribution include both medium (between 600 V and 34.5 kV) and high voltage (above 34.5 kV), high current devices that must automatically disconnect faulted equipment to protect people, prevent damage to upstream equipment, and minimize damage to downstream equipment in two to five cycles. Additionally, the circuit breaker should not damage itself when it operates. To accomplish this task substation circuit breakers are far more complex than the molded case low voltage breakers used in building distribution systems.

Circuit breakers range in size from 125 V, 15 A molded case thermal or magnetic breakers used in buildings to over 800 kV at several thousand amperes. Breakers are classified by voltage, continuous current, interrupting capacity (maximum fault current the breaker can interrupt without becoming dangerous themselves), and methods of extinguishing the arc.

To illustrate the magnitude of the circuit breaker's job let us consider a mechanical analogy. A 100 MVA circuit breaker (they come as large as 20,000 MVA) must stop the flow of power in two to five 60 Hz cycles after it receives a trip signal. Assume the power factor is one and that the power is all mechanical. One hundred MW is 134,048 hp. One hp = 550 ft-lb/s, so 100 MW = $7.37265 \times 10^7$ ft-lb/s. Assume this energy consists of a weight moving at 10 miles/hour, which in ft/s is

$$(10 \text{ mi/hr})(5280 \text{ ft/mi})(1 \text{ hr/3600 s}) = 14.67 \text{ ft/s}$$

The equivalent moving weight is

$$\frac{7.37265 \times 10^7 \text{ ft-lb/s}}{14.67 \text{ ft/s}} = 5.0268 \times 10^6 \text{ lb}$$

A 17-car freight train with an average car weight of 295,695 lb would weigh $5.0268 \times 10^6$ lb. Imagine stopping a 17-car freight train moving at 10 mi/hr in 1 second. Quite a job. A 100 MVA circuit breaker must stop the equivalent energy in 0.083 seconds if it opens in five cycles, or 0.0333 seconds if it opens in two cycles.

### 5.1.1 The Arc

When current carrying contacts open, the initial electric field between the just parted contacts is very high. The high electric field causes any gas between the contacts to ionize and support current flow through it, or arc. The higher the voltage that the contacts are breaking the more severe the arcing.

Inductive loads make the tendency to arc even more severe because inductive loads attempt to keep the current through the opening contacts the same as it was before they opened. Most industrial loads and faulted lines are inductive. The arc reaches very high temperatures because of the current dissipating power in the arc itself. The high temperature at the contact face causes some of the contact metal to burn off and ionize, which worsens the arc. The arc must be extinguished to interrupt the current.

Many methods are used to extinguish an arc. They use one or both of the following two principles. The first is to lengthen the arc until it is long and thin. This causes the arc resistance to rise, thus the arc current to drop, the arc temperature to decrease, and ultimately results in insufficient energy in the arc to keep it ionized. The second is to open the arc in a medium that absorbs energy from the arc causing it to cool and quench. Air, oil, and insulating gas are normally used as the medium.

Direct current (dc) arcs are harder to break than alternating current (ac) arcs because ac goes through a current zero every half cycle, which dc does not. The same principles are used to break dc arcs, but dc rated breakers must use the principles more efficiently. After ac zero the arc will re-establish if the medium between the contacts is still ionized. Even if the arc does not re-establish immediately because the medium is de-ionized, the rise in voltage across the contacts may cause the arc to re-establish if the distance between the contacts is not sufficient to keep the electric field between the contacts below the ionization field of the medium.

Figure 5.1a shows a faulted bus with a circuit breaker. Figure 5.1b shows the arc between the separating contacts, and 5.1c shows the waveforms across the contacts. Recall the $X/R$ ratio of a line or a transformer is high so the fault current lags the voltage. This is typical. Notice that while arc current is flowing the arc voltage is relatively low. At the zero crossing the high $dV/dt$ of the voltage across the contacts may cause stray inductance and capacitance in the circuit to resonate at a frequency higher than 60 Hz, which can cause the voltage to shoot higher than the applied voltage. This voltage can cause re-ignition of the arc as shown at the first two current zero crossings in Figure 5.1c. This can occur several times before the contact separation is great enough that the arc is extinguished. If the medium between the contacts does not de-ionize the high voltage spike does not occur. An arc re-established in the first $1/4$ cycle is called a re-ignition, and an arc re-established after the first $1/4$ cycle is called a restrike. Notice in Figure 5.1c the oscillation riding on the applied voltage after the arc is broken. The peak voltage from oscillation can be high enough to cause equipment damage without transient protection or damping. Figure 5.1d shows an arc on a power line.

The high $X/R$ ratio of most faulted equipment causes the arc to be more difficult to break. Recall that the offset of a fault waveform from symmetrical depends on the time during the waveform that the fault occurs. The offset from zero, as shown in Figure 5.2, is a dc component that is harder to interrupt than ac. The higher the $X/R$ ratio the longer the dc component lasts.

The arc energy transferred to the medium between the contacts causes the medium to reach temperatures as high as 30,000°K and high pressures very quickly. The resulting expansion of the medium is almost explosive. The combination of high mechanical and arc forces can cause the ground to shake around a large breaker when it operates.

A circuit breaker must stretch and cool the arc until it breaks. Circuit breakers are classified by how they accomplish this task.

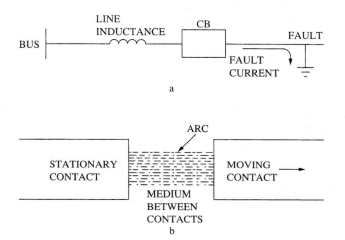

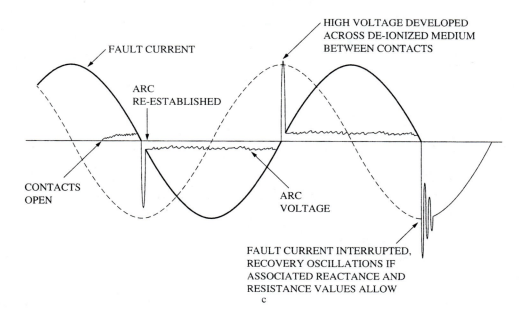

**FIGURE 5.1**   Contact arc (a) Equivalent circuit (b) Contacts and (c) Waveforms

**FIGURE 5.1**   (d) Photograph (courtesy of Lower Colorado River Authority).

## 5.1.2  Air Circuit Breakers

Air circuit breakers use air as the arc interrupting medium. Because air at atmospheric pressure ionizes easily some auxiliary equipment must be used to break the arc except for the very lowest voltage and capacity breakers. Almost all low voltage breakers use air as an interrupting medium. Figure 5.3 shows some low voltage circuit breakers. We will now look at the methods used to break an arc in air.

Convection causes an arc, which is hot, to rise if the contacts are properly oriented. As the rising arc stretches its resistance increases, its current drops, and its increased surface area is exposed to cooler air, causing its temperature to drop until the arc is finally extinguished. The longer an arc can be drawn out the easier it is to extinguish.

Arc tips (also called arcing contacts) break after the main contacts break. This prevents pitting of the main contacts. Because the arc tips travel further than the main contacts they stretch the arc further, thus making it extinguish earlier. Arc tips are shown in Figure 5.4a. Arc horns work on the same principle except convection drives the arc up the spreading horns causing the arc to leave the load current carrying contacts and stretch, as shown in Figure 5.4b.

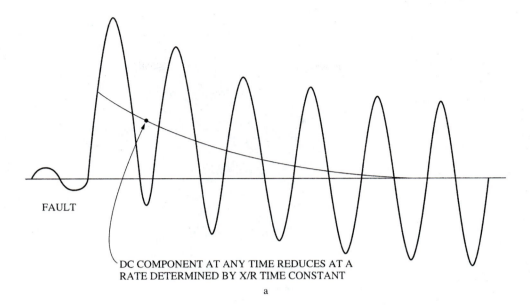

FAULT

DC COMPONENT AT ANY TIME REDUCES AT A
RATE DETERMINED BY X/R TIME CONSTANT

a

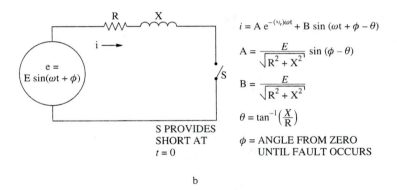

$$i = A\, e^{-(X_r)\omega t} + B \sin (\omega t + \phi - \theta)$$

$$A = \frac{E}{\sqrt{R^2 + X^2}} \sin (\phi - \theta)$$

$$B = \frac{E}{\sqrt{R^2 + X^2}}$$

$$\theta = \tan^{-1}\left(\frac{X}{R}\right)$$

S PROVIDES
SHORT AT
$t = 0$

$\phi$ = ANGLE FROM ZERO
UNTIL FAULT OCCURS

b

**FIGURE 5.2** Asymmetrical fault current waveform. (a) Current waveform (b) Circuit.

Interrupting fins placed in the path of the rising arc will stretch the arc farther, cool it more, and aid in extinguishing the arc. This is shown in Figure 5.4d. Large low voltage breakers will have interrupting fins.

Magnetic blowout refers to the use of a transverse magnetic field near the contacts to stretch and drive the arc into the interrupting fins, as shown in Figure 5.5. The magnetic field interacts with the ions of the arc to provide the driving force. The magnetic field can come from a permanent magnet in small breakers, but is provided by a properly positioned coil through which the contact current flows in larger low voltage breakers. Circuit breakers with magnetic blowout and interrupting fins can even be used for lower medium voltages. This combination can interrupt current as high as 42,000 A at 8 kV per phase. Three-phase circuit breakers

**FIGURE 5.3** Low voltage circuit breakers

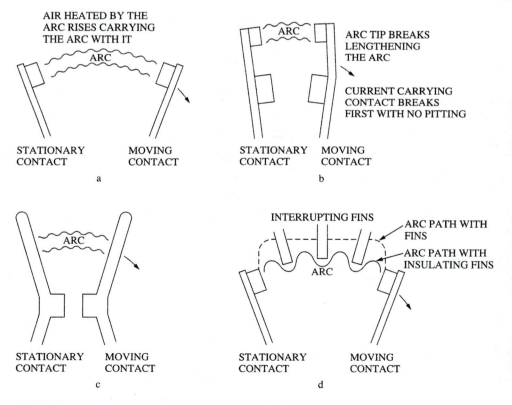

AIR HEATED BY THE
ARC RISES CARRYING
THE ARC WITH IT

ARC

STATIONARY          MOVING
CONTACT            CONTACT

a

ARC

ARC TIP BREAKS
LENGTHENING
THE ARC

CURRENT CARRYING
CONTACT BREAKS
FIRST WITH NO PITTING

STATIONARY   MOVING
CONTACT      CONTACT

b

ARC

STATIONARY   MOVING
CONTACT      CONTACT

c

INTERRUPTING FINS

ARC PATH WITH
FINS

ARC PATH WITH
INSULATING FINS

ARC

STATIONARY      MOVING
CONTACT         CONTACT

d

**FIGURE 5.4** Arc lengthening methods include: (a) Convection (b) Arc tip (c) Arc horn and (d) Interrupting fins.

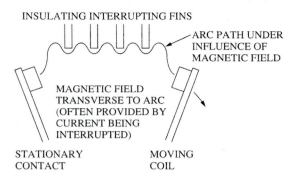

INSULATING INTERRUPTING FINS

ARC PATH UNDER
INFLUENCE OF
MAGNETIC FIELD

MAGNETIC FIELD
TRANSVERSE TO ARC
(OFTEN PROVIDED BY
CURRENT BEING
INTERRUPTED)

STATIONARY
CONTACT

MOVING
COIL

**FIGURE 5.5**   Magnetic blow out

use barriers between the phases made of insulating material. The barriers provide arc chutes to prevent phase-to-phase faults from originating in the breaker.

Air circuit breaker contacts are driven open by strong springs that are compressed (called charging the breaker) by rachet mechanisms or by geared down electric motors. The trip signal is usually originated by a magnetic coil or a thermal element built into the breaker, although some larger, medium voltage air circuit breakers are designed to trip on protective relay signals. Most low voltage breakers have either bolt on terminal connections, or blade springs designed to slide into a mating bus in the cabinet as the breakers are bolted into place. Medium voltage cabinet mount air breakers have spring mounted metal fingers arranged in a circle that slip over a mating tubular contact in the cabinet. Larger medium voltage air breakers are mounted on wheels for ease in installation.

## 5.1.3 Air Blast Circuit Breakers

*Cross air blast circuit breakers* are special purpose medium voltage circuit breakers used where noise is an important factor. They are used primarily at voltages between 14.4 and 34.5 kV. Figure 5.6 shows the principle of cross air blast circuit breakers. A blast of compressed air (to 800 psi) is blown across the circuit breaker contacts as the contacts open. The blast of high pressure air blows the arc into the interrupting fins, stretches the arc, and cools it.

Air blast breaker contact opening mechanisms are powered by powerful springs, as are the opening mechanisms of all high voltage and medium voltage circuit breakers. The contacts are closed by either a motor or compressed air. Most air blast circuit breakers have bushings for connection.

*Axial air blast circuit breakers* blow high pressure air along the axis of the contacts to stretch and cool the arc, as shown in Figure 5.7. Usually the air is blown from a port next to the stationary contact toward the moving contact. Axial air blast breakers are usually high voltage breakers. They can be built to interrupt currents as high as 63 kA at 800 kV. A single interrupting contact such as that of Figure 5.7 can operate reliably to about 70 kV, so higher voltage circuit breakers use several contacts in series. Two sets of contacts can be built into a single package called an interrupter. Figure 5.8 shows a photograph of an axial air blast breaker.

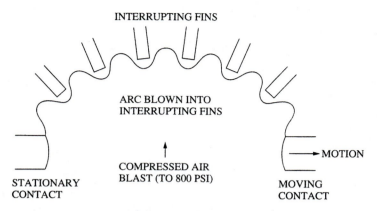

**FIGURE 5.6**   Cross air blast circuit breaker principle

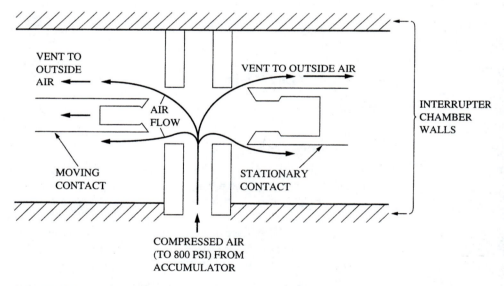

**FIGURE 5.7**   Axial air blast circuit breaker principle

Air blast breakers (assume axial unless otherwise noted) must have a resistor across the arc, breaking contact to dampen possible high voltage oscillations that might re-ignite the arc. Auxiliary contacts break the low current through the resistor after the arc is broken. The contacts are driven open by springs (compressed with compressed air) acting on the actuating linkages after the breaker receives a trip signal from a protective relay.

**FIGURE 5.8** Axial air blast breaker (courtesy of Houston Lighting & Power).

## 5.1.4 Vacuum Circuit Breakers

Vacuum circuit breaker contacts, shown in Figure 5.9, are enclosed in a container with a high vacuum. No significant arcing can occur because there is no air between the contacts to ionize. A small amount of metal on the arcing sleeve around the contact may ionize on each operation, but not enough to sustain an arc. The few metal atoms that arc off on operation are ejected at nearly right angles to the contacts and collected on the metal vapor condensing shield so they do not accumulate around the contacts. A metal bellows allows the moving contact to do so. Vacuum circuit breakers are operated by spring force. Vacuum breakers are available for voltages from 480 V to over 34.5 kV. Many are suitable for cabinet mounted indoor use.

## 5.1.5 Oil Circuit Breakers (OCBs)

Oil circuit breakers use oil as the arc interrupting medium. A photo of a typical high voltage oil circuit breaker is shown in Figure 5.10a, and an outline drawing with some dimensions in Figure 5.10b. Oil has a dielectric strength far in excess of air. When contacts open in oil the arc causes the oil to disassociate which absorbs arc energy. One of the products of disassociation is hydrogen, which has a higher heat capacity than air and is superior to air as a cooling medium (hydrogen is not explosive if no oxygen is present). The hydrogen is ionized by the arc, which

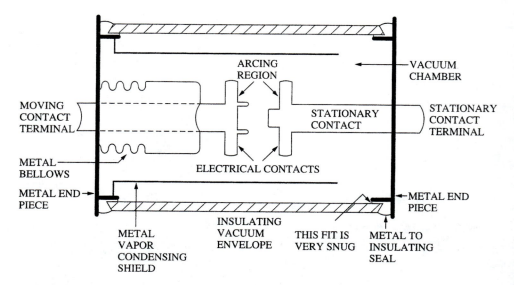

FIGURE 5.9   Vacuum circuit breaker contacts

**FIGURE 5.10**   145 kV 2000 A Westinghouse oil circuit breaker
(courtesy of Westinghouse)  (a) Circuit breaker

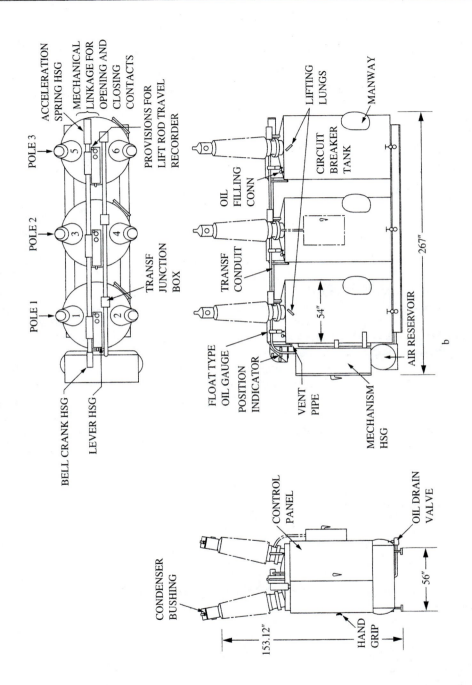

**FIGURE 5.10** (b) breaker outline

removes more energy from the arc. The hydrogen bubbles move upward from the arc so new oil is brought between the contacts, which disassociates, producing more hydrogen, which ionizes and rises, allowing more oil between the contacts. This process continues until the arc is quenched. Bulk oil breakers, in which the contacts are simply immersed in a tank of oil, operate to about 230 kV. Bulk oil breakers have a limited interrupting capacity.

The use of an arcing chamber can increase the capacity of an oil breaker by a factor of 500, to 10,000 MVA and a break voltage of 500 kV. All but very modest capacity circuit breakers use them. An oil arcing chamber is shown in Figure 5.11a. It is an insulating fiber chamber, with arc interrupters on one side (called arc splitters in this application), within which the contacts open. The arcing chamber is also called an oil blast chamber. In operation the arc vaporizes the oil in the chamber very rapidly, evolving high pressure hydrogen (because it is confined in the chamber). The high pressure hydrogen is hard to ionize. The hot, high pressure hydrogen expands with explosive force and blows the arc into the arc splitters causing the arc to stretch and cool. The process could be called *hydrogen blast*. The placement of the chambers in the tank is shown in Figure 5.11b.

There are two tank configurations of oil circuit breakers. Those with large, grounded, tanks are called dead tanks. The pictures of oil circuit breakers in this chapter are dead tank designs. The other design, called live tank or minimum oil, uses less oil. The oil blast chamber is in a small tank supported on insulators like an air blast breaker interrupter. Live tank oil breakers work very well but the dead tank design is preferred in the United States because it is less likely to explode.

Oil circuit breakers are very seldom used above 345 kV even though they can be built for higher voltages because they become very large compared to $SF_6$ and air blast circuit breakers. The contact operating mechanisms are powered by powerful springs that are compressed for operation by compressed air at about 500 psi.

### 5.1.6  Sulphur Hexaflouride ($SF_6$) Circuit Breakers

$SF_6$ gas is a popular interrupting medium for high voltage and extremely high voltage (EHV, above 345 kV) applications. Its voltage withstand rating is about three times that of air and it is extremely electronegative. That means its atoms bind for a considerable time to free electrons, thus becoming negative ions. When free electrons are removed from the arc it is difficult to sustain because no free electrons are available to accelerate and ionize atoms by collision. $SF_6$ breakers are usually live tank designs with the tanks supported on insulators as shown in Figure 5.12a. There are two types in use, two tank and puffer.

Two tank designs have a high pressure tank in which the interrupter is located, and a low pressure reserve tank, as shown in Figure 5.12c. On a fault the high pressure tank is vented to the low pressure tank, creating turbulence to help interrupt the arc. The low pressure $SF_6$ is then compressed and returned to the high pressure tank. The high pressure tank must be warmed at low temperatures to keep the high pressure $SF_6$ from liquefying.

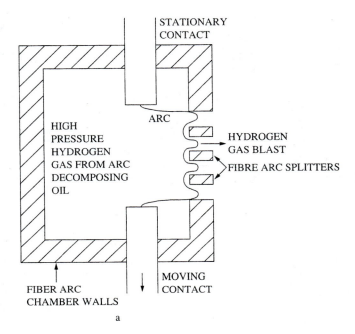

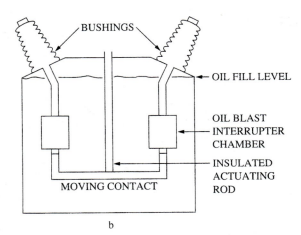

**FIGURE 5.11** Oil blast chamber circuit breaker (a) Oil blast chamber and (b) circuit breaker with two series chambers

Puffer designs are simpler than two tank designs because all of the $SF_6$ is at one pressure, 75 psi, which does not liquify until the temperature drops to -30°C. A puffer type $SF_6$ circuit breaker cut away is shown in Figure 5.12b. The grading capacitors are a capacitive voltage divider to force each of the two interrupters to share the voltage equally. The mechanical actuation is driven by a strong spring.

a

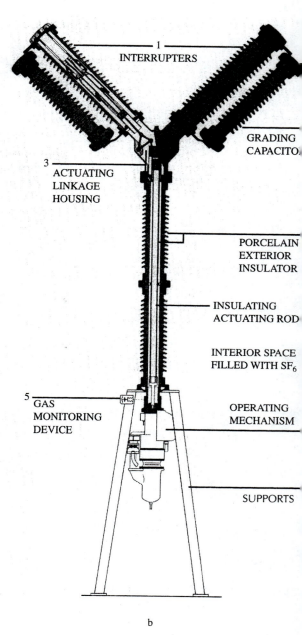

1 INTERRUPTERS

GRADING CAPACITO

3 ACTUATING LINKAGE HOUSING

PORCELAIN EXTERIOR INSULATOR

INSULATING ACTUATING ROD

INTERIOR SPACE FILLED WITH SF$_6$

5 GAS MONITORING DEVICE

OPERATING MECHANISM

SUPPORTS

b

**FIGURE 5.12**   SF$_6$ circuit breaker (a) 362 kV at 4000 A three-phase, interrupting capacity—50 kA (b) Cross section drawing

**FIGURE 5.12**  (c) Dead tank SF$_6$

On actuation a cylinder of 75 psi SF$_6$ is compressed to a high pressure and released across the contacts as they separate. This is illustrated in Figure 5.13.

Figure 5.14 illustrates the use of closing (also called opening) resistors for damping oscillations when the arc is broken. Even though SF$_6$ breakers do not need closing resistors to break the arc, the resistors provide valuable transient suppression in many applications, especially at 345 kV and above.

SF$_6$ breakers are available that can handle up to 8000 A continuous current and they can have interrupt capacities to 63 kA at 800 kV, and 80 kA at 230 kV. Puffer types have somewhat lower interrupt capacities (about 50 kA) than dead tank types because the operating mechanisms become massive when they are built for the same interrupt capacity.

## 5.1.7  Circuit Breaker Ratings

Users of circuit breakers must consider a number of ratings to select the right one. The continuous voltage rating, which may decrease at altitudes above 3300 ft, must be adequate. The rated impulse voltage (BIL) must be considered for insulation coordination, lightning, and surge protection.

The continuous current rating must be adequate for maximum loads and the interrupt capacity must be greater than the maximum fault current the breaker will have to interrupt. This means that the maximum fault current must be known. We will learn to calculate fault currents in later chapters. Additionally the MVA rating must be adequate.

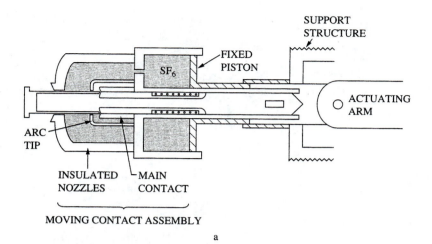

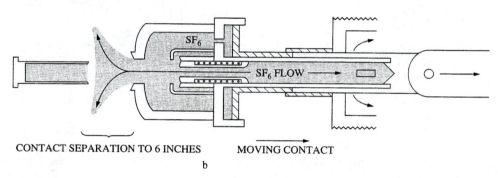

**FIGURE 5.13**   SF$_6$ puffer operation (a) Closed and (b) open

The interrupting time must be fast enough to provide proper protection for the system, and if it is to be automatically reclosed, the reclose time must be known. The current rating must be high enough to allow the system to be energized (recall that each transformer in the system will take some inrush current). Breakers above 250 MVA have no manual closing so the amount of station power needed to close the breaker must be known.

In addition to the electrical parameters there are a number of mechanical considerations such as size, foundation requirements, and space required. Table 5.1 lists a number of OCB ratings, and Figure 5.15 shows the space requirements for a series of SF$_6$ circuit breakers.

## 5.1.8  Circuit Breaker Controls

Circuit breaker controls are quite complex because of the number of functions that must be performed. The controls must include monitoring sensors for the compressed air, and equipment to start the compressor when the pressure is too

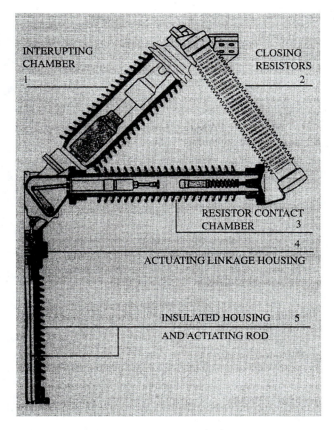

**FIGURE 5.14** SF$_6$ interrupting unit with closing resistor (Modified from ASEA Brown Boveri. With permission.)

low, or send an alarm when the pressure is too high or zero. The interrupting medium level and/or pressure must be monitored. Any auxiliary heating equipment must be monitored and controlled.

The trip signal from the protective relays must be monitored and acted upon, and the status of the breaker (open or closed) must be relayed to the appropriate control station, so must any malfunction information. Figure 5.16 is an oil circuit breaker control panel.

## 5.1.9 High Voltage Circuit Breaker Comparison

Low voltage circuit breakers are usually air with arc interrupting fins, and higher capacity low voltage breakers also have magnetic blowout. Medium voltage breakers are oil, cross air blast, or vacuum typically, although many indoor medium voltage circuit breakers use magnetic blowout with interrupting fins. High voltage and extra high voltage breakers are almost always either oil, axial air blast, or SF$_6$.

■■■■ **TABLE 5.1    Rating Structure for 145 GMB Breaker (Standard)**

**Symmetrical Current Basis of Rating**
**Table 4A, ANSI C37.06-1971**

Breaker Type _145GMB63_    DB No. __33–253__    Standard Outline No. __2069C91__
Breakdown kV ___145___    Voltage Range Factor K = ____1____
Continuous Current Rating __3000__ Amps.  Close & Latch Rating __101__ kA
Rated RMS, 3 Sec. Rating __63__ kA
Rated Short Circuit Current __63__ Sym. kA Max.  Sym. Interrupting Rating __63__ kA
Rated Overhead Line Charging Current Switching __80__ Amp.
Applicable Duty Cycle _CO–15 Sec. –CO_  Special Duty Cycle Rating Factor _____ %
one minute 60 Hertz Withstand __310__ kV  Impulse Withstand

Full wave _650_ kV — Impulse
Full wave _488_ kV — Interrupter
2 μsec. __838__ kV — Chopped wave
3 μsec. __748__ kV — Chopped wave

Open Time to part contacts __1.8__ Cycles  Closing time __16__ Cycles
Reclose Time __20__ Cycles  Rated Interrupting Time 25–100% Rating __3__ Cycles
1– 25% Rating __4__ Cycles
Additional Time for Interruption Resistor Circuit __1.5__ Cycles

Line Spacing ____72____ In.
External Creepage Over Bushing ___92.5___ In.
External Striking Distance Phase-to-Ground __46.31__ In.
Phase-to-Phase __62.62__ In.
Internal Striking Distance Phase-to-Ground __10.16__ In.

Closing Circuit:                                   Tripping Current:
E/Z Current __3__ Amperes at _125_ Vdc.   E/Z Current __20__ Amperes at _125_ Vdc
Cutoff Current __3__ Amperes at _125_ Vdc  Cutoff Current __14__ Amperes at _125_ Vdc

Heater Requirements:     Mechanism Type _AA 10–80_ ; Pneumatic
  Volts ___230___         Stored Operations ___5___
  Watts __700__           Operating Pressure __270__ psi;
                          Range __225__ psi (cutoff) to __270__ psi

Compressor Requirements:
Motor HP ____2____
Voltage __230__         Phase __1__          Start Current __64__ Amperes
Pump Up Time Lock-out to Normal __30__ Minutes  Run Current __12.5__ Amperes
Condenser Bushing Data (not in breaker):
Type _"OC"_ Manufacturer _Westinghouse_ Voltage Rating _145_ kV Current _3000_ Amperes
One Minute Withstand Test RMS Dry — 60 Hertz          _335_ kV
60 Sec. Withstand Test RMS Wet — 60 Hertz            _275_ kV
Impulse Withstand Test 1–1/2 × 40 Full Wave (Crest)  _650_ kV
Safe Cantilever Loading of Bushing Installed         _150_ Lb
Bushing Drawing Number _5621D62 G14_

Space Available for ___18___ Current Transformers No. of CT's Supplied ___6___
CT Type __BYM__  CT Ratio __3000/5__ Amperes  CT Accuracy Class __C800__
Interrupter Type _SB2A_            Contact Resistance Measured Between Bushing
Contact Travel __18__ In.          Top Terminals 127 Microhms
Break Distance Per Pole __33.5__ In.
Head of Oil Above Contacts __45.75__ In. Volume of Air Above Oil __8.8__ Cu. Ft.

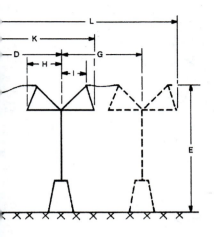

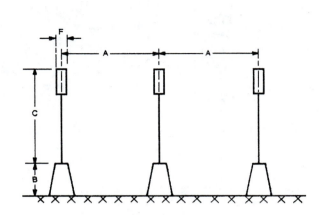

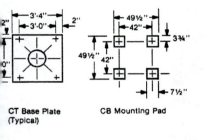

CT Base Plate
(Typical)

CB Mounting Pad

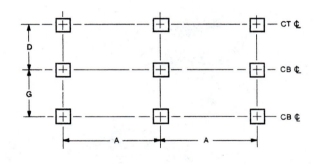

| ating | A | B | C | D | E | F | G | H | I | K | L |
|---|---|---|---|---|---|---|---|---|---|---|---|
| 2 kV | 106 | 98 | 152 | 108 | 250 | 14 | NA | NA | 57 | 188 | NA |
| 2 kV | 161 | 98 | 214 | 174 | 312 | 41 | NA | 93 | 66 | 286 | NA |
| 0 kV | 236 | 98 | 227 | 180 | 325 | 41 | 220 | 87 | 60 | — | 533 |
| 0 kV | 354 | 98 | 286 | 201 | 384 | 41 | 256 | 95 | 66 | — | 576 |

mensions in inches

FIGURE 5.15   SF$_6$ breaker dimensions

**FIGURE 5.16** Circuit breaker control panel

Oil breakers have the advantages of using a very well established technology, low cost, easy installation, low maintenance, and dead tank design. Additionally, oil breakers need no closing resistors to break an arc, although closing resistors may be used to reduce switching transients. The disadvantages of oil breakers are: possible oil fire hazard, slower interrupting times than the other HV breakers, and physical size of breakers for voltages over 345 kV are excessive. Oil breakers are the oldest HV breaker type.

Axial air blast circuit breakers have the advantages of using a well established technology, high interrupt capacity, and fast interrupt times (usually one or two cycles). Their disadvantages are: they are noisy, relatively expensive, need very high pressure air (800 psi), use live tank designs, and must use closing resistors.

Two tank $SF_6$ circuit breakers have high interrupt capacities, a live or dead tank option, and need no closing resistors for arc quenching. Additionally, the live tank design is well known in the United States. The disadvantages are: $SF_6$ tank heaters are needed in cold climates, high pressure $SF_6$ must be stored in the breaker and as a result the breakers have a high $SF_6$ leakage.

$SF_6$ puffer breakers are simpler than two tank types, have low leakage because only low pressure $SF_6$ (75 psi) is stored, and need no tank heaters until the temperature is $-30°C$. They need no closing resistors to break an arc. Currently the interrupting capacity of puffer designs is 20 to 30% less than two tank designs, and there is limited experience with puffer designs in the United States. All puffers are live tank designs. $SF_6$ puffer circuit breakers are rapidly becoming very popular for EHV applications.

## 5.2

## RECLOSERS

Most faults (80-95%) on distribution and transmission lines are temporary, lasting from a few cycles to a few seconds. They are caused by such things as tree limbs falling or blowing across the lines and are removed when the limb burns off or is blown out of the line. Reclosers allow temporary faults to clear and then restore service quickly, but disconnect a permanent fault.

Reclosers are essentially special purpose, light duty circuit breakers. They can interrupt overloads but not severe faults. Reclosers sense an overcurrent, open, then after a preprogrammed time, reclose. They can be programmed to sense an overcurrent, open, and reclose several times (up to five times is typical) and after the preset number of operations remain open (lock out). This process is shown in Figure 5.17. They must be manually reset after lock out. The preset number of

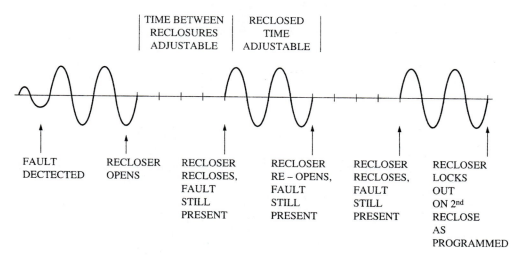

**FIGURE 5.17**   Recloser operation

operations is usually set at three or four. The time between reclosures and the time of the reclose can also be programmed. The time of reclose can be set from a couple of cycles (instantaneous) to around 600 cycles in fixed increments with 2 and 5 second reclosures common, and the time between recloses can be set from about 1 to 10 seconds in fixed intervals. This is necessary to coordinate the reclosers with other devices. If the recloser is not locked out and no further overcurrents are sensed for 120 seconds, the recloser resets to start counting from one again. Reclosers are much lighter, smaller, and cheaper than power circuit breakers. Pole top mounted reclosers are most used. Single-phase and three-phase reclosers are available. If single phasing of three-phase loads is possible on the protected circuit, as in a commercial area feeder, three-phase reclosers must be so that all three phases can be locked out from a single-phase trip. Single- and three-phase reclosers are shown in Figures 5.18a and 5.18b.

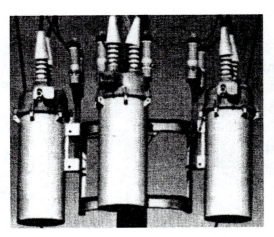

a                                                                      b

**FIGURE 5.18**   Reclosers (a) Single-phase recloser (Reprinted from Cooper Power Systems. With permission.) (b) Three-phase recloser (Reprinted from Cooper Power Systems. With permission.)

Two types of reclosers are currently manufactured. In one type the times are controlled by pistons in hydraulic cylinders, and in the other by electronic circuitry. Figure 5.19 show single phase and three phase hydraulic reclosers. Electronic reclosers are more flexible, accurate, and easily tested than hydraulic reclosers, but are also more expensive so electronic controls are used primarily on heavy duty three-phase reclosers.

Recloser interrupters are usually oil or vacuum. Some hydraulic reclosers use the same oil for timing and interruption. Single-phase reclosers are available to 560 A continuous at 34.5 kV with an interrupting capacity of 8 kA. Three-phase reclosers are available to 560 A continuous at 69 kV with an 8 kA interrupting capacity.

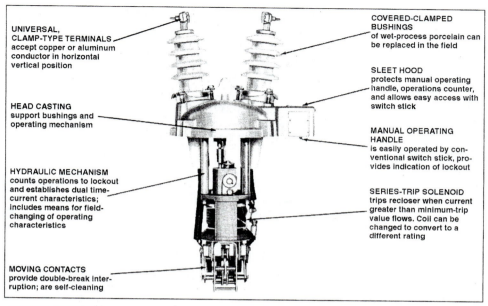

UNIVERSAL,
CLAMP-TYPE TERMINALS
accept copper or aluminum
conductor in horizontal
vertical position

HEAD CASTING
support bushings and
operating mechanism

HYDRAULIC MECHANISM
counts operations to lockout
and establishes dual time-
current characteristics;
includes means for field-
changing of operating
characteristics

MOVING CONTACTS
provide double-break interruption; are self-cleaning

COVERED-CLAMPED
BUSHINGS
of wet-process porcelain can
be replaced in the field

SLEET HOOD
protects manual operating
handle, operations counter,
and allows easy access with
switch stick

MANUAL OPERATING
HANDLE
is easily operated by conventional switch stick, provides indication of lockout

SERIES-TRIP SOLENOID
trips recloser when current
greater than minimum-trip
value flows. Coil can be
changed to convert to a
different rating

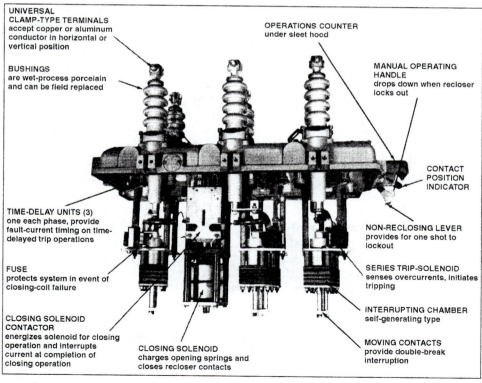

UNIVERSAL
CLAMP-TYPE TERMINALS
accept copper or aluminum
conductor in horizontal or
vertical position

BUSHINGS
are wet-process porcelain
and can be field replaced

TIME-DELAY UNITS (3)
one each phase, provide
fault-current timing on time-
delayed trip operations

FUSE
protects system in event of
closing-coil failure

CLOSING SOLENOID
CONTACTOR
energizes solenoid for closing
operation and interrupts
current at completion of
closing operation

CLOSING SOLENOID
charges opening springs and
closes recloser contacts

OPERATIONS COUNTER
under sleet hood

MANUAL OPERATING
HANDLE
drops down when recloser
locks out

CONTACT
POSITION
INDICATOR

NON-RECLOSING LEVER
provides for one shot to
lockout

SERIES TRIP-SOLENOID
senses overcurrents, initiates
tripping

INTERRUPTING CHAMBER
self-generating type

MOVING CONTACTS
provide double-break
interruption

**FIGURE 5.19** Recloser mechanism (a) single-phase, (b) three-phase (Reprinted from Cooper Power Systems. With permission.)

Reclosers are used to protect feeders leaving substations, for important taps on feeders, and to sectionalize faulted distribution lines to minimize the area out of service. Reclosers must be selected by voltage, continuous current, minimum fault current in the protected zone (it must be enough to operate the recloser), interrupt capacity, ability to coordinate with other protective devices, ground fault sensing capability, and BIL rating for transient protection.

*Coordination* means that the load side device should operate and clear a fault before the source side device operates. That way the source side device serves as a backup for the load side device and the minimum possible area is taken out of service.

## 5.3

## SECTIONALIZERS

A sectionalizer is a device that is used to automatically isolate faulted line segments from a distribution system. It senses any current above its actuating current followed by a line de-energization by a recloser. It counts the number of overcurrent followed by line de-energization sequences and after a preset number of them (usually two or three) it opens and locks out. The sectionalizer must be manually reset after lock out. If normal line conditions continue for a preset length of time after an overcurrent, de-energization sequence below the preset lock out number, the sectionalizer will reset itself to zero count. The delay before reset is usually set between 30 and 90 seconds. Figure 5.20 shows a three-phase sectionalizer.

Sectionalizers can break very modest overcurrents, but are normally not used to do so. Their current ratings in excess of rated load current (only twice rated load on larger units) are to allow the recloser to handle any inrush current on re-energizing a line. Sectionalizers are not designed to break current in normal operation.

Two types of sectionalizers are available. Smaller ones are hydraulically operated in a manner similar to reclosers, and higher capacity ones are electronically operated. Sectionalizers are available to 34.5 kV with continuous current ratings as high as 600 A, interrupt currents as high as 1320 A, and BIL ratings as high as 150 kV. They can be equipped with voltage restraint so that only source side de-energizations are counted, and they can also be equipped to sense ground faults.

Circuit breakers, reclosers, and sectionalizers are used together to provide better protection of lines. Figure 5.21 shows a one-line diagram with a circuit breaker, transformer, recloser, and the sectionalizers for three feeder circuits and a tap from one branch. Assume a fault on feeder three. The recloser senses the overcurrent and opens. After the preset time it recloses. The sectionalizer senses each overcurrent and opening of the recloser. After the third recloser reclose the sectionalizer locks out if the fault is still present and isolates feeder three. It operates when the recloser opens for the fourth time. If the recloser senses a fault after the fourth reclose it means the fault is between the sectionalizers and the recloser. The recloser is set to lock out then. If the circuit breaker opens that means the fault is between the breaker and the reclosers or that the recloser failed to operate. If the

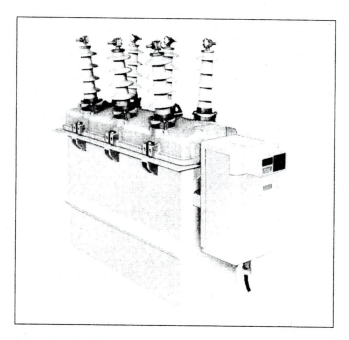

**FIGURE 5.20** Three-phase electronically controlled sectionalizer (Reprinted from Cooper Power Systems. With permission.)

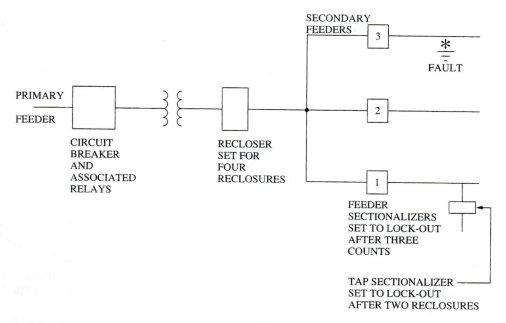

**FIGURE 5.21** Circuit breaker, recloser, and sectionalizers together

fault had been on the tap circuit of feeder one, the tap recloser would have locked out upon the third opening of the recloser (after its second reclose). If the fault is temporary in nature, such as a tree branch fallen across the line, the recloser and the associated sectionalizer would reset to count equal zero after a short while.

Variations of the scheme outlined above are in wide use with the preset counts and actuating currents set to meet the particular application. A circuit breaker might cost $200,000 for a particular 138 kV 2000 A application with another $15,000 to $20,000 in protective relaying and instrument transformers. A recloser would cost about $80,000 and a sectionalizer about $20,000 for use in the protection scheme. By combining circuit breakers, reclosers, sectionalizers, and fuses (see the next section) the distribution system protection can be good while the cost is minimized.

## 5.4

### FUSES

Fuses are one-time devices that must be replaced each time they open a fault. They use a metallic element that melts when an overload current passes through it. The melted element separates breaking the circuit.

### 5.4.1 Low Voltage and Current Limiting Fuses

Low voltage fuses use zinc, copper, or silver as the metallic element, while medium and high voltage fuses typically use tin, cadmium, or silver. Figure 5.22a is a cutaway drawing of a fuse. Notice that the narrow fuse element points, which melt, are in series so that the voltage is divided between two or more breaks, making the arcs easier to extinguish. Two element fuses (often called slow blow) are available for loads with high inrush currents, such as motors. One element is a normal fault interrupting element to open if a short circuit causes current to be several times the continuous current rating, and the other element is an overload sensing element that opens only if a current of about five times the continuous rating remains for over a second. Lower current overloads take longer to open the fuse.

Current limiting fuses are fuses that limit the peak fault current to less than it would be without a current limiting fuse, and break the circuit (clear) in less than one-half cycle. A current limiting fuse cartridge is filled with sand. The sand melts but does not disassociate so it absorbs heat energy cooling the arc, plus the sand filling leaves little air to support an arc. Silver elements are used in the best low voltage current limiting fuses because silver is a good conductor, so very little silver is required to carry the continuous current. Thus the arc contains fewer metal ions to be quenched. The response of a current limiting fuse to a fault is shown in Figure 5.22b.

Current limiting is valuable because both the heating and mechanical damage caused by a fault is proportional to the square of the current. The heating to the rms current squared, and the mechanical, caused by magnetic forces, to the peak current squared. Current limiting fuses reduce both the rms and peak fault current.

Current limiting fuses are available with voltage ratings to 69 kV, continuous current ratings to several thousand Amperes, and interrupt current capacities to 200 kA.

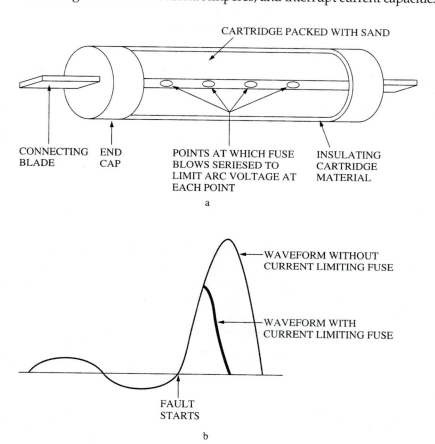

**FIGURE 5.22** Current limiting fuse (a) Mechanical configuration and (b) fault current waveform

## 5.4.2 Expulsion fuses

Expulsion fuses, when blowing, use the heat generated by the melting element to decompose a material on the inner fuse wall. The decomposed material produces a high pressure, turbulent gas that blows the arc out the end of the fuse tube. They sound like a shotgun when they blow. The material is usually the interior fiber of the fuse tube, boric acid powder on the interior of the fuse tube, which melts, then turns to a turbulent steam, then blows the arc out, or a nonflammable gas such as carbon tetrachloride. A number of expulsion fuses are shown in Figure 5.23. Expulsion fuse links are usually tin.

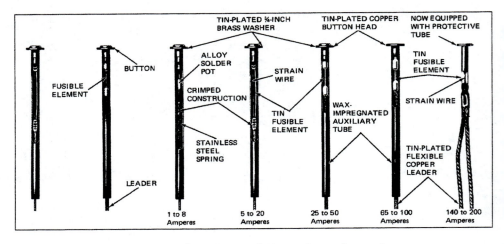

**FIGURE 5.23**  Expulsion fuse links (courtesy of Cooper Power Systems)

Expulsion fuses are not independent of their holders (called cutouts). The holders are designed to help break the arc by pulling the fuse element ends away from the melted point, and to provide an indication, visible from a considerable distance, that the fuse is blown. To accomplish this the cutout fuse holders are spring loaded. The fuse is attached to the spring-loaded fuse clip support. The fuse has a fine, high resistance strain wire running through it to support the force from the spring. The strain wire softens enough to break when the fuse blows. The spring-loaded part of the cutout then flies apart with great force. The fuse cutouts are designed so that they cannot be closed without a good fuse in them. Figure 5.24 shows some expulsion fuse cutouts.

Expulsion fuses are available with voltage ratings to over 34.5 kV, continuous current ratings to over 1000 A, and interrupt capacities to over 12.5 kA.

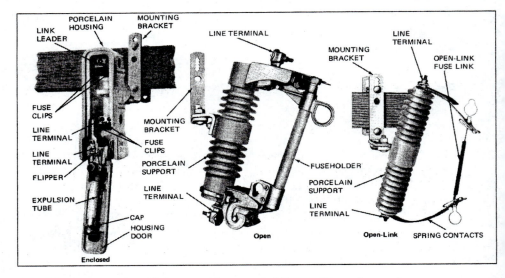

**FIGURE 5.24**  Expulsion fuse cutouts (courtesy of Cooper Power Systems)

## 5.4.3 Fuse Application Considerations

The six items to be considered for fuse applications are also items that must be considered for any overcurrent protection device.

1. Voltage: The voltage rating of the fuse must be greater than the system voltage. The fuse holder or cutout voltage rating must also be higher than the system voltage, and the BIL of the medium and high voltage fuse cutouts must be acceptable for the application. Higher voltage fuses are usually larger than lower voltage fuses to allow sufficient room to break the arc.

2. Continuous current: The continuous current rating should be 125% of the maximum load current for low voltage fuses, with the exception of bolt-on fuses in which they can be equal. Imprecision and variables such as ambient temperature can lead to fuse operation when the maximum overload current has not been exceeded if the fuse continuous current rating is exactly equal to the maximum allowed current. Medium or higher voltage fuses should have a continuous current rating equal to or higher than the maximum load current.

3. Interrupt Capacity: The fuse should be able to interrupt the highest fault current available in the zone to be protected.

4. Current Limiting: The desirability of using current limiting fuses should be considered. Often a current limiting fuse can limit the maximum current of a severe fault to a value that allows lower interrupt capacity devices to be used downstream for overload protection. An example of this is the use of current limiting fuses to protect downstream low voltage breakers in lighting panels. High interrupt capacity in a low voltage molded case breaker is very expensive, so the higher cost of the current limiting fuse is easily justified. Current limiting fuses may allow the use of equipment with lower fault current withstand ratings, such as bus structures, in some applications.

5. Time-Current: The time-current characteristic of a fuse is the time it takes the fuse to blow with different size fault currents. Fuses blow more quickly at high currents than at low currents. Figure 5.25 shows the time current curves for EEI-NEMA type T and K links, slow and fast links, respectively. The slow links allow normal motor and transformer inrush current to occur without blowing the fuse. Low voltage slow fuses use dual elements to obtain slow-blow characteristics.

6. Fuse Coordination: Coordination is accomplished by making sure the downstream fuse clears before the upstream fuse element melts. Figure 5.26 shows the melt and clear curves of a fuse. This principle is applied to the time current curves to select fuses. Coordination charts are supplied by fuse manufacturers to simplify the task of coordination. The time current curves to be coordinated need not all be fuse curves. Fuses can be coordinated with circuit breakers and reclosers. The curves for each are superimposed for evaluation. Figure 5.27b shows a group of time current melt-clear curves for coordination.

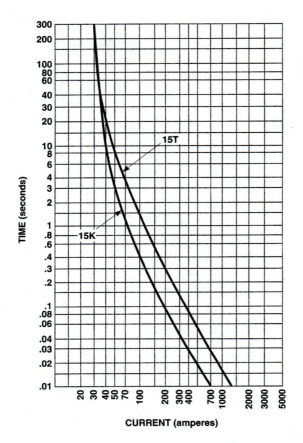

**FIGURE 5.25**   Fuse time-current curve (courtesy of Cooper Power Systems)

**Example 5.1:**

Select the proper fuses from the time current curves of Figure 5.27b, and continuous current ratings in Table 5.2, for proper overcurrent protection coordination of the circuit shown in Figure 5.27a. The maximum and minimum fault currents for each segment of the circuit is shown in the circles (we will learn to calculate fault currents a bit later).

**Solution:**

An interrupt current rating of 2400 A is sufficient for any of the circuit fault currents, so we will select fuses with this rating or higher. This is an easily obtained rating.

Starting at point C, we find from Table 5.2 that the 15T designated fuse carries 23 A continuous, so it is usable at point C. At the maximum fault current at point C it melts in 0.008 s and clears in 0.03 s.

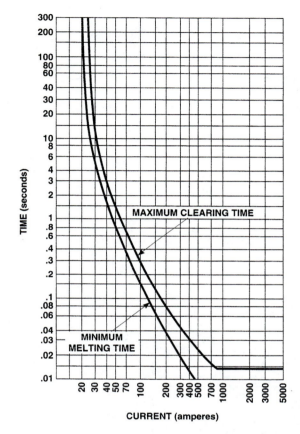

**FIGURE 5.26**   Melt-clear graph for 10K fuse link (courtesy of Cooper Power Systems)

Next to point B. The next fuse on the time current chart of Figure 5.27b, the 25T, cannot be used because it cannot carry 40 A continuously. The 30T can carry 45 A continuous. Its melt time is 0.033 s, and its clear time is 0.056 s at the maximum fault current. It meets the criteria of not melting until the downstream fuse has cleared.

The 80T is able to carry 100 A continuously, and at the maximum fault current its melt time is 0.16 s, which is longer than the clear time of the 30T. Its clear time is off the chart, but we can estimate it at about 0.3s. It coordinates with the 30T, which coordinates with the 15T.

The fuses should also coordinate at the minimum fault currents. The method is the same. As long as the time-current curves do not cross, there is usually no problem at the minimum fault currents, but the coordination should be checked anyway. Once again, fuse manufacturers coordination tables simplify the job of fuse coordination.

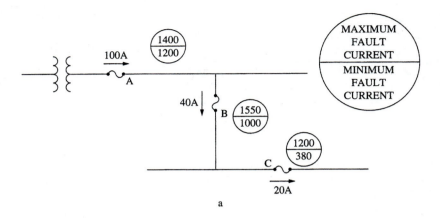

a

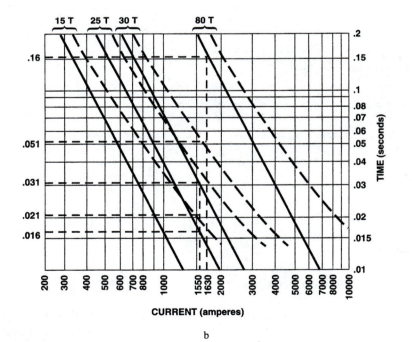

b

**FIGURE 5.27**   Fuse coordination (a) Circuit Fuse coordination (b) Time-current curves, minimum melting curves are solid, maximum clearing curves are dashed (courtesy of Cooper Power Systems)

■■■■ TABLE 5.2   Fuse Ratings, Tin Fuse Links

| IEEE-NEMA K or T rating | Continuous current (amperes) |
|:---:|:---:|
| 6 | 9 |
| 8 | 12 |
| 10 | 15 |
| 12 | 18 |
| 15 | 23 |
| 20 | 30 |
| 25 | 38 |
| 30 | 45 |
| 40 | 60* |
| 50 | 75* |
| 65 | 95 |
| 80 | 120† |
| 100 | 150† |
| 140 | 190 |
| 200 | 200 |

*Only when used in a 100 or 200 A cutout
†Only when used in a 200 A cutout

## 5.5

## LIGHTNING PROTECTION

People have always been fascinated by lightning. The immense power and beauty of a lightning bolt is awe inspiring. Based on recorded tower strikes the current in a lightning strike can be as high as 200,000 A, but only about half of the strikes exceed 15,000 A. Lightning strikes power lines somewhere between 59 and 232 times per 100 miles each year on the average. The number varies greatly from year to year, and for different geographical locations, with Naples, Florida, having the greatest number of lightning storms per year in the United States. The National Weather Service keeps maps showing the number of strikes by location. The number of lightning strikes per year on electric power lines is sufficient to provide unreliable service in most areas if lightning protection is not used.

### 5.5.1  Lightning

Cloud Charge Distribution is the key to lightning formation. There are many theories on the formation of the cloud charges that lead to lightning. The most accepted ones are those of C. T. R. Wilson and G. C. Simpson with some modifications. Both relate ascending air currents and raindrop size to charge accumulation. They are summarized briefly in the following paragraphs.

The atmosphere is filled with ions that attach themselves to dust particles and small water droplets. The larger ions have low mobility (ease with which they move) and smaller ions have higher mobility. Many larger ions are positively charged. Mackey found that water droplets in a high electric field become elongated because water is easily polarized; it has a dielectric constant of about 80. At very high electric fields, around 10,000 volts/meter, drops with an average diameter greater than 0.15 cm set up a glow discharge that evaporates water until the droplet size is reduced. Wilson posited a normal atmospheric electric field of approximately 1 V/cm at the Earth's surface that decreases to about 0.02 V/cm at 30,000 ft. Relatively large drops of water polarize in the field, positive at the bottom and negative at the top. Large drops fall fast to the cloud bottom sweeping up negative ions on their bottom as they fall, becoming negatively charged. Smaller drops fall more slowly or are held near the top of the clouds by air currents. Thus the smaller drops pick up the more slowly moving positive ions. In this fashion the top of the clouds become positively charged and the cloud bottoms become negatively charged. Ascending air currents greater than 8 m/second keep the cloud together.

Cloud movement causes an incoming high air flow at the front of the cloud that excludes larger water droplets and breaks up any that do get through. This causes the lower front part of the cloud to be positively charged near the bottom but most of the cloud bottom is negatively charged. Figure 5.28 shows a cloud whose charge distribution incorporates the modified theories of Wilson and Simpson. The theories are only partly verified but they are fairly consistent with balloon and other measurements. There is room for argument though.

*The lightning stroke* is not a simple phenomenon. Measurements indicate that an average of 90% of the lightning strokes lower negative charge to the ground and 10% raise negative charge to the cloud. Lower current amplitude strokes are as much as 37% positive (negative to cloud) and 63% negative. No one is sure why the difference in the percentage of positive strokes with stroke size. The fact that most of the cloud bottom is negatively charged may account for most of the strokes being negative.

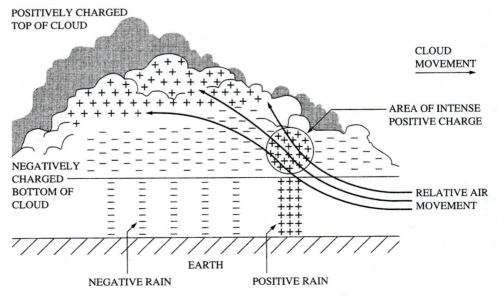

**FIGURE 5.28**   Charge conditions in a thundercloud from Simpson's revised theory.

Figure 5.29 shows the sequence of a lightning stroke. The initial streamers, or pilot streamers, are only a few amperes in amplitude. It has been posited that the initial streamers are following the paths of cosmic particles as they pass through the Earth's atmosphere. They move in jumps or steps of about 50 meters with each step being quite straight, but each one propagating in a new direction. The steps each propagate at about one-sixth the velocity of light, but the overall streamer propagation is only about 0.05% of the velocity of light. The leader propagates in a zigzag path.

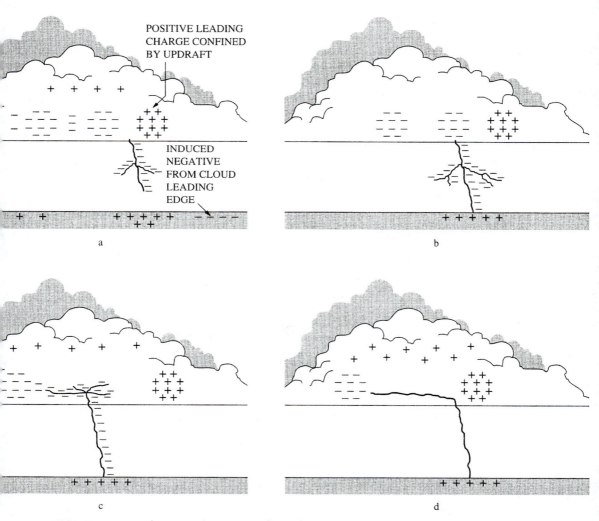

**FIGURE 5.29** Lightning strike to ground (a) Charge centers in cloud, pilot streamers propagate towards Earth (b) Streamer touches ground, heavy return streamer propagates toward cloud (c) First charge center discharged to Earth, interior streamers developing between charge centers, second pilot leader along same ionized path about to touch Earth (d) Heavy return streamer to Earth, charge movement between charge centers in interior return streamer.

When the leader strikes the Earth, usually at the highest grounded point, the bright return stroke occurs following the same path as the leaders. It propagates at about 10% of the speed of light. The return stroke consists of the discharge at a high current of the charge distributed along the leader to ground. When the return stroke reaches the charge center of the cloud, more charge is lowered to the ground through the luminous ionized path of the return stroke. It takes about 10 ms for the step leader to reach the ground, but only about 50 to 100 $\mu$s for the return stroke and discharge. Most lightning flashes consist of a number of strokes that follow the same path discharging more than one cloud charge center.

The lightning current waveform (an average shape) is shown in Figure 5.30. The initial very high current associated with the return stroke lasts only a short time but the subsequent charge center discharge takes much longer. The voltages induced on a line by a stroke can reach 5 million volts without lightning protection and low impedance grounds. A lightning stroke hitting a power line will cause a high voltage wave front to travel along the line that can damage any unprotected equipment by breaking down the insulation. Notice the similarity of the lightning stroke waveform to the BIL test waveform. This is not a coincidence.

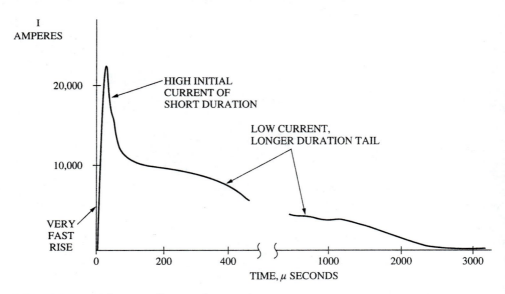

**FIGURE 5.30**   Lightning strike waveform

## 5.5.2 Lightning Arresters

The job of the lightning arrester is to clip the induced voltage transient caused by a lightning strike at a level below the BIL, but above the normal operating voltage, of the protected equipment. The lightning arrester should be an insulator at any voltage below the protected voltage, and a good conductor at any voltage above to pass the energy of the strike to ground.

The first lightning arresters were spark gaps such as the one in Figure 5.31. A spark gap breaks down at a voltage governed by the distance between the electrodes and the condition of the air between the electrodes—wet, dry, clean, or dirty. Enclosed, sealed electrodes with controlled interior atmospheres are often used to remove the variability of the spark breakover voltage caused by the uncertainty of the atmospheric condition. After a spark gap breaks down its impedance is very low, and the normal working voltage of the line must be sustained by auxiliary impedance, or the gap must become nonconducting after the strike quickly enough that the lightning arrester itself does not act as a fault. Careful selection of lightning arresters is required to accomplish this. Spark gap, as well as other types of arresters, are equipped with isolators to remove the arrester from service if the arrester becomes damaged. The isolator is essentially a fusing structure that can handle high short-time current, but cannot handle long-term current such as would occur if the arrester break over voltage dropped below the peak line voltage.

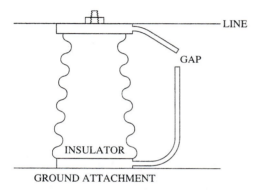

**FIGURE 5.31** External spark gap lightning arrester

Newer lightning arresters use enclosed spark gaps plus a ceramic valve block. The ceramic valve block is a tubular ceramic holder filled with discs of an appropriate material that has a predictable voltage breakdown and sustains the applied breakdown voltage while conducting. For example, if the breakdown voltage is 400 V, when the device is shunting current to ground its voltage remains near 400 V (usually a bit higher because of the IR drop across the device internal resistance), then when the applied voltage drops below 400 V the device stops conducting. Silicon Carbide was a popular low voltage transient suppression material, and zinc oxide is popular for medium and high voltages. Zinc oxide has become popular at low voltages also. A drawing of a typical distribution lightning arrester is shown in Figure 5.32a. The ceramic valve block is filled with discs of solid zinc oxide stacked until the correct voltage breakdown voltage is reached. The spark gap prevents any current flow until a transient occurs. The zinc oxide material does not have an infinite nonconducting impedance so it will draw a quiescent current without the spark gap. A lightning induced voltage transient will break down the spark gap, the spark gap impedance will drop very low switching the ceramic valve block between the line and ground, and the valve block will prevent the line voltage from rising higher by shunting the stroke current to ground.

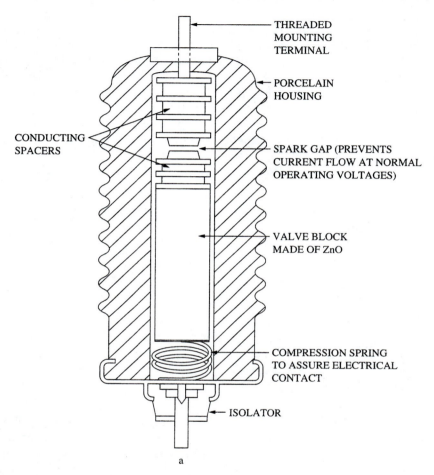

a

**FIGURE 5.32**   (a) Distribution lightning arrester

The lightning arrester should be connected as close as possible to the terminal of the equipment being protected to reduce the voltage build up from $IX_L$ on the connecting lines before the spark gap breaks down. Figure 5.33 shows the lightning arrester connection for one phase of a distribution transformer; one set of arresters is used for each phase. Note the separate ground wires for the lightning arresters. The grounded tank is not considered a reliable enough ground for the arrester. The ground wire is solidly connected to the buried metal stake or wire mesh that serves as the station ground. Only the line side of a building distribution transformer has a lightning arrester because the line is most likely to be struck, although a tall building will very likely have a series of lightning rods for its protection.

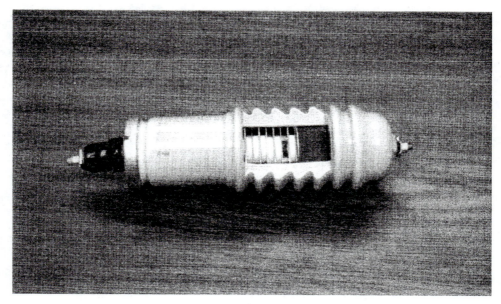

b

**FIGURE 5.32** (b) Distribution lightning arrester

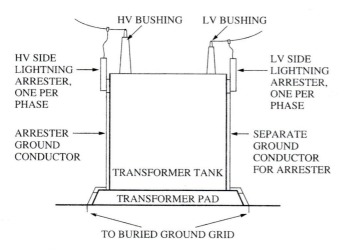

**FIGURE 5.33** Lightning protection of distribution substation power transformer

Figure 5.34 shows a typical connection for a single-phase lightning arrester and for a three-phase wye with neutral interconnection. A delta would not have the neutral interconnection. The margin of protection (PM) by a lightning arrester is

$$PM = \frac{BIL - \text{arrester discharge voltage}}{BIL} \qquad (5.1)$$

Where the arrester discharge voltage is the arrester break down voltage plus the arrester internal voltage drop at the expected stroke or transient current. The discharge voltage may be 5% higher than the break down voltage. The margin of protection should not be lower than 0.2.

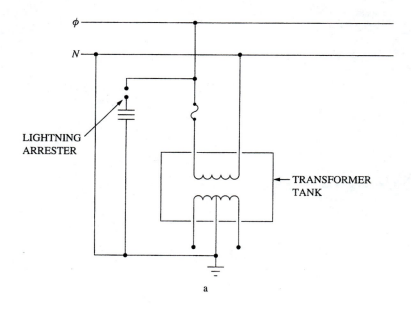

a

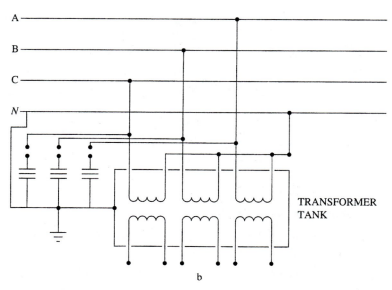

b

**FIGURE 5.34**  Distribution transformer lightning and surge protection diagrams
(a) Single-phase transformer and (b) three-phase wye, neutral interconnected transformer

### 5.5.3  Shield Wires for Lightning Protection of Lines

Shield, or static wires, are conductors strung above the load carrying conductors on transmission and distribution towers and poles to protect the load carrying conductors from lightning strikes. The shield wires provide a place for lightning strokes to terminate instead of the power carrying conductors, thereby protecting the power conductors. Almost all lines 34.5 kV and above use shield wires.

Figure 5.35a shows the shield wire as it is strung on a pole. Shield wires provide a 30° zone of protection on either side of a vertical line drawn from the ground to the wire, as shown in Figure 5.35a. Towers in which the power carrying conductors do not fit within the zone of protection of a single shield wire use two, as shown in Figure 5.35b. Equipment in station yards can be protected by placing it within the 30° protection zone of a tall mast with a conductor running from the tip to ground, as shown in Figure 5.35c. Shield wires must be grounded to provide a path for the lightning current. The lightning current is very high but of short duration so the conductors, sized to carry the peak current for only a few microseconds, can be relatively small.

## 5.6

## PROTECTIVE RELAY INTRODUCTION

A relay is an electromechanical- or microprocessor-controlled electronic system that senses an abnormal or fault condition, such as an overcurrent, under or over voltage, or low frequency, and sends a trip signal to a circuit breaker. They are used to protect generators, transformers, motors, and lines.

Other types of relays exist as well as protective relays. They are systems that perform functions other than fault sensing. The major types are listed below.

*Monitoring relays* verify conditions in the power system or power system protection system and send an alarm when the conditions are abnormal. Monitored conditions include network phasing and voltage levels. Monitoring relays often are used in conjunction with protective relays.

*Programming relays* sequence events or detect sequences of events. They are used to control and monitor synchronization and reclosing sequences.

*Regulatory relays* are used to determine if a parameter, such as line voltage, is between programmed limits, and send a control signal to force the parameter to return to within the limits, such as a tap change signal, if the parameter leaves the limits. They send an alarm signal if the parameter will not return to within the limits.

*Auxiliary relays* provide miscellaneous functions within other relaying systems. Timers are an example of an auxiliary relay function.

Relays must operate reliably and quickly, be economical, and selective, operating only on the desired input. Entire books have been written about protective relays and their applications. The remarks in this section are introductory. We will consider relay types as needed throughout the text, but in this section we will discuss only overcurrent relays, under voltage relays, and differential relays.

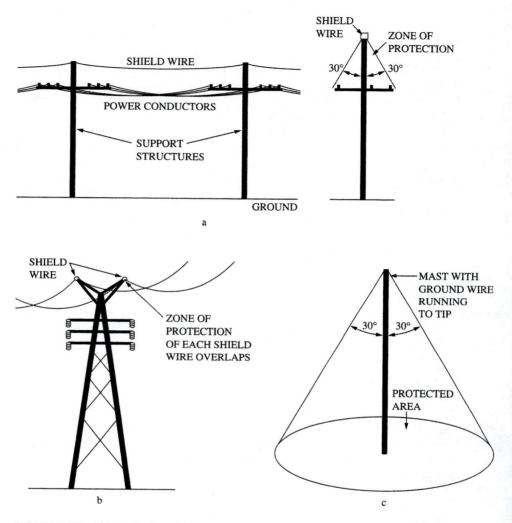

**FIGURE 5.35** Shield (static) wire (a) Narrow tower or pole (b) wide tower and (c) mast for substation yard protection

## 5.6.1 Microcomputer Controlled Relays

The use of microprocessors in microcomputer relay systems has allowed relay systems to perform several relaying functions with a single central relaying package in a very economical manner. The multifunction capability of microprocessor-controlled relay systems has resulted in a drop in the cost per function of such relays when compared to electromechanical relays.

A one-line diagram of a microcomputer-controlled relay system is shown in Figure 5.36a. The current and potential transformers provide current and voltage information to the relay from which the relay microcomputer calculates any additional parameters needed, such as impedance, VAR and power quantity and flow direction, trends over a fixed time, and running averages of quantities as needed. In addition the current entering and leaving the protected machine or section can be compared, called *differential relaying* (covered later in this chapter), because a difference in these currents indicates a fault within the protected section. The relay can also make use of other parameters, such as temperature and vibration sensor outputs, to monitor more conditions than electromechanical relays are able to monitor. The relay will react to out of limit parameters by sending a trip signal to a circuit breaker and an alarm signal to a central monitoring point via a telecommunication system. In addition to responding to out of tolerance parameters, a microcomputer-controlled relay may have the capability of sending monitored data, such as current, voltage, temperature, vibration, or other information, back to a central control room via a data communication system. Thus a single microcomputer-controlled relay can monitor and respond to abnormal conditions while gathering data for use in control and trend analysis.

Figure 5.36b is a block diagram of the microcomputer-controlled relay. The data acquisition system collects the transducer information and converts it to the proper form for use by the microcomputer. Information from current transformers (CTs), potential transformers (PTs), and other systems is sent through an isolation transformer or optically-coupled amplifier, amplified if needed, and sampled at a frequency well above the power line frequency (at least several kilohertz). The signal samples are digitized with an analog to digital converter and fed to registers in the microprocessor system. The microprocessor then compares the information directly with preset limits for over/under voltage, overcurrent, over/under

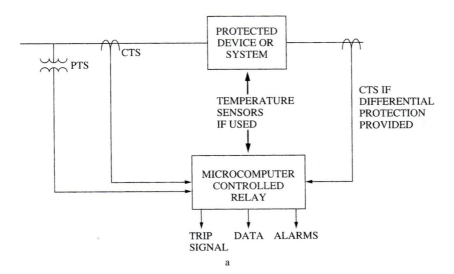

**FIGURE 5.36** Microprocessor controlled relay system (a) System

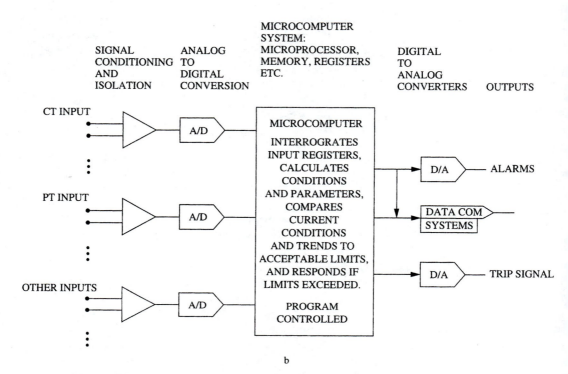

b

**FIGURE 5.36** Microprocessor controlled relay system (b) microcomputer controlled relay

temperature or pressure, or any other directly measured parameter. The microprocessor then calculates parameters necessary for other functions such as difference current, effective impedance, frequency measurement, or compares present measured or calculated quantities with previously measured (and stored in memory) values to establish trends and checks to see if the trends are within allowable values. The output data will be in the form of data or alarm signals to be sent to a remote control room, trip signals for such conditions as overcurrent, or possibly control changes such as a voltage adjust signal to a transformer with an automatic tap changer from an over/under voltage condition.

The microprocessor-controlled relays work under program control. The programs are stored in read only memory. The more complex a particular relay function is the more complex the program to control that function is, and the longer the amount of time that is needed for the relay to perform that function. Thus, when a relay has many complex functions, such as impedance measurement, to perform it must use more time to perform them than if it had only a few relatively simple functions, such as instantaneous overcurrent measurement to perform. A trade-off must be made in system protection between complexity of monitoring and speed with which the relay responds. The faster the microcomputer and data acquisition systems, the more functions a microprocessor-controlled relay can perform in a given time. Multiple microcomputers sharing the same data acquisition system can be used to handle functions that are too complex or too numerous for one microcomputer to handle alone. An optimized system can respond more rapidly to some conditions than others if the microprocessor system and program are sufficiently sophisticated.

Microcomputer relays are being designed for specific protection applications. For example, specialized generator and motor protection relays are commercially available that provide the specific protection that these devices need.

Microcomputer-controlled relay systems are being designed into most new electrical systems and retrofitted into older systems as relay replacement is needed because of changes in the system or protection needs. However, an enormous amount of electromechanical relaying equipment still exists, and it may be more than a decade, or maybe much more time, before it is all replaced by solid state microprocessor-controlled protective relays. Thus, electromechanical relays are also covered in the sections that follow.

## 5.6.2 Basic Current Relay Electromechanical Mechanisms

The *instantaneous hinged armature (clapper)* is illustrated in Figure 5.37. Current through the coil magnetizes the core, the frame, and by magnetic induction, the armature. The magnetic attraction then pulls the armature toward the coil, closing the contacts. The core plunger adjusts the magnetic circuit to adjust the pull-in current of the relay. The closing action is instantaneous (2–4 ms operating time) once the pull in current is reached. The armature movement is usually used to release a catch on a spring or gravity-driven indicating flag to show the relay has operated. Thus it is called an *indicating instantaneous trip relay* or IIT. The relay can be manufactured with many poles, normally open contacts, normally closed contact, or both.

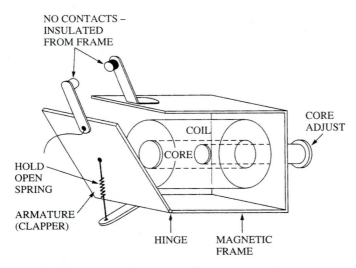

**FIGURE 5.37**   Hinged armature relay

The indicating contact switch (ICS) is a close cousin of the IIT. It is wound so as to reduce its operating voltage and current. It has one normally open contact set and an indicating trip unit. Its coil is in series with other relay contacts, and its contacts in parallel with other relay contacts to help carry trip current. Its primary function is to indicate a relay operation.

The *solenoid (plunger) relay,* see Figure 5.38, centers the plunger by magnetic induction when the coil current is high enough to cause it to operate. The plunger operates against a spring that holds it out when the relay is not energized. The solenoid relay is usually used as a high current instantaneous trip relay. The pull in current can be adjusted by the magnetic core air gap adjust, which varies the reluctance of the core. The contacts can be in a variety of arrangements, but the moving contact disc, which bridges the stationary contacts on operation, is popular. The operating time varies from 5 to 50 ms, depending on the overcurrent magnitude (slower with lower currents), and the relay size.

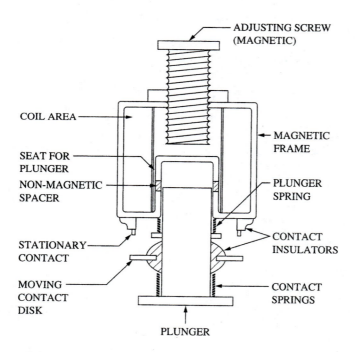

**FIGURE 5.38**   Solonoid or plunger relay

*Induction disc* relays are used as inverse time overcurrent relays, where high overcurrent causes faster operation than low overcurrent. The relay consists of an aluminum induction disc suspended between bearings with a magnetic structure held in place on one side of the disc shaft, as shown in Figure 5.39a. The disc shaft has a moving contact mounted on it that makes contact with a stationary contact, mounted on the relay frame, if the disc rotates far enough. A spiral spring is

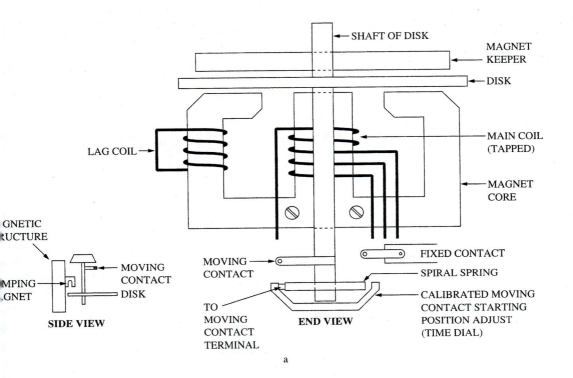

SHAFT OF DISK

MAGNET KEEPER

DISK

MAIN COIL (TAPPED)

LAG COIL

MAGNET CORE

GNETIC
RUCTURE

MOVING CONTACT

DISK

MOVING CONTACT

FIXED CONTACT

SPIRAL SPRING

MPING
GNET

CALIBRATED MOVING
CONTACT STARTING
POSITION ADJUST
(TIME DIAL)

TO
MOVING
CONTACT
TERMINAL

**SIDE VIEW**                          **END VIEW**

a

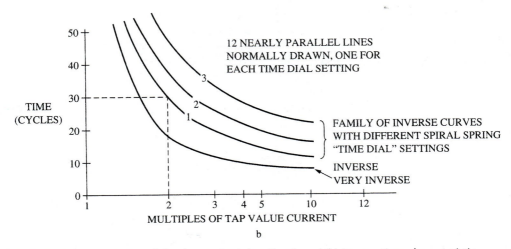

12 NEARLY PARALLEL LINES
NORMALLY DRAWN, ONE FOR
EACH TIME DIAL SETTING

TIME
(CYCLES)

FAMILY OF INVERSE CURVES
WITH DIFFERENT SPIRAL SPRING
"TIME DIAL" SETTINGS

INVERSE
VERY INVERSE

MULTIPLES OF TAP VALUE CURRENT

b

**FIGURE 5.39**   Induction disk relay (a) Drawing, E unit and (b) inverse time characteristic

mounted on one end of the shaft to provide both a counterforce against the disc rotation and provide a current path for the moving contact. A dial, called a time dial, is mounted over the spring to adjust the starting position of the moving contact. The higher the time dial setting, the further the moving contact must rotate before reaching the stationary contact.

The magnetic structure consists of an E core in this model with a magnetic keeper bar on the other side of the disc to lower the reluctance of the magnetic path. A damping magnet, to assure smooth operation and bounceless contact closure, is mounted near the disc with its pole facing toward the disc. The E core has a main coil through which the monitored current from the CT passes. This is tapped to adjust the trip current of the relay. A lag coil around one of the other core legs completes the magnetic circuit. The lag coil is shorted. The magnetic plugs vary the saturation characteristic of the coil to vary the inverse time characteristic of the relay.

In operation the current through the main coil induces a flux in the core. The main coil flux induces a voltage in the lagging coil, which causes current in the lagging coil to flow. As the flux from the main coil grows the lagging coil flux opposes it on the lagging coil core leg. When the main coil flux is decreasing the lagging coil flux aids the main flux in the lagging coil leg. The resultant flux sweeps from the right core leg to the left core leg, the core leg with the lagging coil, in the Figure 5.39a end view. The sweeping magnetic field induces current in the disc under the field. The magnetic field of the disc current interacts with the core field producing torque on the disc, just like a shaded pole induction motor. The torque will cause the disc to rotate toward the lag coil. The torque on the disc acts so as to rotate the moving contact toward the stationary contact while the spiral spring torque acts to keep the contacts apart. When the closing torque exceeds the opening torque long enough for the disc to rotate the moving contact to the stationary contact the relay has operated.

The inverse time characteristic, shown in Figure 5.39b, illustrates the results from high currents rotating the disc faster. The time dial moves the curve up and to the right with higher numbered settings because higher number settings move the starting point of the moving contact farther away from the stationary contact so it has to rotate more to make. Dial numbers normally range from 0.5 to 11. All 12 time dial lines are normally shown on a time-current chart. The time to make is given in cycles, and the current to make is given in multiples of the tap value. Typical tap values range up to five amperes.

### Example 5.2:

If the tap value of the relay whose inverse time curves are shown in Figure 5.39b is 3 A and the time dial setting is 1, how long does it take the relay to trip from a CT current of 6 A?

### Solution:

The tap value is 3 A, so 6 A is two times the tap value. Moving vertically from 2 on the multiple of tap value scale to the curve with time dial setting 1, and from that intersection horizontally to the time axis, we find the operate time is 30 cycles, or 0.5 second.

Many alternate designs of the induction disc relay are available, as well as solid-state and microprocessor-controlled systems that perform the same function at a somewhat higher price. The microprocessor-controlled systems can perform the function of several relays using the same data acquisition system, which makes them very economical on a per function basis. Microprocessor-based relay systems are discussed later in this chapter.

The lag coil terminals can be connected to an auxiliary relay to give the induction disc directional characteristics. If the induction disc is to sense only load current, a fault on the source side could cause power flow toward the source relative to the relay location, an auxiliary relay that senses power flow can be connected to the lag terminals. If, in the event of power flow toward the source, the auxiliary relay opens the lag winding no torque can be developed on the disc, with the result that only current associated with power flow toward the load will be sensed.

The *induction cup relay* operates similarly to the induction disc relay. The lag and main coils are arranged on poles around the interior of a magnetic frame, as shown in Figure 5.40. The tubular cup fits over a circular fixed magnetic core in the center of the frame. The rotating magnetic field set up by the main and lag coils causes the cup to rotate against a spring to operate the relay. The light, low inertia cup allows induction cup relays to operate faster than induction disc relays. Like the induction disc the lag coil can be connected to give the relay directional characteristics.

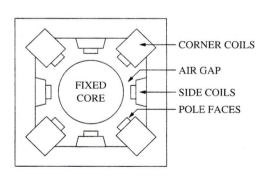
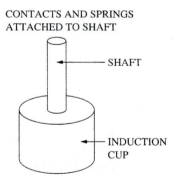

**FIGURE 5.40**   Induction cup relay

## 5.6.3 Overcurrent Relays (Type 50/51)

Type 50 relays are instantaneous overcurrent relays and type 51 are inverse-time overcurrent relays, so a 50/51 relay is a combination of both. A relay usually consists of more than one relay mechanism. The 50/51 relay, shown in Figure 5.41, on the left consists of three relay mechanisms. It has the following parts: an instantaneous unit (IIT) for high value, close in faults; an induction disc (type CO) or cup inverse time unit for coordination with other protective devices; an indicating unit (ICS) to indicate a time trip and relieve the time unit contacts of the trip burden; a tap block to set the overcurrent that will cause the disc to turn; a time dial to set the inverse time characteristic so that a given current takes longer to activate the relay; and if the relay is directional, a directional unit.

**FIGURE 5.41**   50/51 overcurrent relay

An elementary diagram, or schematic, of a basic 50/51 relay is shown in Figure 5.42. The diagram is usually shown in two parts, the sensing coils and the contacts. Note that the current from the CT flows through both the CO and IIT coils, although the relay could be connected for either one to be used alone. Also note that either the CO or the IIT contacts can connect the 125 V dc circuit breaker trip voltage to the breaker trip coil.

The operating sequence is as follows. A high overcurrent causes the IIT to trip, connecting 125 V dc to the breaker trip coil tripping the breaker. The IIT indicator drops the red flag indicating an instantaneous trip. A low value overcurrent starts the CO disc turning at a rate dictated by the relay settings and the value of the overcurrent. If the overcurrent is removed before the contacts make, the disc resets. Otherwise the contacts make. The CO contacts allow current through the ICS coil causing it to pull in. The ICS contacts then carry the bulk of the trip current to the circuit breaker trip coil. The ICS also drops a flag indicating that there has been a time trip.

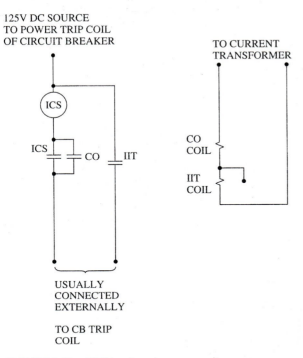

125V DC SOURCE
TO POWER TRIP COIL
OF CIRCUIT BREAKER

TO CURRENT
TRANSFORMER

ICS

ICS      CO      IIT

CO
COIL

IIT
COIL

USUALLY
CONNECTED
EXTERNALLY

TO CB TRIP
COIL

**FIGURE 5.42**   50/51 relay elementary diagram

Figure 5.43a shows the connection of 50/51 relay coils to provide overcurrent protection for a transformer secondary. Note the polarity of the current transformers. The polarities can be reversed but all three CTs must be the same polarity. The connection, which uses three 50/51 relays, provides a trip signal for an overload or fault on any one of the lines. The current tap and time dial settings are set to coordinate with downstream protective devices and are normally the same for all three of the 50/51 relays. Downstream overcurrent devices are set to operate faster than the upstream devices. Thus the downstream device clears before the upstream device operates.

The dc trip circuit is shown in Figure 5.43b. The 125 V dc trip bus is usually supplied by batteries that are continuously trickle charged. The batteries provide trip power even when the fault is so severe that the station voltage is very low. Notice that each relay trip circuit is in parallel with the rest so that each relay can trip the circuit breaker independently. A transformer would have more protection than we have shown here. The trip contacts for the other protective relays would be in parallel with the overcurrent relays as shown.

With a balanced load the CT neutral current will be zero, assuming matched CTs. A fourth 50/51 relay can be placed in the neutral to sense ground faults. Sensitive overcurrent relays are often used for this function. Another parallel set of contacts would be needed in the trip circuit. We will study ground fault protection more thoroughly in a later chapter.

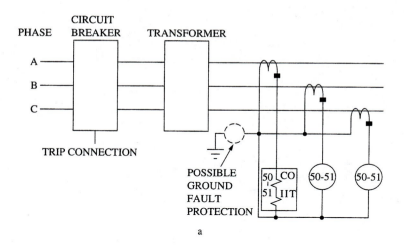

a

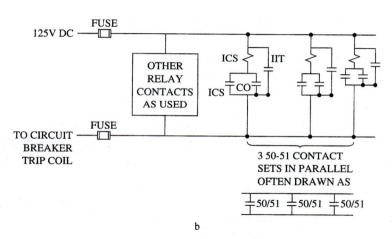

b

**FIGURE 5.43**   Transformer secondary circuit protection with 50/51 relay: (a) AC circuit and (b) DC trip circuit

Suppose the overcurrent relay is to trip in 30 cycles at a line current of 3000 A. Suppose further that the relay time current curve is that of the time dial setting equal one shown in Figure 5.39b. From the time current curve we see that a 30-second trip on the time dial one setting occurs at two times the tap setting. Recall that CT ratios are XXX A:5 A. Some common CT ratios are 300:5, 600:5, and 2000:5. If we choose the 2000:5 ratio (400:1) our CT current is 3000 A/400=7.5 A. The tap on the relays should be set to 3.75 A because the 2 times tap value provides the proper trip time, or the tap setting nearest to 3.75 A (perhaps 3.8 A). Microcomputer-controlled relay programs can provide the time-current functions available on electromechanical relays.

## 5.6.4  Under-Voltage Relays

Under- and over-voltage conditions can seriously damage electrical equipment. Instantaneous low voltage relays can be constructed from plunger relays with normally closed contacts, closed when the relay is de-energized. The relay is energized at normal line voltage holding the contacts open, but cannot hold in at low line voltage, thus it drops out when the line voltage is low causing the circuit breaker to trip. Instantaneous relays are not normally used because most line faults are transient.

An induction disc with voltage coils (more turns of higher resistance than current coils) can be used as under-voltage relays. In this application the relay is designed so the coil torque holds the moving contact away from the stationary contact. When the voltage is low the spiral spring torque is sufficient to rotate the moving contact into the stationary contact causing a trip. The normal coil voltage for an under- (or over-) voltage relay is 120 V, supplied by a potential transformer. A one-line diagram for application of under-voltage protection is shown in Figure 5.44a, and the time-voltage curves for an under-voltage relay in Figure 5.44b. Notice that the curves are the inverse of the overcurrent relay curves. The vertical axis is time in seconds, and the horizontal axis is in percent of tap value. There are 12 nearly parallel curves, one for each time dial setting between 0.5 and 11 (just as in CO overcurrent curves), even though only five are shown in Figure 5.44b. An example demonstrates the use of the curves.

### Example 5.3:

The under-voltage protection circuit of Figure 5.44 is set to trip the breaker if the voltage drops to 90% of normal, or 3744 V. The time dial setting is three. The PT ratio is 4200:120 V. What relay voltage will trip the breaker in 6 seconds?

### Solution:

First the normal relay voltage must be found to determine the proper tap setting. The PT ratio is $4200/120 = 35$, so the normal relay voltage is $4160 \text{ V}/35 = 118.5$ V. The tap is set as close as possible to $3744 \text{ V}/35 = 106.97$. Assume a tap setting of 107 V. From the relay curves we see that the relay will send a trip signal in 6 seconds if the voltage drops to 67% of the tap value, or $(0.67)(107 \text{ V}) = 71.7$ V. The relay will operate at 90% of the tap setting in 18 seconds. The relay will operate at 100% of tap value, but the curves are not plotted to 100% because they are not predictable with precision beyond 90% of tap value.

An induction disc relay can be used as an over-voltage relay by connecting the voltage coils so the coil torque rotates the moving contact toward the stationary contact, and the spiral spring torque rotates it away.

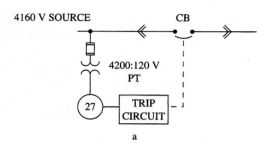

a

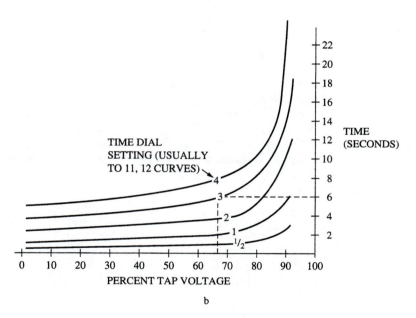

b

**FIGURE 5.44**  Under voltage relay use (a) One-line diagram and (b) typical curves for CV under voltage relay

## 5.6.5  Differential Relays (Type 87)

A differential relay responds to the difference in two currents. If the two currents are the same, even if they are excessive, the relay does not operate. If their difference is above a preset limit the relay operates.

Any overcurrent relay can be operated in a differential mode but in most applications differential relays are more versatile and precise. The principle of differential operation is shown in Figure 5.45. When the current in both CTs is the same, as in Figure 5.45a, the current out of each CT is equal and opposite because the current is entering each CT from opposite directions with respect to the polarity markings. Thus the current from the CTs bypasses the overcurrent relay, or equal and opposite currents pass through it and cancel at any instant of time. Either

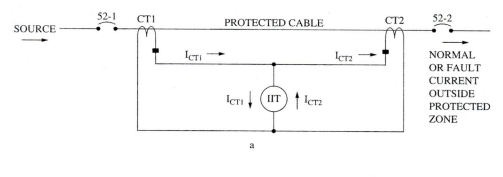

a

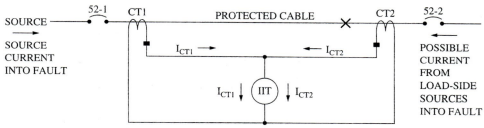

Note: $I_{CT1}$ can be less than $I_{CT2}$, of same polarity, of opposite polarity, or zero.

b

**FIGURE 5.45**   Differentially connected relay (a) $I_{CT1} = I_{CT2}$, polarity opposite, currents cancel in IIT, no relay operation (b) $I_{CT1} \pm I_{CT2}$, IIT current not zero, relay operates when $I_{CT1} - I_{CT2} = $ IIT operating current, circuit breaker then receives trip signal.

concept works so pick the one you are most comfortable with. One could also say that the equal and opposite phase voltages from the CTs result in zero potential across the IIT so no current flows through it. The area between the CTs is called the *protected zone*. If a fault occurs inside the protected zone, as shown in Figure 5.45b, the CT currents are no longer the same and the current through the IIT no longer equals zero. In extreme faults the current through $CT_2$ changes directions and the two CT currents add at every instant of time. When the IIT current is above the actuating current the relay operates.

A differential relay provides better protection than an IIT connected differentially because it has provisions for balancing the effect of the CTs if their currents are different. Differential relays use restraining coils to prevent accidental operation in response to a fault outside of the protected zone because of CT mismatch. CTs that match well at lower currents do not match as well at current levels near or at saturation.

A *type CA induction disc differential* relay is shown in Figure 5.46a. The restraining coils are wound so that when the current through them is the same and in the same direction at any instant of time they produce opening torque, and the operating coil with no current produces no torque at the same time. If a fault occurs in the zone of protection the current through the restraining coils is no longer the

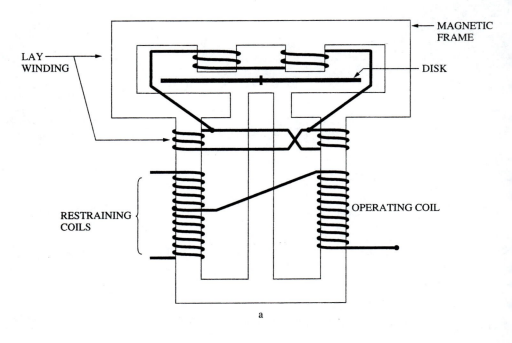

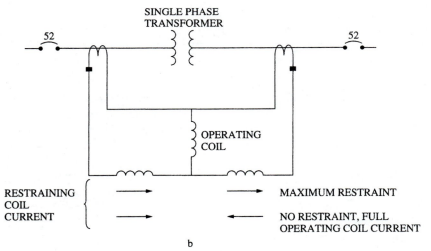

**FIGURE 5.46** Type CA differential relay (a) Magnetic configuration and (b) elementary diagram, transformer protection

same and the opening torque weakens. At the same time current through the operating coil produces closing torque. When the closing torque is greater than the opening torque the relay operates. A severe fault in the protected zone causes the current through the restraining coils to be opposite in direction at any instant and the restraining coils cancel. The currents through the operating coil add and the relay operates more quickly.

The restraining coils are tapped to allow for mismatch in the CT currents, which can be as high as 10 A. The operating coil is also tapped to set the operating current, which can be set to between 2.75 and 5 A. The type CA differential relay is fairly insensitive to transformer inrush current, which can cause a differential set up like that of Figure 5.45 to operate nearly every time if used for transformer protection. Type CA relays operate as fast as two to six cycles with severe faults. The connection for the protection of a single-phase transformer is shown in Figure 5.46b.

*Differential relays with harmonic restraint units (HRU)* are required for transformers. Large transformer inrush current, which has no corresponding outrush current, can cause a type CA relay to operate every time the transformer is energized. Harmonic restraint units prevent operation from transformer inrush current. Figure 5.47a shows a harmonic restraint unit. The dc relay mechanism has a restraining coil to prevent it from operating, and an operate coil to cause it to operate. The principle of the mechanism is shown in Figure 5.47c. The operate coil current provides closing force but if the restraining coil current is high enough the operate coil cannot overcome the opening force and operate. The relay coil voltage is obtained by rectifying ac from the HRU transformer.

The HRU makes use of the high second harmonic content of transformer inrush current to restrain the HRU relay. A second harmonic band pass filter passes the second harmonic to the restraining circuit where it is rectified and applied to the restraining coil of the HRU relay. A second harmonic content of about 16% is enough to prevent the relay from operating. Another filter blocks the second harmonic from the operate coil of the HRU so that the HRU relay is sure to close on any fundamental current that operates the differential unit (DU). Note in the trip circuit, shown in Figure 5.47d, that the HRU and DU contacts are in series. Even if the DU operates on inrush current the HRU contacts remain open preventing a breaker trip. On a fault in the protected zone the second harmonic content is low and both the DU and HRU contacts close to send a trip signal to the breaker.

A simplified diagram of a differential relay with an HRU is shown in Figure 5.47b. Note that the restraining current from the CTs is transformed through transformers with air gaps in their cores to make saturation difficult, rectified, and applied in parallel to the DU operating relay mechanism restraint coil. The necessary additional transformer-rectifier connection is shown for the tertiary of a three winding transformer. The operate transformer core has no air gap so that it more efficiently couples low amplitude operate current to the rectifiers that supply the operate coil. When the current through the CTs is the same and opposite in direction with respect to the polarity markings, the restraint voltages are highest and the operate coil cannot operate the DU relay. A fault in the zone of protection causes at least one restraint transformer primary current to lessen as current is diverted to the operate transformer primary and the HRU transformer primary. This current is transformed and rectified, and when high enough it operates the DU coil. The same primary current passes through the HRU primary, is rectified, and passed to the HRU operate coil so it operates at the same time as the DU. The

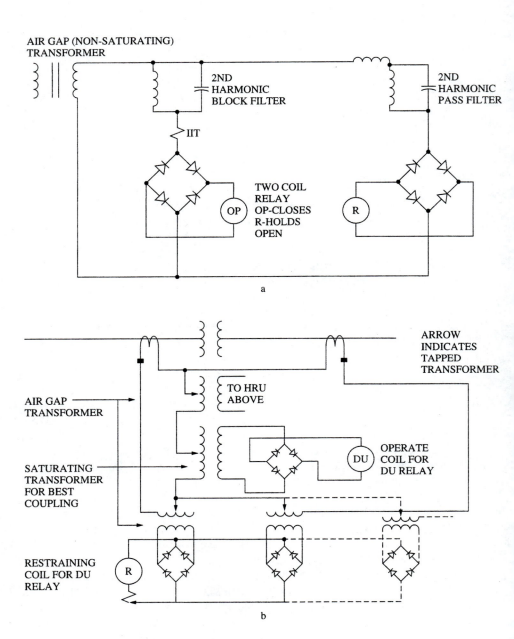

**FIGURE 5.47**   Differential relay with harmonic restraint unit (HRU) (a) Harmonic restraint unit (b) differential relay organization (c) two coil relay, balanced beam type and (d) trip circuit

restraining transformer continuous current rating is 10 to 22 A depending on the tap, and the operating coil pick up current is 30% of the tap value.

The trip circuit is shown in Figure 5.47d. Notice that the HRU differential relay makes use of an ICS relay to indicate a timed differential trip. It also has an IIT

CONTACTS

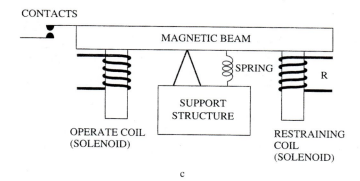

c

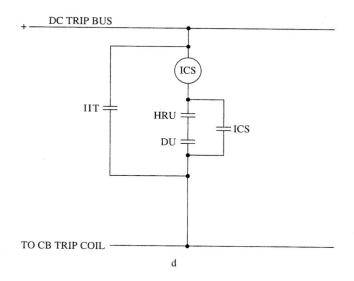

d

unit to send a trip signal in response to massive faults that require instantaneous tripping. The trip circuit for a CA type differential relay is the same except the HRU contact is missing.

The CA type differential relay is used in smaller substations where the inrush current is smaller, and the more expensive HRU type is used in generating stations and large substations where the inrush current is high.

*The Y-Δ CT connections for relaying* are critical if the differential relaying is to operate properly. For Y-Y and Δ-Δ connections the CTs can be connected in a Y configuration. A Y-Δ or a Δ-Y connected transformer has a 30° phase shift between the primary and the secondary. CTs connected in the same configuration on both the primary and secondary side of the transformer have the same 30° phase shift in their currents. This acts as differential current even if the CT currents are the same. To compensate for the 30° phase shift the CTs are connected Δ on the Y side and Y on the Δ side, as shown in Figure 5.48. Additional tertiary winding CTs are connected oppositely to the tertiary connection, CTs connected Y if the tertiary is Δ and vice versa.

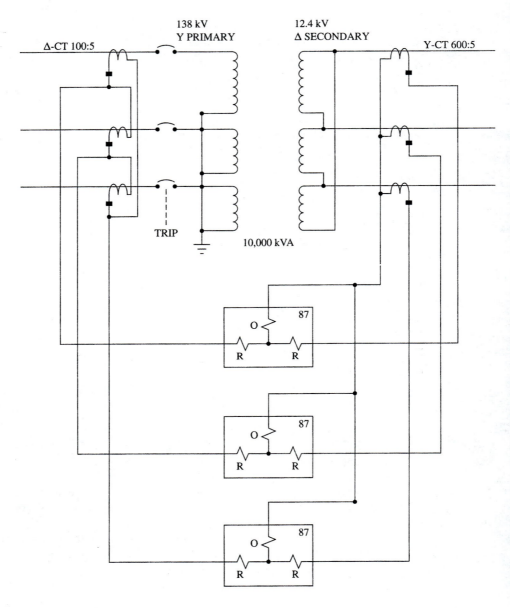

**FIGURE 5.48**   Current transformers in Δ-Y configuration for differential protection of a Y-Δ connected three-phase transformer

**Example 5.3:**

Calculate the relay currents and the tap settings for the circuit of Figure 5.48.

**Solution:**

First calculate the primary and secondary currents.

$$I_P = \frac{kVA}{kV(\sqrt{3})} = \frac{10,000 \text{ kVA}}{138 \text{ kV }(\sqrt{3})} = 41.84 \text{ A}$$

$$I_S = \frac{10,000 \text{ kVA}}{12.4 \text{ kV}(\sqrt{3})} = 465.6 \text{ A}$$

The CT ratios are 100:5 = 20:1 and 600:5 = 120:1. From these the relay currents are:

Primary relay current = $(41.84 \text{ A}/20)(\sqrt{3}) = 3.62 \text{ A} = I_{RH}$

The $\sqrt{3}$ is because the primary CTs are $\Delta$ connected.

Secondary relay current = $465.6 \text{ A}/120 = 3.88 \text{ A} = I_{RL}$

Where the subscripts RH stands for high voltage side relay quantity and RL stands for low voltage side relay quantity.
    The tap ratios are set as close to the relay current ratios as possible.

$$\frac{I_{RL}}{I_{RH}} = \frac{3.88 \text{ A}}{3.62 \text{ A}} = 1.07$$

CA type relay taps are 5-5, 5-5.5, 5-6.6, 5-7.3, 5-8, 5-9, 5-10 for tap ratios of 1, 1.1, 1.32, 1.46, 1.6, 1.8, and 2. If our relays are CA type differential relays the 1.1 ratio is the closest. HU relay taps are 2.9, 3.2, 3.5, 3.8, 4.2, 4.6, 5, and 8.7 on both the primary and secondary side. Table 5.3 shows the ratios. The 1.094 ratio, corresponds to the 3.2 and 3.5 primary and secondary taps respectively. If the mismatch ratio between the CT current ratio and tap ratio is less than 35% for CA relays, and 30% for HU relays, normally our relaying circuit will work. The allowed mismatch varies with the relay subtypes, so check for the allowed mismatch.
    Mismatch is found from the equation

$$M = \frac{\dfrac{I_{RL}}{I_{RH}} - \dfrac{T_L}{T_H}}{S}(100) \tag{5.2}$$

Where $I_{RL}$ and $I_{RH}$ are from the same kVA base, $T_H$ and $T_L$ are high and low side relay tap settings respectively, and S is the lower of the tap and relay current ratios. For our example and an HU relay well within the most stringent requirement (10% for the CA-26 type).

**TABLE 5.3   Harmonic Unit (HU) Taps and Ratios**

| Taps | Ratio | | | | | | | |
|------|-------|-------|-------|-------|-------|-------|-------|-------|
|      | 2.9 | 3.2 | 3.5 | 3.8 | 4.2 | 4.6 | 5 | 8.7 |
| 2.9 | 1.000 | 1.103 | 1.207 | 1.310 | 1.488 | 1.586 | 1.724 | 3.000 |
| 3.2 |       | 1.000 | 1.094 | 1.188 | 1.313 | 1.438 | 1.563 | 2.719 |
| 3.5 |       |       | 1.000 | 1.086 | 1.200 | 1.314 | 1.429 | 1.486 |
| 3.8 |       |       |       | 1.000 | 1.05 | 1.211 | 1.316 | 2.289 |
| 4.2 |       |       |       |       | 1.000 | 1.095 | 1.190 | 2.071 |
| 4.6 |       |       |       |       |       | 1.000 | 1.087 | 1.890 |
| 5 |       |       |       |       |       |       | 1.000 | 1.740 |
| 8.7 |       |       |       |       |       |       |       | 1.00 |

$$M = \frac{\dfrac{3.88\,A}{3.62\,A} - \dfrac{3.5}{3.2}}{\dfrac{3.88\,A}{3.62\,A}} \times 100 = \frac{1.071 - 1.093}{1.071}(100) = 2.04\%$$

If the CT current mismatch is too large for the tap ratio, current balancing autotransformers are used to bring the CT ratios close enough for the relay taps. For example, if a high side CT current of 6 A has to be brought down to 4.6 A to match the low side current close enough for the relay taps to take care of, the autotransformer ratio would be (6 A)/(4.6 A) or 1.33:1 turns ratio. The autotransformer is connected between the CT line to the relay and the relay, as shown in Figure 5.49. The autotransformer is Y connected regardless of the CT connection. Auxiliary CTs are available with a turns ratio of 3/2, which is often enough to obtain relay currents within the tap range.

## 5.7

# DISCONNECT SWITCHES

Disconnect switches are designed to open and close a circuit. At high voltages the switches must have a large gap when open. An air gap of about 11 feet is required at 230 kV. Disconnect switches cannot open a fault.

## 5.7.1 Non-load Break Disconnect Switch

High and medium voltage disconnect switches are designed to isolate a section of a circuit after the protective device has de-energized the circuit. They allow disconnecting the faulted circuit or equipment for repair while the rest of the circuit in the zone of protection is put back into service and provide for personnel safety

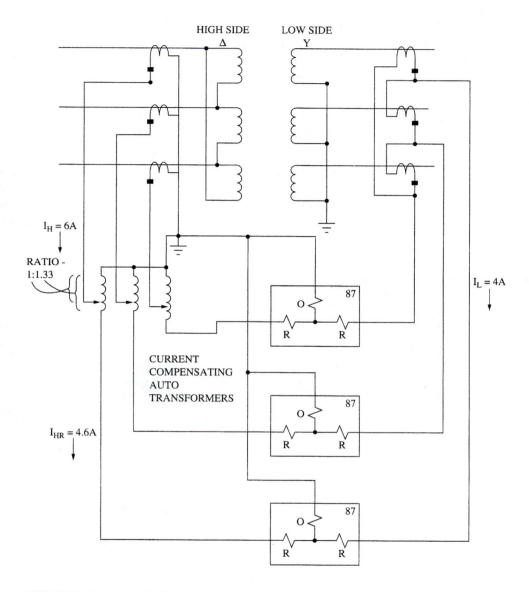

**FIGURE 5.49** Current balancing autotransformer connection

while the malfunction is being repaired. Disconnect switches can open very small charging currents to unloaded equipment but that is all. The low current arc is broken by swinging the moving arm in a 90° arc, providing a long gap.

Disconnect switches can be operated by motors, as most high voltage switches are, by an insulated lever connected to a actuating arm that moves the switch blade, as many medium voltage pole top switches are, or manually by using a hook stick (a long insulated pole with a hook attached to one end). Disconnect switches are classified as to voltage, continuous current rating, and function such as distribution pole top. They are available in single phase, single pole, and three phase, three pole. Unless a disconnect switch is specifically specified as load break it is a non-load break switch. Figure 5.50 shows a non-load break disconnect switch.

**FIGURE 5.50**   Non-load break switch (courtesy of Houston Lighting & Power).

## 5.7.2 Load Break Switches

*Load break disconnect switches* can interrupt normal load currents, but not large fault currents. The wall switch is the most common load break switch. Load break switches for medium and high voltages use interrupters built into the switch to break the load current before the switch disconnecting arm is swung open. The interrupters are often vacuum (medium voltage), oil, and $SF_6$. Some newer high voltage load break switches equipped with $SF_6$ interrupters can break significant fault current and accept protective relay trip signals.

Most load break switches use motors to open and close the switch blades but the interrupters are actuated by strong spring pressure. Figure 5.51 shows a $SF_6$ load break switch.

**FIGURE 5.51**   Load break switch (courtesy of Gulf States Utilities).

## 5.8

## METERING EQUIPMENT

Power metering equipment records the amount of power used in a particular area, sent down a particular line, and used in a particular structure. Metering equipment provides power use information for planning for future needs from power use trends, and of course billing for revenue.

Power meters use a current coil and a potential coil to turn and an induction disc, which then turns accumulating devices. The disc rotation is proportional to the power passing through the metered line. Electronic circuitry with microprocessor control can perform the same function and also transmit the accumulated power use data to a central point or record it in a digital memory for retrieval at the meter site. Most meters for industrial use meter demand, peak power use over a given period of time (usually 15 or 30 minutes), as well as power use. High demand requires lines and equipment with capacities in excess of normal loads and the demand charges amortize a portion of the extra cost as well as encourage more consistent power use.

Low voltage meters can be hooked directly to the monitored line voltage and current, but meters for station monitoring and large commercial and industrial customers must be fed through instrument transformers. Figure 5.52 shows some power meters. Reactive power meters are also used to monitor the VARs in a system.

**FIGURE 5.52**   Power meters (courtesy of Houston Lighting & Power).

## 5.9

### SUMMARY

Distribution equipment covered in this chapter includes circuit breakers, reclosers, sectionalizers, disconnect switches, metering and protection equipment. This equipment along with transformers, lines, and bus equipment are necessary to distribute power to utility customers or within large industrial concerns or multibuilding commercial establishments.

The purpose of a circuit breaker is to open a circuit both as a switch and during a fault. Thus, a circuit breaker has to operate during a fault when abnormally high currents are present. To interrupt the current the circuit breaker must extinguish the arc, the path of ionized gas, that occurs when the circuit breaker contacts are opened. Many means of breaking the arc have been successfully used in circuit breakers. Air circuit breakers stretch and thin out the arc until it extinguishes by using a long contact opening path and special baffles to further stretch the arc. Air circuit breakers are suitable for low and lower medium voltage applications.

Oil circuit breakers (OCBs), in which the contacts are immersed in a tank of oil, use the cooling and energy absorbing properties of oil to quench the contact arcs. Oil circuit breakers are suitable for medium and high voltages.

Circuit breakers in which the contacts are immersed in $SF_6$, an insulating gas that efficiently absorbs arc energy, are used for medium and high voltage circuit breakers.

Forced air circuit breakers called air blast breakers are suitable for high voltages. They interrupt an arc by forcing a high pressure air blast onto the arc to stretch and cool it.

A circuit breaker interruption method that is rapidly gaining in popularity is the use of vacuum bottles to contain the circuit breaker contacts, one bottle per phase. The bottles have bellows to allow the contacts to move for opening and closing. There are few atoms in a high vacuum to ionize, thus no arc forms. Vacuum circuit breakers have been made that are suitable for low, medium, and lower high voltage applications. To allow vacuum circuit breakers to be used at higher voltages the vacuum bottles containing the contacts are enclosed in a container of $SF_6$.

Most faults on a distribution line are temporary, having been caused by a falling tree branch that falls to the ground after causing a brief short circuit, or by some other brief event. Reclosers are basically smaller circuit-breaker type devices that open on a fault, but reclose after a set time. The process can be repeated for several times, but the limit is typically three recloses, after which, if the fault still exists, the recloser stays open, called *locking out*. Thus, brief faults are allowed to clear, then reclosing establishing service with minimum loss of service, but a fixed fault results in the removal of power from the circuit.

Fuses are alloy strips in a housing that melt if exposed to a current above that for which they were designed. They are an economical means of interrupting overcurrent faults. Current-limiting fuses are filled with sand that quickly absorbs the energy from the arc that forms at the point(s) that the fuse element melts. They limit the maximum amount of current that can pass during a fault. Fuses must be manually replaced after they operate so that service is restored only after someone looks for the cause of the fault. Dual element fuses have an element that operates quickly for massive faults, and one that operates slowly for overloads. Dual element fuses allow transformer energizing and motor starting current to occur without operating, but operate rapidly for large faults, and after a time for moderate overloads.

Lightning occurs nearly everywhere, just in varying frequency. Lightning induces current and voltage transients in electrical equipment. Equipment that is to be used for electrical power transmission or distribution must be protected against the effects of nearby or direct lightning strikes. Lightning arresters are devices that absorb the energy from the lightning induced transient while limiting the voltage from the transient to less than the BIL of the equipment. The most commonly used arresters in current use are made of zinc oxide. Shield (also called static) wires are strung above distribution and transmission lines to shield the lines from direct lightning strokes. They provide a zone of protection of 30° on either side of the shield wire.

To minimize the damage to a system from a fault condition devices called protective relays are used to sense and react to abnormal conditions on the line. Sets of protective relays are used together in systems to protect the line or equipment from a variety of line disturbances such as sudden and long duration overcurrents, over- and under-voltage, and/or over and under frequency. Relay systems are designed to react to faults within an area (or zone) of protection and ignore faults outside that zone, which are detected by other relays. Instrument transformers for current and voltage transform the line currents and voltages to levels that are convenient for use by relay systems to monitor the status of the protected equipment. Microprocessor-based relays, in which data is acquired from the instrument transformers and changed to digital form for processing under program control, are rapidly replacing electromechanical relays because they are more flexible than electromechanical relays and can provide several functions in a single package. Specialized relays for motor, generator, and transformer protection are commercially available.

## 5.10

### QUESTIONS

1. State the function of circuit breakers.
2. What is an arc and how does it originate at the breaker contacts?
3. State the two principles used in breaking an arc.
4. Why are dc arcs harder to break than ac?
5. List the four major arc extinguishing mediums.
6. List five arc breaking aids used in air circuit breakers and explain how they work.
7. What is the source of power for the opening mechanisms of almost all medium and high voltage circuit breakers?
8. How do air blast breakers extinguish an arc, and why must they always use opening resistors?
9. State the principle of the vacuum circuit breaker.
10. Describe the process by which bulk oil breakers quench an arc.
11. Describe the oil blast method of breaking an arc.
12. What property of $SF_6$ makes it such a good arc interrupting medium?
13. State the function of closing resistors.
14. List the factors that must be considered when selecting a circuit breaker for a given application.
15. List two advantages and two disadvantages of each of the following high voltage circuit breakers: air blast, oil, two tank $SF_6$, and puffer $SF_6$.
16. State the function of a recloser and its operating sequence.

17. Define coordination with reference to protective devices for electrical power equipment and lines.
18. Describe the function of a sectionalizer.
19. Describe the relationship of circuit breakers, reclosers, and sectionalizers in circuits that use all three.
20. State the principle of fuse operation.
21. List and briefly explain the six considerations for fuse selection.
22. Describe the arc breaking mechanism of both current limiting and expulsion fuses.
23. State the principle of fuse coordination.
24. State the purpose of lightning arresters.
25. State the difference between the terminal characteristics of spark gap and ceramic valve block types of lightning arresters.
26. State the function of a shield wire on a transmission or distribution line.
27. What is a protective relay?
28. The induction disc relay whose characteristics are shown in Figure 5.39b has a time dial setting of 3, a tap value of 4A, and a CT current of 8 A. How long will it take to operate?
29. List the major parts of a 50/51 overcurrent relay.
30. The circuit of Figure 5.44a is to be disconnected by the breaker if the source voltage falls to 80% of the nominal value of 4160 V. The under voltage relay whose characteristics are shown in Figure 4.44b is to be used. The relay time dial setting is four. Find the proper tap value and the operate time if the PT voltage is 50% of tap value.
31. State the operating principle of a differential relay.
32. State the function of an HRU in a differential relay, and the principle of HRU operation.
33. State the method of connecting CTs for Δ-Y and Y-Δ connected transformers, and the reason the method is used.
34. State the major advantages of microprocessor-controlled relays over electro-mechanical relays.
35. What is the reason for the trade-off of response time and complexity in micro-computer-based relay systems?
36. For the circuit of Figure 5.48 calculate the relay currents, find the HRU type restraining coil tap settings, and calculate the mismatch if the transformer is 230kV/34.5kV of the same connection and kVA rating.
37. State the difference between load break and non-load break switches in terms of operation.

# CHAPTER 6

# Distribution Substations

The purpose of an electrical power distribution substation is to lower the voltage from the high voltage transmission and subtransmission levels to a voltage that can be safely installed on the streets and alleys that are adjacent to the customer's premises. The smaller right of ways available near the customer's premises make the lower voltage necessary. Shapes, configurations, capacities, and voltages of distribution substations vary widely, but they all have the same purpose.

Substations for distribution consist of equipment (items that use electric power to perform a function such as motors and transformers) and devices (which help control but do not use electric power such as switches, lightning arresters, and fuses). The devices are often referred to as switch gear. Circuit breakers, though a switch, are so costly and complicated that they are often classed as equipment. Figure 6.1 illustrates a distribution substation.

## 6.1

### SUBSTATION LAYOUT SELECTION CRITERIA

The design or layout of a substation depends on many factors, including: voltage; load requirements, density, and expected growth; available site space; location of the site; protection system sophistication and cost; and other stations in the general area.

The substations of any given power company are not alike because the items listed above vary from place to place, but substations do have as much in common as the design criteria allow. Standardization in substation design is desirable where

**FIGURE 6.1** Distribution substation (courtesy of Houston Lighting & Power).

possible because of the savings in design time and the simplification of construction and maintenance. Standardization simplifies and usually simplification reduces cost, even in such things as stocking spare parts. Over standardization can lead to the selection of a substation design that does not economically meet the needs of the area to be serviced, so there should be a happy medium in standardization.

The design of a substation depends on both the subtransmission and/or transmission voltages available and the low voltage requirements. The trend is toward the highest distribution voltage that can be safely and economically installed in the available distribution right of ways. The higher the voltage, the lower the line resistive losses are. This is generally 34.5 kV for new distribution systems. Lower voltages must be used if a substation is to serve a lower voltage distribution line system, of course. If the substation must serve distribution lines at more than one voltage, 34.5 kV and 12.5 kV for example, it will be more complex than a single voltage substation.

## 6.1.1 Voltage and Spacing

The voltages determine the spacing between conductors, and the conductor clearance to ground or grounded structures. The minimum clearances are specified by the National Electric Safety Code (NESC), and usual clearances (which are larger than the minimum) are published by the Institute of Electrical and Electronic Engineers (IEEE) electrical power related divisions and the Edison Electric

Institute (EEI). The IEEE is a professional organization of individual engineers and technical people and the EEI is the professional association of electric power companies. For example the minimum clearances are 22.3 inches for 34.5 kV, and 13.2 inches for 12.5 kV for rigid buses with fixed supports.

Often the substation conductor ground clearance is dictated more by safe working and walking distance than by minimum safety code clearance. The safe walking and working distances, which are greater than the minimum safety code clearance, are reduced if only qualified personnel have access to the substation. To reduce the cost of substation by reducing the safe clearances, substations are enclosed in a fence to limit access to qualified people, who by education and experience understand the hazards involved and safety procedures to be followed when working around high voltage. Some substations, called unit substations, are completely enclosed in a metal structure. Rigid bus, copper or aluminum pipes or solid bars designed as substation conductors, can be installed in less space than required to install wire bus because wire bus sags between insulators. The sag varies with temperature and load current, which heats the conductor. Additionally, blowing wind can displace wire conductors. Rigid bus clearance can be equal to the minimum deflected wire bus clearance, which is usually considered too close for reliable wire bus installation.

## 6.1.2 Load

The size of the load to be served determines the capacity of the substation. Dense loads such as commercial and industrial districts are served by larger substations than loads spread over a wide area. The load must be distributed such that it can be served with reasonable feeder loss or more, smaller substations closer to the load may be used. Critical loads, such as industrial districts, are served by more complex substations, designed for maximum reliability and speed of power restoration, than residential areas where a short time power loss is usually not a disaster. Substations for critical loads usually use more than one transformer so that the load is served even if one transformer is out. Otherwise a single large three-phase transformer is used because it costs less per kVA of capacity, and requires less room, bussing, and simpler protective relaying. The expected growth of an area also influences substation design. An area that is expected to grow uses an easily expandable substation design.

Other substations in the area influence the design of a new substation. The ability of the other substations' capacities to be increased to serve growth in the area between the stations (instead of expanding the capacity of the new substation) affects the design of the new substation. Additionally, it is advantageous from a maintenance point view to use similar new and old substation designs in the same general area if possible.

### 6.1.3 Space and Location

More space available for substation construction allows more construction options. Densely populated areas may contain few locations suitable for substation construction, and those locations may be unsuitable for some substation designs. In some areas land costs may be so high that one substation design or layout provides considerable savings by virtue of needing less space, even if the construction cost is more for the more compact layout.

The visual sensitivity of the residents of an area can dictate substation construction. Some people do not like to see substations and insist that their neighborhood be served by a low profile substation enclosed in a fence that hides it. This costs more and the added cost for the same revenue base has an effect on the substation layout. The aesthetic considerations are the same ones that have led to increased underground distribution.

### 6.1.4 Distribution Substation Protection Needs

Above a minimum protection needed to avoid injury to people and damage to equipment, the level of protection of a substation is determined by how critical the loss of power is to the load. For example, the loss of electrical power to a hospital is very serious while the loss of power to a residence is inconvenient. In the event of a fault the hospital electricity must be restored in the shortest amount of time possible while the residence can be without electricity several hours without serious consequences. Industrial and commercial areas cannot be without power for any length of time without serious economic consequences. Equipping a substation with automatic switching to restore power when it is lost and to assure the least possible damage and repair time after a fault is expensive. It would be nice, but hard to justify to public utility comissions (PUCs), to provide each substation with the maximum possible protection. Thus judgment must be used in substation design and protection to assure a safe, reliable, economical substation with protection that is adequate for the load served. Figure 6.2a shows a small substation at the end of a radial subtransmission line that might be used to serve a small group of residences. It consists of two dead end poles to terminate the lines. It has two manual non-load break switches, and primary fusing. Together with the customer distribution transformers primary fusing, and fuses at taps with a 30% or more decrease in wire size, it is adequately protected to perform its function. The substation in Figure 6.2b might be used to serve a small commercial area. It has a circuit breaker as well as a primary fuse for back up, and more disconnect switches for isolation during maintenance. Additionally the circuit breaker will operate from relays that require a metal clad enclosure, instrument transformers, and a dc power supply for the trip circuit. The increased speed of fault removal supplied by the circuit breaker for this substation has substantially increased its cost over the substation shown in Figure 6.2a. Both substations are simple single-source, single-transformer, single-feeder types. The cost differences increase with the size of the substation, and the size and number of transformers used.

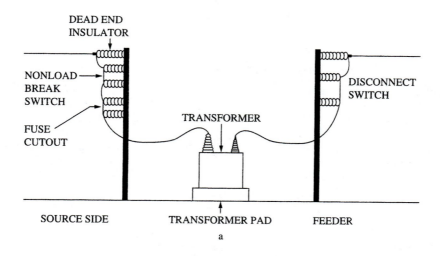

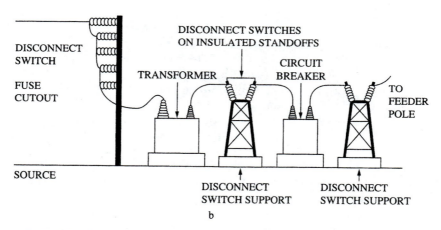

**FIGURE 6.2**   Simplest substations (one phase drawn) (a) Fuse protected transformer and (b) circuit breaker protected

## 6.2

## DISTRIBUTION SUBSTATION CONSTRUCTION METHODS

Four basic methods exist for substation construction: wood, steel lattice, steel low profile, and unit. Each has advantages and disadvantages. These are shown in Figure 6.3.

Wood pole substations are inexpensive, and can easily use wire bus structures. Wood is suitable only for relatively small, simple substations because of the difficulty of building complex bus and switch gear support structures from wood. Very few wood substations are built anymore.

**FIGURE 6.3**   Substation construction methods include: (a) steel lattice (courtesy of Gulf States Utilities) and (b) low profile (courtesy of Gulf States Utilities).

Lattice steel provides structures of low weight and high strength. Complex, multilevel bus structures can be fabricated relatively easily from lattice steel, as can switch gear support structures. Lattice steel is reasonably economical and is the preferred material for substation construction whenever possible.

Solid steel low profile substations are considered by many to be aesthetically superior to lattice or wood constructed substations. However, low profile construction is more expensive than either wood or lattice steel, and requires more land because multilevel bus structures cannot be used.

The unit substation is a relatively recent development. A unit substation is one that is factory built and tested, then shipped in modules that are bolted together at the site. They are completely enclosed in metal cladding to deny access to unqualified individuals, and protect the equipment. Unit substations usually contain high and low voltage disconnect switches, one or sometimes two three-phase transformers, low voltage breakers, high voltage fusing, bus work, and relaying with overcurrent relaying the minimum. Many unit substations have provisions for underground high voltage lines to be brought in. Unit substations are usually 3000 kVA at 34.5 kV or less. Smaller unit substations can be installed in underground vaults or utility access holes. Unit substations are very expensive, but they are also very reliable and compact.

## 6.3

## TRENDS IN DISTRIBUTION SUBSTATIONS

While large substations of 10,000 kVA or more are necessary and economical in areas of dense load, the trend is toward smaller substations where possible. There are a number of reasons for this trend. It is difficult to add small increments of capacity to large substations economically. Maximum economy is realized when low voltage feeder mileage is kept as low as possible, and all line possible is at higher subtransmission voltages. Smaller substations allow a lower loss mix of high and low voltage lines in lower density load areas. Growth is very hard to predict and has been known to be away from substations designed to serve an area making long, lossy, low voltage feeders necessary. Smaller substations also allow simpler layouts that make design, construction, relaying, and maintenance simpler and more economical.

The trend toward smaller substations is complemented by a trend toward three-phase, as opposed to three single-phase, transformers in substations. Three-phase transformers cost less per kVA of capacity, are smaller, lighter, require fewer fittings to install, and are highly reliable. Three single-phase transformers can be operated in the open delta configuration at reduced capacity should one transformer fail, but that is their only advantage. There is also a trend toward smaller transformers, in the 600 to 3000 kVA range where possible, because a spare transformer in this size range can fairly easily be kept at a central location for emergency replacement. This is the size range in which a mobile transformer can be made available for emergency service while substation repairs are made. When mobile transformers are equipped with disconnects and protection they are often called mobile substations.

## 6.4
## INSULATION COORDINATION

A typical method of insulation coordination is to establish a definite common level for all insulation in the station and bring all equipment and device insulation to that level. Lightning protection is set at least 20% below this level, but above normal peak operating voltages. The basic impulse level (BIL) standards have been established as a result of insulation flashover tests. For example the BIL standard for 34.5 kV is 200 kV, and for 8.7 kV is 75 kV for distribution class equipment and 95 kV for power class equipment when tested with the standard 1.5 us $\times$ 40 us impulse crest waveform. Solidly grounded systems are easier to protect against overvoltage.

## 6.5
## VOLTAGE REGULATION

When the feeder load is such that a fixed tap cannot keep its voltage between the allowed limits some form of voltage regulation is required. Automatic tap changing under load is the most popular voltage regulation scheme, but the voltage level of all feeders from the transformer are changed simultaneously. If this is not acceptable, it usually is, then capacitors or regulating transformers must be used on the individual feeders involved, as shown in Chapter 4.

## 6.6
## DISTRIBUTION SUBSTATION LAYOUT

This section discusses some common distribution substation layouts or designs. These are a representative sample of the layouts in use. The variations on the main themes are many. The designs vary from utility to utility and within the same utility. The nomenclature used for substation designs also varies from utility to utility, so the names we used here are as generic as possible.

### 6.6.1 The One Feeder Substation

Figure 6.4 shows the one-line diagram of a single-source, single-feeder substation. Note the minimum equipment used. A bypass switch is often provided so service

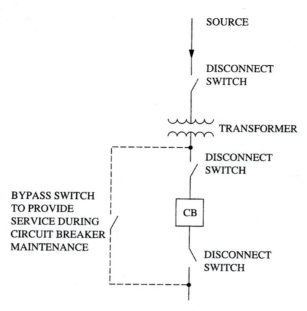

**FIGURE 6.4**   One-line diagram of single source, single feeder substation

can continue during circuit breaker maintenance. The probability of a fault during circuit breaker maintenance is small, but still there. The transformer usually has a primary fuse to back up the breaker, and provide some protection for internal transformer faults. The minimum relaying is overcurrent on the secondary of the transformer. Frequently differential relaying is provided for the transformer. The switches, as in all substations in this section, can be manually or motor operated. They are normally manual in a substation this simple.

## 6.6.2 Single Bus Substation

Figure 6.5a shows the one line of a single bus substation fed by a single radial subtransmission line. Each feeder must have its own overcurrent protection. The primary side switch should be interlocked with the feeder breakers so it cannot be opened under load. The primary switch must be able to break the transformer excitation current. The high voltage side breaker is often omitted because the probability of a fault between the transformer high voltage bushings and the feeder breakers is small in a small substation. The transformer rating is often so small compared to the subtransmission circuit that a transformer fault will not operate the subtransmission circuit breaker, so the transformer primary is protected by at least a fuse whose clear time is longer than the feeder breaker operate times. The transformer may have differential relaying that trips all of the feeder breakers in the event of a fault. Each distribution voltage the substation supplies must have its own bus.

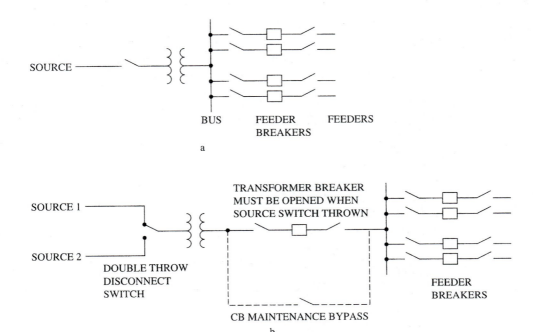

**FIGURE 6.5** Single bus substation (a) Single source and (b) two source

The possibility of a subtransmission circuit fault is much higher than a transformer fault. Two sources allow service to be restored more quickly upon a subtransmission circuit fault. The double throw switch on the primary side, shown in Figure 6.5b, allows the transfer to be made quickly from one subtransmission circuit to another. The switch should have an open position to remove the high voltage completely. The switch is interlocked with the transformer breaker so it cannot be opened under load. The switch can be replaced by two manual high voltage breakers that can break the load, and expected fault, current. If high voltage breakers are used the transformer secondary breaker is usually omitted. The transformer secondary breaker makes possible very effective differential bus protection to detect faults internal to the bus. The bus relays then trip all of the circuit breakers connected to the bus upon a bus fault.

Please note that all of the circuit breakers probably have a bypass switch to allow service during circuit breaker maintenance. Figure 6.5b shows only the bypass switch for the transformer breaker to avoid cluttering the one-line drawing. The reader can assume a bypass switch around all feeder breakers henceforth.

## 6.6.3 Two Transformer Distribution Substations

More critical loads may be out of service an unacceptable length of time in the event of a transformer failure. Two transformers used at a substation can eliminate this eventuality. Normally the transformers are rated at 75% capacity when self cooled and equipped with automatic air cooling that is used when one transformer must handle the entire substation capacity. Sometimes three transformers are used to reduce the spare capacity the others must have when one of them is out of service. A two transformer substation is shown in Figure 6.6.

Notice the tie switch between the two transformer connections to the bus. This is open when both transformers are in use to prevent the transformer secondaries from operating in parallel. Even small differences in the two transformer secondary voltages can cause large currents to circulate between them and cause excessive heating, as well as make automatic tap changers malfunction. Special reactors must be used to force equal current sharing if transformers are to operate in parallel. Momentary parallel operation during switching is often permissible. The primary side switching is arranged so that either or both transformers can be fed by either subtransmission line.

In a substation of this complexity the transformers normally have primary fuse protection as well as differential relaying, and secondary overcurrent protection. The bus often has differential relay protection to disconnect the bus if it faults and each feeder has overcurrent relaying.

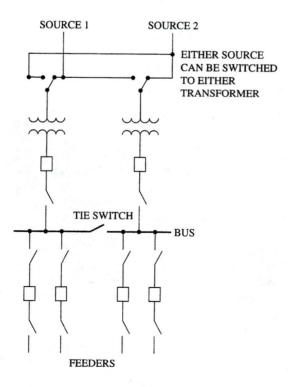

**FIGURE 6.6**  Two transformer substation

### 6.6.4 Automatic Switching (Throw-over)

All of the substation layouts we have looked at so far require manual switching to restore power after a fault. Service outage time can be reduced considerably by using circuit breakers to automatically, or on command from a central control station, disconnect the faulted source or bus and connect the substation so that power can reach all of the feeders. Figure 6.7a shows a substation connected for automatic switching, also called throw-over or roll-over. Assume sources 1 and 2 are connected as radial lines. If source 1 should be lost, breaker 1 would open under relay control disconnecting source 1, and breaker three would close connecting transformer 1 to source 2, and vice versa. If transformer 1 should fail breakers 1, 3, and 4 would open to disconnect it. The low voltage bus tie breaker, would then close to connect all of the feeders to transformer 2. The low voltage tie breaker is interlocked with transformer secondary breakers 4 and 5 to prevent parallel transformer operation. If even momentary parallel operation cannot be tolerated, relaying can assure that the tie breaker opens as the transformer being reconnected to the bus is energized. This station is said to have both high and low side throw-over. Once the breakers have operated on a fault they must be manually closed to assure that the fault has been corrected.

If the sources were part of a loop the high side tie breaker would be normally closed instead of normally open as it is when the sources are radial lines ends or taps. Circuit breaker 1 would remove source 1 in the event of a fault, and circuit breaker 2 would remove source 2. A bypass switch, shown in the dashed line, would close the loop in the event circuit breaker 1, 2, or 3 were down for maintenance. A preferred automatic switching scheme for loop connected supply lines is shown in Figure 6.7b. Circuit breakers A and B remove a faulted line from service, while circuit breakers C and C' and D and D' remove the transformers in the event they fault. Upon a transformer failure the low voltage tie breaker connects all of the feeders to whichever transformer is working.

Notice that each circuit breaker has disconnects on either end to isolate it from the circuit during maintenance. Each feeder breaker probably has a bypass switch so the feeder can operate during breaker maintenance.

A substation of this complexity is well protected with overcurrent protection on the transformer secondaries, and the feeders. The transformers and the bus have differential relay protection with overcurrent backup.

### 6.6.5 Double Bus Substation

In densely populated areas, whether with people or business, and high load areas, like downtown, 10,000 kVA and larger substations are economical. A bus fault will bring the entire substation down interrupting service to all of the customers served by the substation. The solution is to have a spare bus to use in the event one faults. This solution should be familiar by now. The spare bus is often called an inspection bus because it is in service during maintenance in a single-source substation. Each bus is protected with differential relaying that connects the feeders

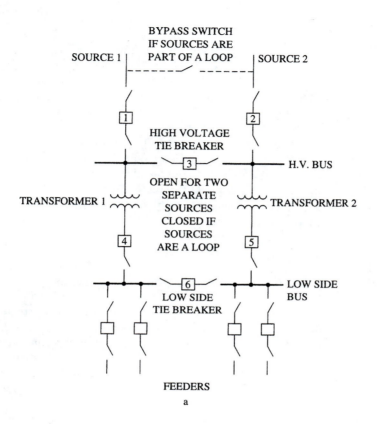

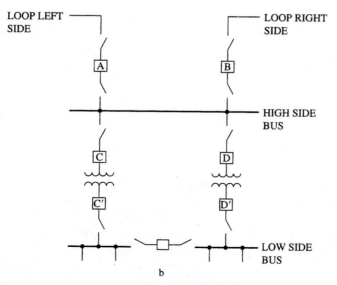

**FIGURE 6.7**   Automatic switching two source, two transformer substation (a) Two source, radial or loop and (b) better high side loop arrangement

served by one bus to the other bus upon a fault. The automatic switching also removes power from the faulted bus.

A double bus substation is shown in Figure 6.8 that has double bus on both the high and low voltage sides. The reader may enjoy determining the breakers to be switched in the event of a fault in various parts of the substation. Recall that the transformers are not operated with their secondaries in parallel. The relaying for this substation is quite complex and thorough. Notice the feeders are connected to the busses by disconnect switches that must be manually thrown, even if a motor does the work. It is much quicker to connect automatically the feeders to the desired bus.

### 6.6.6 Bus Arrangements

The bus arrangements in this section can be used for high or low voltage side busses.

The *ring bus,* shown in Figure 6.9 requires only one circuit breaker per line so it is economical. Power can reach any feeder from two directions so no feeder need be disconnected when one breaker is down for service. Two breakers are operated when a feeder fault occurs, one on either side of the feeder.

If a second breaker trips when one feeder is down the entire bus may be disconnected from a source. The ring bus is therefore seldom used if the feeders exceed the source lines by more than a factor of two. Additionally the ring bus is hard to expand. The current capacity of the circuit breakers must be large enough to handle the load of all feeders that may be connected through it. The ring bus is normally used in smaller substations.

The *breaker and a half,* shown in Figure 6.10, is a reasonably economical, versatile, reliable, and easy to expand bus arrangement. It is very popular. A bus or feeder fault can easily be isolated while the station remains in service. The name is from the fact that three circuit breakers are required for every two feeders, or 1.5 breakers per feeder. The breaker and a half is usually used in substations with more than four feeders.

The current carrying capacity of the circuit breakers must be equal to the maximum load of the two feeders they serve because the feeders may be fed from either bus.

The *double bus-double breaker* arrangement, shown in Figure 6.11, is the ultimate in versatility, protection, and the highest in cost. Each circuit has full protection with any breaker out of service for maintenance and no breaker carries the load for more than one circuit. If the bus switches of Figure 6.8 were replaced with circuit breakers, which would replace the individual feeder breakers on the low side and the transformer primary breakers on the high voltage side, it would be a double bus-double breaker substation. This arrangement is only used at important locations because of the expense.

### 6.7

## FAULTS

A fault, recall, is a malfunction in the system. Most faults are, or result in, short circuits. Most faults are the result of lightning and wind of storms, with lightning causing the greatest number. The wind may cause lines and poles to break, or tree

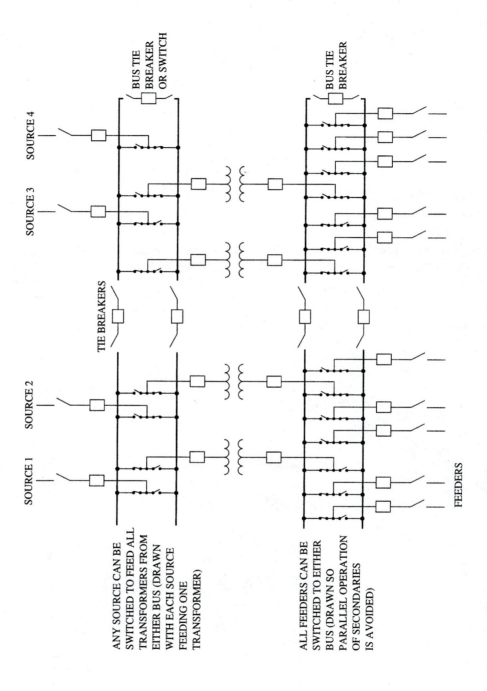

**FIGURE 6.8** Double bus distribution substation (high and low voltage double bus)

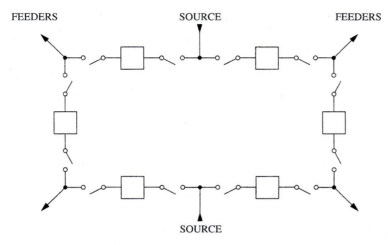

**FIGURE 6.9**　Ring bus

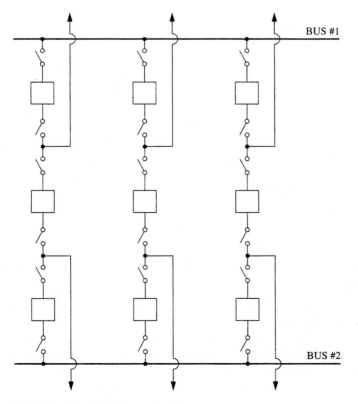

**FIGURE 6.10**　Breaker and a half

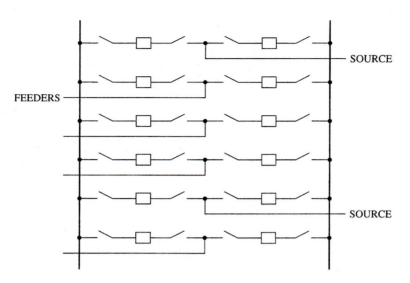

**FIGURE 6.11**   Double bus-double breaker

limbs to blow across lines, or in the winter ice and snow loading may cause lines to break.

The major categories of faults are: line to ground, about 70%; line to line, about 15%; double line to ground, approximately 10%; three-phase faults to ground, less than 1%; and open circuits not accompanied by a short, very few. The first three are illustrated in Figure 6.12. The percentages given vary from system to system but the order of frequency does not. People and equipment must be protected from system faults by disconnecting the faulted system segment with circuit breakers, reclosers, sectionalizers, and fuses.

Prompt removal of electrical supplies from a faulted circuit is essential to prevent several unpleasant consequences. The sooner the fault is isolated the less damage is done to the equipment, lines, and devices in the faulted circuit. A sustained fault may melt lines, cause transformers windings and bus to twist from mechanical forces caused by the fault current magnetic field, overheat and ignite the oil in transformers causing dangerous explosions and fires, and extend the service outage because of the extensive repairs required. Another problem a fault can cause, we will discuss why later, is to reduce the stability margins of the system. This can result in generators being dropped off line by their protective relaying.

The maximum and minimum values of fault current must be known to provide proper protection for a circuit. An approximate method of fault current calculation is available for building distribution systems that assumes the voltage does not drop during a fault (which it does), but it is not adequate for distribution and transmission faults. In the transmission section we will study *symmetrical components*, which is an accurate method of calculating fault currents. In fact, symmetrical components are most of Chapter 10.

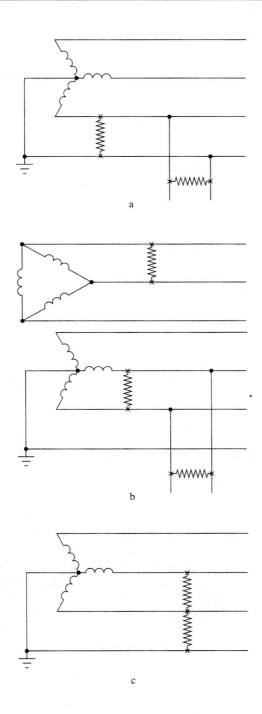

**FIGURE 6.12**    Major fault types include: (a) Line to ground (b) Line to line and (c) Double line to ground.

## 6.8

### DISTRIBUTION SUBSTATION PROTECTION

Circuit breakers tripped by protective relays are used to protect the equipment within a substation, with primary fusing used to protect the transformers in some smaller substations. Transmission substation protection is discussed in the transmission section of the book.

### 6.8.1 Zones of Protection

Each relay set and circuit breaker is set to protect a certain portion of the substation and restrict the amount of the substation removed from service for a given fault. The portion of the substation removed from service by a given relay set is its zone of protection. In Figure 6.13 the feeder breakers zone of protection is the individual feeder. The bus zone of protection is the bus and connections to the bus. Should a catastrophic close in fault occur on a feeder the overcurrent relays for the transformer secondary breaker should eventually operate if for some reason the feeder breaker does not trip. This breaker is then a backup for the feeders and for the bus too. Each protective element normally has a backup in this manner to provide protection if the first line protection fails to operate. More of the substation, in Figure 6.13 the whole substation, is taken out of service by the back-up protection, but most of the station can be put back into service quickly, which would not be the case if the back-up protection was not used. Notice that the zones of protection overlap.

The relays operate from current and potential transformers, but their trip circuits operate from 125 V dc normally supplied by a substation battery bank. The batteries, usually lead-acid, are trickle charged so they are fully charged when needed.

The overcurrent protective devices must be coordinated to assure that only the faulted substation section is taken out of service unless the first line protective device fails. Thus the upstream protective element must take longer to operate on any fault current than the downstream protective element. The inverse-time characteristic of fuses, reclosers, and inverse time (CO) type overcurrent relays aid in their coordination. The procedure for coordinating CO overcurrent relays is identical to that for fuses, except for the necessity of selecting a tap. Fuse coordination was illustrated in Chapter 5. Relay, recloser, and fuse manufacturers have detailed literature on coordination for their protective devices, usually with application examples.

Instantaneous trip overcurrent devices (IT and IIT) must be coordinated strictly by operating current. The upstream IT must trip at a higher current than the downstream IT. Usually this is no problem because ITs are set for fault currents instead of overload, and upstream fault currents, both minimum and maximum are higher in a distribution system. If ITs cannot be coordinated on current trip points, then the upstream device will have a fixed time delay to prevent operation until the downstream IT has time to operate. If the downstream IT fails to operate then the upstream IT will operate after the fixed delay.

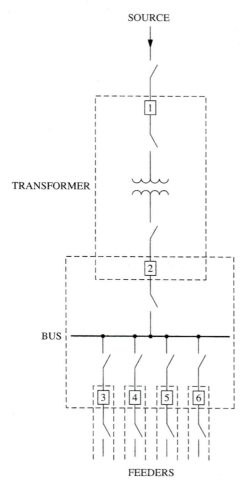

**FIGURE 6.13** Zones of protection

Coordination by inverse-time devices (fuses, CO relays, and reclosers) along a distribution feeder is illustrated in Figure 6.14. The solid curves are the CO relay inverse time characteristic. Recall that the further from a source that a fault occurs, the more line impedance is in the circuit, and the lower the fault current is. Usually the maximum fault current is available from a fault just downstream from the protective device, and the minimum fault current is available from a fault at the downstream end of the protective device zone of protection. The inverse-time characteristic means that the protective device operates faster for high fault currents than for low. Thus, as shown in Figure 6.14, the time current curve of an inverse-time protective device correlates with distance along the line. The higher the fault current, the closer the fault, the faster the protective device operates, and vice versa. The instantaneous IT devices used to back up the inverse-time

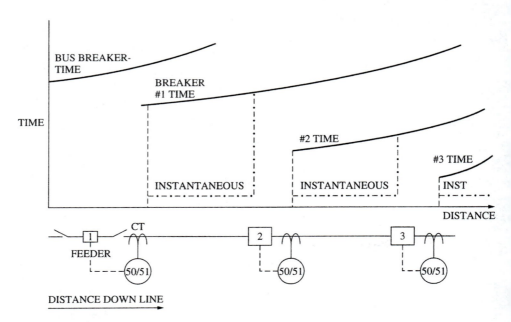

**FIGURE 6.14**    Coordination of inverse-time over current relays

overcurrent devices are set to trip quickly in the event of very high fault currents, usually from close in faults, and thereby minimize damage caused by the fault.

Coordination is easiest if all inverse-time devices have similar characteristics. For example use all CO-9 relays. In CO overcurrent relays the higher the number the more inverse the characteristic, with CO-2 called short (not very inverse but not an IT characteristic), CO-5 is long (meaning more inverse), CO-6 is definite (for definitely inverse), CO-7 is moderately inverse, CO-8 is inverse, CO-9 is very inverse, and CO-11 is extremely inverse. Figure 6.14 shows all CO relays and breakers but the protective devices might be a circuit breaker, recloser, and fuse in that order.

## 6.8.2 Transformer Protection

Transformers are, as we noted before, very expensive, and so are well protected. A substation transformer has a minimum of: secondary overcurrent relaying, primary fusing, and a sudden pressure relay. Except for very small substations transformers also have differential relaying with back-up overcurrent relaying. The basic transformer relaying scheme is shown in Figure 6.15a. The delta winding is shown on the source side because the distribution side is normally wye to make the phase to neutral taps needed by the customers. Not all subtransmission transformer connections are delta, but if they are then no tertiary is needed for harmonic control. Note that a circuit breaker is a type 52 device.

The sudden pressure relay (SPR) is a relay, attached to a valve on the transformer tank, that will detect the sudden rise in pressure caused by internal arcing

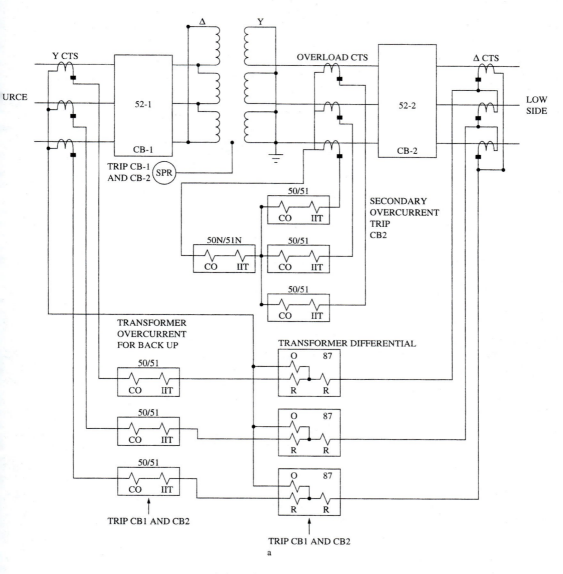

**FIGURE 6.15**   Transformer protection (a) Simplified circuit

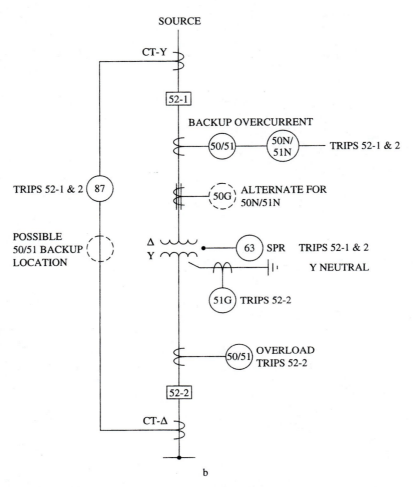

**FIGURE 6.15**   Transformer protection (b) one-line diagram

and trip the secondary breaker, and primary breaker if there is one. The SPR will not respond to the gradual rise in pressure caused by an increase in loading. The SPR is classified as a type 63 devise, and operates in 2 to 37 cycles, depending on the rate of pressure rise.

Notice in Figure 6.15a that the differential relays also have back-up 50/51 overcurrent relays. The CO relays used in this application must usually be set well above the maximum allowed overload current (200 to 300%) to coordinate with downstream devices. The IIT will be set at 125% of the maximum through fault. Primary overcurrent relays alone can protect against internal transformer faults, and are used on many small installations, but to minimize internal transformer damage from internal faults differential relays are needed.

The secondary overcurrent relaying of Figure 6.15a includes a 50N/51N relay to detect ground faults. The 50N/51N relay is a sensitive overcurrent relay designed for this purpose. If a fault to ground occurs on one of the phases, even a fairly high impedance fault, the phase currents will no longer sum to zero, so the CT currents will no longer sum to zero, and CT neutral current will flow through the 50N/51N relay. If enough flows, the relay trips CB-2. We could have connected a 50N/51N in the CT neutral line of the differential/back-up overcurrent relays.

Figure 6.15b shows a one-line drawing of a well-protected transformer. Notice that the back-up primary overcurrent protection can use separate CTs than the differential relays. A number of different ground fault protection possibilities are shown in the figure. Starting at the top, the 50N/51N in the primary overcurrent relaying is connected exactly the same as shown in Figure 6.15a in the secondary overload relays.

An alternative method of detecting ground faults in delta systems is shown next. A closer look is shown in Figure 6.16a. This is a current transformer with an aperture of sufficient size to allow all three (four including neutral if used with a grounded wye) phase conductors to pass through. The current transformer is called a zero sequence CT for reasons we will discover in the symmetrical components section. The CT output should be zero as long as the phase currents sum to zero, but in the event of a ground fault the phase current will no longer sum to zero and relay current will flow causing an alarm or a trip. A 50G/51G ground fault relay is normally used with a zero sequence CT.

A good place to detect a ground fault in a grounded wye system is on the neutral to ground conductor. This is shown in Figure 6.15. No current should flow in this conductor unless a ground fault exists. Resistors are used in the ground conductor to limit the transient voltage from line capacitance charging on long subtransmission and transmission lines, and to limit current in the event of a fault from a phase to ground. Reactors, which limit sudden changes in current, can be used instead of resistors but the common practice in the United States is to use resistors, which cost less and perform as well as reactors. When possible, solidly grounded neutrals are preferred. If a neutral resistance is used a type 59G overvoltage relay provides positive and sensitive ground fault detection, as shown in Figure 6.16b. Ground fault neutral current causes a voltage across the resistor proportional to the fault current through it.

A more sensitive ground fault detection method using standard CTs is shown in Figure 6.16c. The 50N/51N relay obtains all of the current from a mismatch because the phase CTs have no relay burden.

Ground fault relaying, especially on delta systems, may only provide an alarm, allowing the system to continue operating while the fault is found, instead of tripping a circuit breaker. The phase overcurrent relays then trip on a sufficient fault to cause a serious problem. A combination can be used with a CO providing an alarm and an IIT providing a trip if the fault is serious enough to warrant it. The ground fault relaying must not be so sensitive that normal phase load mismatch causes operation, though excess sensitivity is seldom a problem. There are many ground fault protection schemes, but these four are the most common. Their use is not limited to transformer ground fault detection.

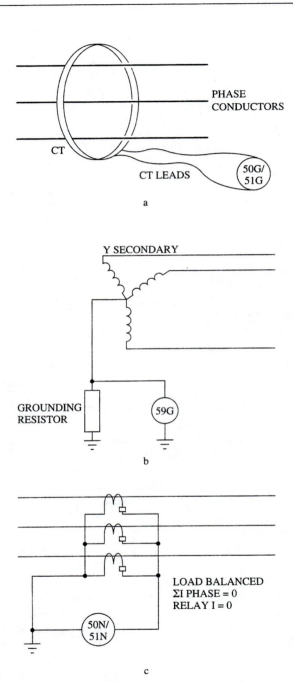

**FIGURE 6.16**  Ground current detection (a) Ungrounded system ground fault detection using zero sequence CT (b) 59G over voltage relay for ground fault detection (c) Three CT method for ground fault protection

### 6.8.3  Bus Protection

Differential protection is effective for bus faults because the current leaving the bus on feeders and the current entering the bus from sources should be zero at any instant. Additionally, differential protection can distinguish between internal bus faults and external feeder faults. A feeder fault can result in the CTs on the feeder saturating, and the dc offset of a fault worsens the situation. Thus special care must be taken in bus differential relaying to prevent external faults from causing a trip on the circuit breakers supplying the bus. Three major systems are used:

1. Linear coupler (LC) system, which works by eliminating the iron core of the CTs.
2. Multi-restraint, variable percentage relays (CA-16).
3. High impedance voltage operated differential relays (KAB).

*Linear coupler systems* use air core transformers that do not saturate. Those used for LC relaying look similar to normal iron core CTs, and can be specified for bushing mount at the time of purchase of the substation circuit breakers. The LC system is fast, less than 16 mS for LC-1 and 32 mS for LC-2 relays, easy to apply, and easy to adapt to bus changes. The LCs are actually designed to provide 5 V secondary voltage for every 1000 A primary current. The relay detects voltage.

The LC connection for a very small bus is shown in Figure 6.17. Notice that the LCs are connected in series. At least one of the lines is a source. In normal operation or during an external fault the sum of the LCs voltages is zero because the LC voltages are proportional to the line current and the source and feeder currents sum to zero. During a bus fault at least one of the feeder currents approaches zero and its LC voltage drops. The voltage sum of the LCs is no longer zero and the relay operates.

The LC-1 relay uses a solenoid connected through an impedance matching autotransformer. The impedance matching autotransformer is to allow the relay impedance to match the total impedance of the LCs. The LC-1 and LC-2 relays both have taps of 30, 40, 60, and 80 Ohms. The LC-2 relay uses a polar dc relay mechanism energized by the rectified output of an impedance matching transformer. The transformer source is, of course, the LCs. The LC-1 is designed for use where high fault currents are available, and the LC-2 for lower available fault current applications, though it can be used with high fault currents.

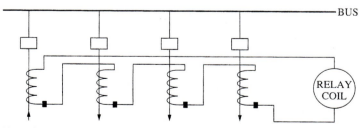

Normal operation or external fault – voltage sum zero
internal fault, feeder current drops – voltage sum not zero

**FIGURE 6.17**  Linear coupler bus protection (one-phase of connections shown)

The *multi-restraining CA-16* uses conventional CTs that saturate on heavy fault currents. The relay is more complex to compensate for CT saturation. The CA-16 consists of two sets of three restraining coils and one operating coil each on an induction disc. The two discs are mounted on a common shaft, which carries the moving contact. The restraining coils provide opening torque to the disc, and the operating coils provide closing torque. Each restraint coil is paired with another giving a total of six restraining coils. More than one relay is used if more restraining coils are needed.

The relay is said to be variable percentage because on light faults, where CT performance is good, its sensitivity is high and a very small current difference causes the relay to operate. On heavy faults, where CT performance is poor, a higher difference current is needed to operate the relay. This characteristic is provided by a saturating autotransformer across the operating coil, as shown in Figure 6.18. Heavy fault currents saturate the autotransformer causing current to be shunted away from the operating coil. The autotransformer also shunts the dc component from asymmetrical fault currents away from the operate coil. The dc in the restraining coils has a braking effect on the induction disc.

The connection for one phase of a four circuit CA-16 bus protection relay system is shown in Figure 6.18a. A circuit consists of from one CT, especially on a source, to several paralleled CTs from different feeders. Preferably each circuit provides

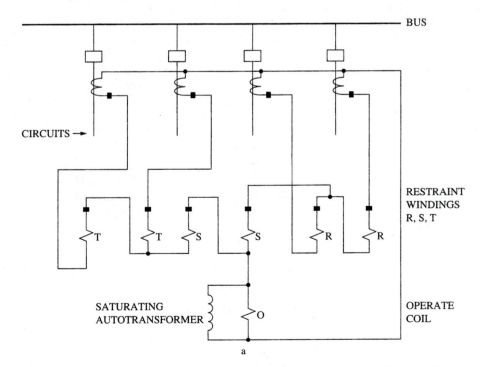

**FIGURE 6.18** Multi-restraint bus protection (one-phase connections shown). Each circuit may be two or more paralleled feeder CTs or two or more parallel source CTs. (a) Four circuit connection, one CA-16 used per phase

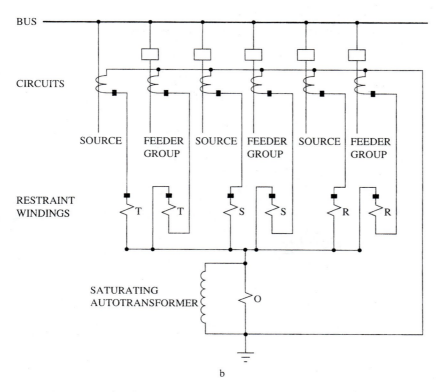

BUS

CIRCUITS

SOURCE | FEEDER GROUP | SOURCE | FEEDER GROUP | SOURCE | FEEDER GROUP

RESTRAINT WINDINGS

T   T   S   S   R   R

SATURATING AUTOTRANSFORMER

O

b

**FIGURE 6.18** (b) Six circuit protection, one CA-16 relay per phase.

about the same total CT current. The total parallel CT current should not exceed 14 A through any restraining coil. The restraining coils are not tapped. Figure 6.18b shows the relay connections for one phase of a six-circuit bus. Notice the source and feeder group polarity connections. For a three-circuit connection use the R, S, T windings of the same polarity and do not connect the other three restraining coils.

The *KAB* is a *high impedance differential relay.* The KAB relay uses conventional current transformers, but loads them with a high impedance overvoltage relay, as shown in Figure 6.19. The high relay impedance forces false difference currents from external faults that saturate a CT through the lower impedance CTs. The KAB has an instantaneous overvoltage cylinder unit ($V$), a tuned circuit to pass the power line frequency and block dc from fault current asymmetry, an IT over-current unit connected in series with a low voltage varister (blocks current until threshold voltage is reached), and an ICS unit. The IT provides faster action on severe internal faults. The KAB operating speed is about 25 mS.

The voltage can rise on a CT loaded with a high impedance so use of taps on the CTs should be avoided. If taps must be used the voltage across the CT from auto-transformer action must be safe at the highest available fault current. The CTs par-alleled connections should be connected in the station yard as close to the CTs as possible and the lead lengths should be as even as possible.

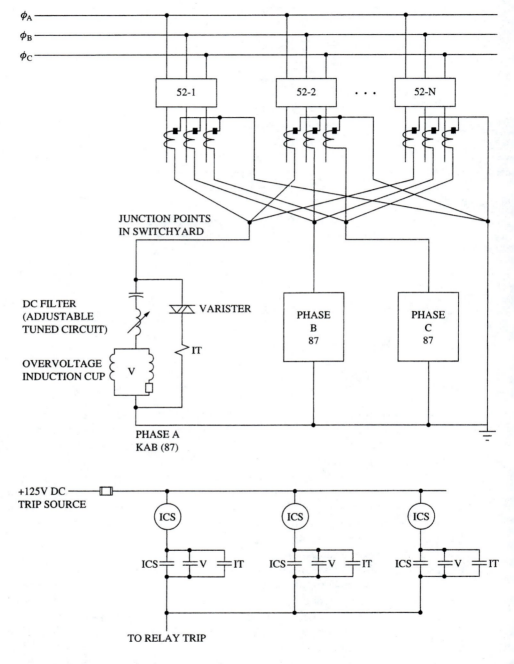

**FIGURE 6.19**    KAB relay bus protection

### 6.8.4 Combined Protection of a Bus and a Transformer

Combining bus and transformer protection is generally done on small substations. Ideally the bus and transformer have separate protection. Separate bus and transformer protection provides better fault isolation and easier fault location. If the substation location and/or economy require that the bus and the transformer be protected together the CA-26 or HU type relays must be used. The CA-26 is more inverse than the CA-16, and has a higher pick up current. It has an operating time of about 3 cycles. The HU relays are preferred because the HRU provides more immunity to false operation from inrush current.

The HU-1, with three restraining coils, connection is illustrated in Figure 6.20. The HU-4, with four restraint coils, must be used for a transformer with a tertiary or a two winding transformer with three feeders. Beyond that level of complexity separate protection is almost always used.

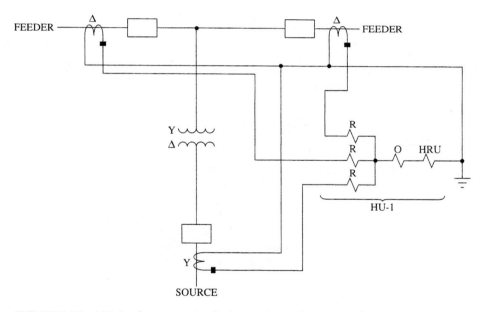

**FIGURE 6.20**  HU-1 relay protection for bus and transformer together

### 6.8.5 Feeder Overcurrent Protection

Basic feeder overcurrent protection is shown in Figure 6.21. The basic and now familiar 50/51 with 50N/51N ground fault sensing is the standard relaying scheme for feeders. If the relays must be coordinated with downstream fuses CO-8 or 9, very inverse, or CO-11, extremely inverse relays are usually needed. Recall that a tap off of a feeder in which the conductor size decreases by 30% or more must be fused.

The feeder, bus, and transformer protection must all be coordinated using the principles already discussed.

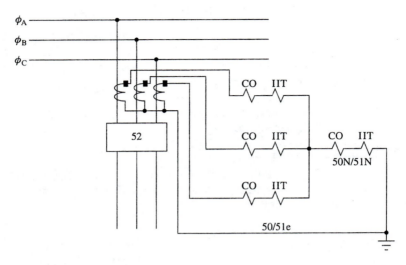

**FIGURE 6.21**   Distribution feeder overcurrent protection, one per feeder needed

## SUBSTATION GROUNDING

Substation grounding is first for safety, and secondly to provide a stable reference voltage for protection systems. The grounding system of a substation consists of a ground mat made of substantial size (4/0 to 600 MCM) bare conductors, connected in a grid pattern, and buried beneath the substation. The perimeter of the grid is connected to metal rods driven about 30 feet into the ground. The grid wires are about 20 feet apart but the spacing varies with the conductivity of the soil. More highly conductive soil can use larger grid wire spacing. The soil must be tested at each location and the soil characteristics are used in designing the ground mat. All substation structures are to be constructed within the perimeter of the grid.

The fence around a substation has two buried ground wires connected to the fence every few feet. One runs about 3 feet outside the fence, and one runs about 3 feet inside the fence. Both wires are connected to grounding rods every 50 feet. This is to protect the public from shocks due to voltage induced in the fence by the substation electric and magnetic fields.

## SUMMARY

Among the factors that effect the design of a substation are: voltage, load requirements, available site space, site location, expected load growth, aesthetic considerations, and protection requirements. The substations of a large industrial concern or utility vary because of changes in these factors, but as much substation standardization as is practical is used because standardization reduces cost and eases maintenance.

The National Electric Safety Code contains the minimum clearance between conductors and conductors to ground, with the spacing increasing as voltage increase. Usually clearances are larger than those specified in the code to allow additional safety margins. The usual clearances are published by the IEEE.

Distribution substations support structures are constructed of wood, steel lattice, steel low profile, or unit. Wood is inexpensive but suitable only for small substations. Steel lattice construction is flexible, strong, low weight, and reasonably expensive. Solid steel low profile stations are considered by many to be aesthetically superior to other construction materials, but are expensive and normally require more space because multilevel buses are not used. Unit substations are factory built and tested, and shipped as modules that are connected in the field. Unit substations are normally totally enclosed. They are very expensive, but compact and reliable.

Single-source single-feeder substations provide maximum economy for noncritical loads. The single-source single-bus substation provides a bus so that several feeders can tie into the substation. More critical loads have two sources available to power the substation in case one feed is lost. Two transformer substations are used for critical loads that cannot tolerate down time in case of a transformer failure. Automatic switching (throw over) is used to change sources or transformers in the event of a failure of either in substations with critical loads.

Densely populated areas with critical loads that require many feeders are served by double bus substations so that service can continue in the event of a single bus failure. Double bus substations are expensive but provide a high degree of fault tolerance. The standard substation bus arrangements are ring, breaker and a half, and double bus-double breaker in order of simple to more complex, flexible, and expensive.

On the average, phase to ground faults account for about 70% of electrical system faults, line to line faults cause about 15%, double line to ground faults account for about 10%, and three-phase to ground faults are responsible for somewhat less than 1%. Substation protection includes both time and instantaneous overcurrent, transformer and bus differential, and ground fault. Substations must be well grounded for reliable protection operation and personnel safety.

## 6.11

## QUESTIONS

1. State the purpose of an electrical power distribution substation.
2. List five criteria that must be considered in substation design.
3. What are the advantages of standardization in substation design?

4. What two electrical factors directly influence substation site size? What non-electrical factor?
5. Why are some substations more heavily protected than others of the same capacity?
6. List the four major substation construction methods along with a major advantage and disadvantage of each.
7. List and explain two major trends in substation design.
8. What is meant by insulation coordination?
9. Why is substation voltage regulation often needed?
10. The two single-source single-feeder substations in Figure 6.2 are quite different. Why?
11. Why are two sources desirable for a substation?
12. List two reasons for using two transformers in a substation.
13. Automated throwover is very expensive. Under what conditions is its use justified?
14. State the purpose of double bus substation layouts.
15. List the advantages and disadvantages of each of the following bus arrangements: ring, breaker and a half, and double bus-double breaker.
16. What changes are needed to convert the substation in Figure 6.8 to a double bus-double breaker?
17. Define zone of protection.
18. What is meant by back-up protection?
19. List and state the function of each relay normally used with a well-protected distribution substation power transformer.
20. List and briefly explain the principle of the three major methods of bus protection.
21. Why is it best to protect the bus and transformer separately?
22. State two purposes of substation grounding.
23. Draw three methods of ground fault protection.
24. How are IT units coordinated?
25. Explain how time-current curves can correspond to distance along a line.

# CHAPTER 7

# Distribution Line Construction

Electrical power distribution lines connect the distribution substation to the ultimate customer. They are an important part of the electrical power system. They must be reliable, low in maintenance, economical, and long lasting. Successful distribution line construction principles and techniques have become standard practice in the industry. This chapter briefly describes the major construction practices.

## 7.1

### DISTRIBUTION CONSTRUCTION SPECIFICATIONS AND STANDARDS

Distribution lines are normally connected grounded wye. The most common distribution voltages are 12,470 V line to line, 7200 V phase to ground, 34.5 kV line to line, and 20 kV line to neutral. The grounded wye configuration aids in loading all three phases proportionately and any accidental phase grounding shows up as a phase to ground fault, which is easily detected by the overcurrent relays.

Distribution lines are normally constructed along easements on streets, alleys, or along property lines between or behind buildings. The easements are granted by the governing body of a locality to providing utility service to its citizens. Often the easements are called *right of ways*. The close proximity of distribution lines to buildings and dwellings gives the public a very real and legitimate interest in the voltages and construction methods used for distribution systems. Thus these have been codified by the National Bureau of Standards (Standards Handbook C-2) and designated as the National Electric Safety Code (NESC).

Work on the NESC was begun by committees of the Association of Public Utilities, the Inspection Societies of Municipalities, and the IEEE (Institute of Electrical and Electronic Engineers) in 1913. The work was completed in 1938 and has been periodically revised and updated since then. In 1973 the IEEE was designated as the administrative secretariat for the National Bureau of Standards Handbook C-2 (NESC). Since then revisions have been published approximately every 3 years. The revisions are studied and recommended to the IEEE by committees composed of representatives from the IEEE, EEI (Edison Electric Institute), Association of Inspectors, and National Fire Underwriters. One committee was formed for each of the four sections of the NESC.

The four parts of the code are:

Part I:   Rules for the Installation and Maintenance of Electrical Supply Stations and Equipment.
Part II:  Safety Rules for the Installation and Maintenance of Overhead Electrical Supply and Communication Lines.
Part III: Safety Rules for the Installation and Maintenance of Underground Electric Supply and Communication Lines.
Part IV:  Rules for Operation of Electric Supply and Communications Lines and Equipment.

Anyone may address a proposed change in the NESC to the IEEE, which will refer it to the appropriate committee for consideration. The work of the committees is periodically reviewed and the proposed changes accepted and endorsed or rejected. Endorsed changes are printed in the next published edition of the code.

Legal acceptance of the NESC comes by its adoption in the building codes of municipalities and cities for projects within their jurisdiction, and state and federal government for contracts supported with their funds. In areas in the United States in which there are no building codes (primarily county land outside cities and towns), the courts in judicial proceedings involving liability for accidents in which electricity is a factor use the NESC as the standard of proper electrical practices for the installation and maintenance of electrical systems. This practice makes the NESC the de facto standard throughout the United States.

Most electric utilities have developed company standards that incorporate the provisions of the NESC as well as local practice and good judgment concerning local conditions. For example, a locality with frequent high winds may need line spacing considerably in excess of the NESC requirements. Less stringent requirements than the code are not an option. One of the most widely used standards is the one developed by the REA (Rural Electrification Administration) as a set of construction specifications for use by all borrowers of REA funds. Appendix B contains sample REA specifications.

One of the more critical specifications contained in the NESC is the safe minimum line spacing. Table 7.1 gives a partial listing from tables 1, 4, and 8 of the

NESC. NESC table 1 deals with line clearance over specified items, table 4 is for clearance of supply lines from buildings, and table 8 concerns line to line clearances, which vary with the line sag between conductor supports. The NESC tables and standards cover nearly every conceivable construction situation.

■■■■ **TABLE 7.1   Sample Spacings from the NESC Tables**

| From Table 1   Open Supply Lines | | | |
|---|---|---|---|
| | **Voltage** | | |
| | **0 to 750** | **750 to 15,000** | **15,000 to 50,000** |
| | **Required clearance over specified items** | | |
| Railroad tracks | 27' | 28' | 30' |
| Public streets, alleys, or roads | 18' | 20' | 22' |
| Driveways to residential garages | 10' | 20' | 22' |
| Areas accessible to pedestrians only | 10' | 15' | 17' |
| Parallel to streets | 18' | 20' | 22' |
| Parallel to rural roads | 15' | 18' | 20' |

| From Table 4   Clearance of supply lines from buildings | | |
|---|---|---|
| **Voltage of supply conductors** | **Horizontal clearance** | **Vertical clearance** |
| 300 to 8,700 | 3' | 8' |
| 8,700 to 15,000 | 8' | 8' |
| 15,000 to 50,000 | 10' | 10' |

| From Table 8   Separation in inches required for line conductors #2 AWG or larger versus span sag | | | | | |
|---|---|---|---|---|---|
| **Voltage between conductors** | **Span sag in inches** | | | | |
| | **36** | **48** | **72** | **96** | **120** |
| 2,400 | 14.5 | 16.5 | 20.5 | 23.5 | 26.0 |
| 7,200 | 16.0 | 18.0 | 22.0 | 25.0 | 27.5 |
| 13,200 | 18.0 | 20.0 | 23.5 | 26.5 | 29.5 |
| 34,500 | 24.0 | 26.5 | 30.0 | 33.0 | 35.5 |
| 69,000 | 36.6 | 36.5 | 40.5 | 43.5 | 46.0 |

**7.2**
■■■

# DISTRIBUTION CONDUCTOR SUPPORTS

Distribution systems are usually designated as either overhead or underground. Overhead distribution consists of conductors, structures to support conductors, and insulators. Underground systems consist of ducts, vaults, and cables. Underground systems are discussed a bit later in this chapter.

The overhead structures consist of wood or steel poles, cross arms, brackets, insulator supports, and miscellaneous hardware for device and conductor attachment. By custom the cross arms, brackets, and related hardware are referred to as *framing*. Distribution circuits very rarely use steel lattice towers, so we will discuss wood and steel poles here. Figure 7.1(a–d) shows a variety of distribution poles.

**FIGURE 7.1**    Distribution pole types include: (a) Wood (courtesy of Gulf States Utilities)

(b) Wood dead end (courtesy of Gulf States Utilities)

(c) Steel (courtesy of Houston Lighting & Power)

(d) Steel dead end (courtesy of Houston Lighting & Power).

## 7.2.1 Wood Poles

The simplest and cheapest support structures are wood poles. These were the first extensively used electrical conductor support structures. Wood poles have been made from larch, spruce, cedar, pine, and fir trees, selected for height and straightness. The most commonly used trees for poles are the Southern Yellow Pine (SYP), about 70%, and Douglas Fir, about 25%. All of the others comprise only about 5 percent of the wood poles used. Poles from 25 to 65 feet are generally SYP while poles over 65 feet are generally Douglas Fir. Wood poles are available in heights from 25 to 130 feet (or more on special order) in 5-foot increments.

Wood pole heights and strengths have been codified from tests, and tables prepared to make distribution design easier. A reliable average for maximum longitudinal fiber stress in both pine and fir is 8000 psi. That is the psi at which the wood fibers will start to split or slide past each other. So all design strengths have been calculated from this value. The loading profile from a force at the top of a pole is shown in Figure 7.2. As force is gradually increased, a pole will fail first by splitting along the pole and then by breaking.

Poles are designated by strength and degree of straightness as class 1 (best) through 5 (worst). Larger poles have their own classification as extra heavy duty classes 0, 00, 000, and 0000, with more zeros being heavier duty. The zeros are usually written as H1, H2, H3, H4. Figure 7.3 shows where to find the pole classification on a pole.

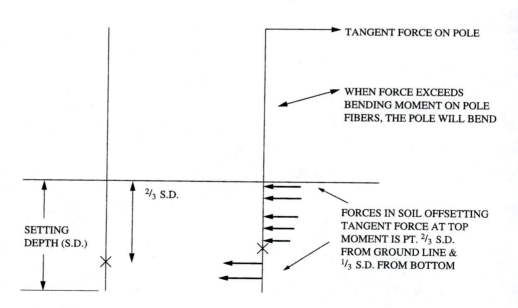

**FIGURE 7.2**   Pole loading profile

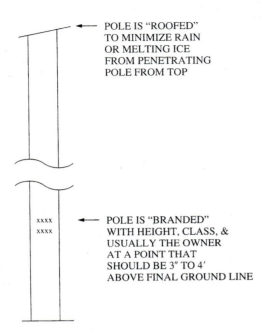

POLE IS "ROOFED"
TO MINIMIZE RAIN
OR MELTING ICE
FROM PENETRATING
POLE FROM TOP

POLE IS "BRANDED"
WITH HEIGHT, CLASS, &
USUALLY THE OWNER
AT A POINT THAT
SHOULD BE 3" TO 4'
ABOVE FINAL GROUND LINE

**FIGURE 7.3**   Wood pole details

Table 7.2 gives some sample pole strengths. The table gives the pole classification and its weight first. The first column gives the distance from the top of the pole that the force is applied, the second column gives the diameter of the pole at that point, and the third column gives the moment (essentially torque) in units of feet-klb, usually abbreviated ft-k. In other words the maximum allowable force perpendicular to a class 1 pole at the top of the pole, 50 feet above the pole attachment point, is $(51.5 \times 10^3 \text{ ft-lb})/50 \text{ ft} = 1034 \text{ lb}$. At 10 feet below the top, 40 feet up, it is $(77.3 \times 10^3 \text{ ft-lb})/40 \text{ feet} = 1932 \text{ lb}$. The tables are based on an 8000 psi ultimate pole bending stress. The force on the pole is called tangent, see again Figure 7.2, because it is tangent to the line attached to the pole. A perpendicular pole force is taken to mean a force perpendicular to the direction that the line is strung.

Wood poles have the flexibility to bend and give under sudden severe loads, and resume their shape when the load is removed. This property makes wood a very desirable conductor support material when it can be used. Under severe force, such as storm winds perpendicular to the line, the force on the pole may be dissipated in the ground. Usually the pole can be straightened or, after extreme loads, reset. Wood poles seldom break in two.

After they are cut and cured, wood poles are pressure treated with a coal tar derivative, usually creosote, until the wood has been completely saturated. This treatment gives the pole an expected lifetime of 40 to 50 years unless attacked by ground rot or woodpeckers, the major enemies of wood poles.

## TABLE 7.2    Wood Pole Strength Tables

### 50 Ft. Pole, Douglas Fir or SYP

| Class H1, Wt. = 2,200 lb. | | | Class 1, Wt. = 1,970 lb. | | |
|---|---|---|---|---|---|
| Distance from top (feet) | Diameter (inches) | Moment (ft.-k) | Distance from top (feet) | Diameter (inches) | Moment (ft.-k) |
| 0 | 9.23 | 51.5 | 0 | 8.59 | 41.5 |
| 10 | 10.57 | 77.3 | 10 | 9.90 | 63.4 |
| 30 | 13.28 | 152.1 | 30 | 12.50 | 127.9 |
| 40 | 14.58 | 203.0 | 40 | 13.80 | 149.6 |
| 50 | 15.92 | 264.2 | 50 | 15.11 | 179.2 |

### 50 Ft. Pole, Douglas Fir or SYP

| Class 2 Wt. = 1,700 lb. | | | Class 3 Wt. = 1,220 lb. | | |
|---|---|---|---|---|---|
| Distance from top (feet) | Diameter (inches) | Moment (ft.-k) | Distance from top (feet) | Diameter (inches) | Moment (ft.-k) |
| 0 | 7.96 | 33.0 | 0 | 7.32 | 25.7 |
| 10 | 9.19 | 50.8 | 10 | 8.48 | 39.9 |
| 30 | 11.65 | 103.4 | 30 | 10.79 | 82.3 |
| 40 | 12.88 | 139.4 | 40 | 11.95 | 111.7 |
| 50 | 14.11 | 183.7 | 50 | 13.11 | 147.4 |

### 70 Ft. Pole, Douglas Fir or SYP

| Class H1, Wt. = 3,640 lb. | | | Class 1, Wt. = 3,225 lb. | | |
|---|---|---|---|---|---|
| Distance from top (feet) | Diameter (inches) | Moment (ft.-k) | Distance from top (feet) | Diameter (inches) | Moment (ft.-k) |
| 0 | 9.23 | 51.5 | 0 | 8.54 | 41.5 |
| 30 | 12.96 | 142.5 | 30 | 12.18 | 118.1 |
| 50 | 15.45 | 241.3 | 50 | 14.56 | 202.1 |
| 70 | 17.93 | 377.6 | 70 | 16.95 | 318.7 |

### 70 Ft. Pole, Douglas Fir or SYP

| Class 2, Wt. = 2,840 lb. | | | Class 3, Wt. = 2,470 lb. | | |
|---|---|---|---|---|---|
| Distance from top (feet) | Diameter (inches) | Moment (ft.-k) | Distance from top (feet) | Diameter (inches) | Moment (ft.-k) |
| 0 | 9.23 | 33.0 | 0 | 7.32 | 25.7 |
| 30 | 11.00 | 96.7 | 30 | 10.60 | 78.0 |
| 50 | 13.68 | 167.5 | 50 | 12.79 | 137.0 |
| 70 | 15.97 | 266.3 | 70 | 14.98 | 220.0 |

Poles can be protected against woodpeckers by wrapping the pole with a ¾ inch galvanized wire mesh. This is not completely effective, but it is the best available protection that does not harm the birds.

Ground rot occurs in areas where the pole base is alternately wet and dry—almost everywhere. It occurs from just above the ground line to a foot or so beneath the ground, and severely weakens the pole if not treated. Tamping the fill dirt well and mounding it around the base of the pole, so the water runs away from the pole, helps prevent ground rot. Treatments to prevent ground rot, or kill it if it has started, are applied to the wood from 6 inches above ground to 18 inches below ground. Several treatments are available including some in which the chemical is forced into the wood under pressure through small holes drilled in the wood. If ground rot has caused pole weakening often the pole can be repaired after the rot has been killed. Repairs can be as simple as bolting split pipe to the base of the pole across the damaged section. Several commercial repair methods are available in which a loose metal sleeve is put around the damaged area (usually from 18 inches under ground to a foot or so above ground) and the space between the sleeve and the pole are filled with high density epoxy or foam.

## 7.2.2 Steel Poles

Steel poles have been in use now for a long time, and are well proven. They do not have the elasticity of wood poles, nor do they have the lifetime. The life of a steel pole is governed primarily by the quality and thickness of the galvanizing. There are special paints to make a steel pole look better longer, but it is impractical to paint the inside of the pole, so the galvanizing sets the pole life. Steel poles typically last from 25 to 30 years.

Steel poles are much more expensive than wood poles, but become more economically favorable in lengths from 90 to 130 feet. The principal reason for using steel poles is their looks. Many people think steel poles are more attractive than wood poles. They are extensively used in highway and street lighting, sports arena lighting, and in commercial and residential areas where the developers believe that steel poles are preferable. Ninety foot steel poles for tangent use (line extends both ways from the pole) cost about $2600 as opposed to about $1600 for a 90 foot wood pole. The reason that steel poles become competitive in the longer lengths is shipping costs. Long wood poles are grown in the Northwest United States, and require two train cars to ship because of their length. Steel poles can be manufactured in convenient shipping lengths that can be slip fitted together like fishing rods at the construction site, as shown in Figure 7.4. Also, steel poles are manufactured at several places throughout the United States. The shipping savings can offset much of the higher price of long steel poles, particularly in the Eastern half of the United States.

There are no standard steel poles. They are manufactured to order for each job so the steel thickness varies with the expected load. This is part of the reason the price of steel poles is so high. The typical weight for a 50 foot tangent steel pole is 2500 lb, and 4500 lb for a 70 foot pole. Dead end poles (wires attached to only one side) are built heavy enough to take the strain without bending so typical weights for steel dead end poles are 5000 lb for a 50 foot, and 9000 lb for a 70 foot pole.

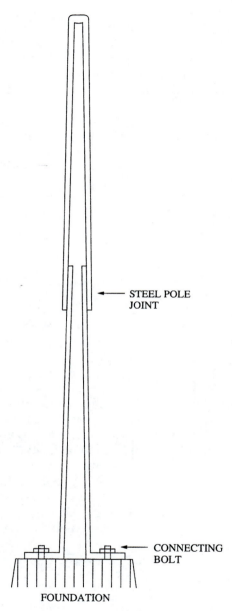

STEEL POLE
JOINT

CONNECTING
BOLT

FOUNDATION

**FIGURE 7.4**   Steel pole construction (cut away view)

## 7.2.3  Pole Foundations

Wood poles are set by direct imbedment as shown in Figure 7. 5. Steel poles may be set by direct imbedding also, or may be set on a foundation and attached with anchor bolts. Direct imbedment for either steel or wood poles is the same and

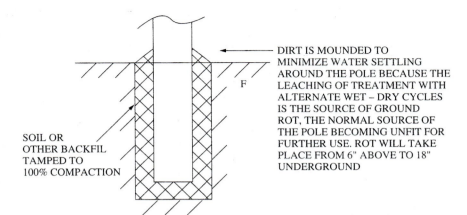

DIRT IS MOUNDED TO
MINIMIZE WATER SETTLING
AROUND THE POLE BECAUSE THE
LEACHING OF TREATMENT WITH
ALTERNATE WET – DRY CYCLES
IS THE SOURCE OF GROUND
ROT, THE NORMAL SOURCE OF
THE POLE BECOMING UNFIT FOR
FURTHER USE. ROT WILL TAKE
PLACE FROM 6" ABOVE TO 18"
UNDERGROUND

SOIL OR
OTHER BACKFIL
TAMPED TO
100% COMPACTION

**FIGURE 7.5**   Wood pole imbedment

requires an imbedment depth of 10% of the height plus 2 feet for standard soil. Thus a 50 foot pole has an imbedment of 7 feet while a 70 foot pole has an imbedment of 9 feet. The soil must be tamped to 100% compaction, meaning to the density it was before the hole was dug. Specifications usually require two tampers for each shoveler when the soil is filled in around the pole. A pole set in soil can be loaded immediately.

Imbedment can be reduced if concrete stabilized sand, rock, or concrete are used for fill. Concrete stabilized sand is a mixture of $3/4$ sand and $1/4$ concrete used for fill where the soil has too much give to support a pole well. The concrete stabilized sand essentially increases the size of the pole end so that the loading per unit area on the soil is lower. The concrete must set up before the pole can be loaded.

Reinforced concrete foundations for steel poles can be very expensive, costing more than the pole for some dead end poles. They are necessary in soils that are particularly corrosive to steel. Figure 7.6 is a sketch of a reinforced concrete steel pole foundation. Anchoring is much cheaper and is used on wood and directly imbedded steel poles.

Poles and imbedments are designed to support forces that are balanced on either side of the pole. Down guy wires and anchors are used to balance the forces on the pole in situations in which balanced pole loading is not possible. Anchoring for a dead end pole is shown in Figure 7.7. Extra loading in the tangent direction is absorbed by the conductor, guy wire, and anchor as long as the ultimate strength of these components is not exceeded. Perpendicular forces, such as storm winds, are absorbed by the pole and transferred to the ground. As mentioned earlier, if the pole is not damaged from excess perpendicular loading it can be straightened or reset. Usually the conductors, anchor, and guy wire are not severely damaged and can be reused. Pole anchors are needed any place that the normal pole loading changes from tangent, so both bends and dead ends need to be anchored. Bend anchoring is illustrated in Figure 7.8.

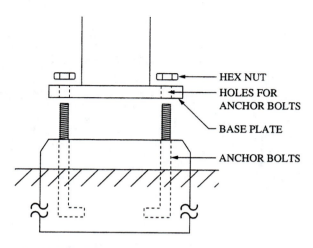

**FIGURE 7.6**   Reinforced concrete steel pole foundation

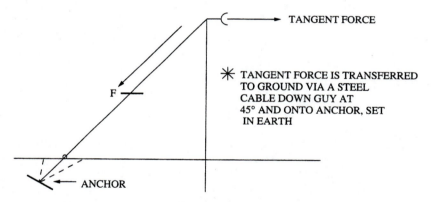

**FIGURE 7.7**   Pole down guy and anchor

## 7.2.4 Framing

The structural member attached to the pole to support the insulators and conductors, including all of the bolts, washers, and other associated hardware, is called *framing*. The framing holds the conductors in the proper position with the proper spacing. Both wood and fiberglass are used for the cross arms. Steel or fiberglass is used for the brackets. Figure 7.9 shows some typical framing.

## 7.2.5 Insulators used for Distribution

Insulators separate the current carrying conductors from grounded structures. They are usually made of glazed porcelain or glass, but fiberglass is becoming increasingly popular. Insulators must be clean to perform properly so they are

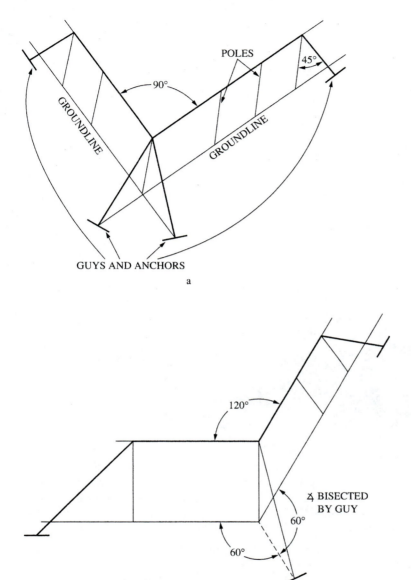

**FIGURE 7.8** Pole anchors on line direction changes (a) 90° and (b) 120°

designed to use rain to wash away dirt and sediment from areas on which they can settle.

The crucial electrical factor for insulation is the length of the path from the conductor to the insulator support. Insulator skirts are fluted to increase the path length, as shown in Figure 7.10. The conductors are heavy so glass and porcelain

**FIGURE 7.9**   Framing

may seem to lack the strength to support them. However, both materials have great compression strength, and insulators are designed so that the glass or porcelain is loaded only in compression. The compression loading is easy to see with the pin insulator shown in Figure 7.10, but is less obvious in the ball and socket and clevis insulators shown in Figure 7.11. In these the steel supporting structures overlap, with glass or porcelain (usually porcelain in these types) between the overlapping parts. Thus the porcelain is loaded in compression only. These parts are totally enclosed within the insulator so the skirts can provide the proper path length.

The clevis and ball and socket insulators are used only as dead end insulators in distribution, although they are the normal tangent insulators in transmission. The clevis insulators use a clevis on one insulator across a mating structure on the next one with a pin holding the two insulators together. The ball and socket insulators are held together by a ball on one insulator fitting into a socket on the other. The ball goes into the socket from the side of the insulator and slips forward a bit so that a perpendicular force cannot cause the ball to come out of the socket when the insulators are under load. The ball and socket is the most expensive but it can be

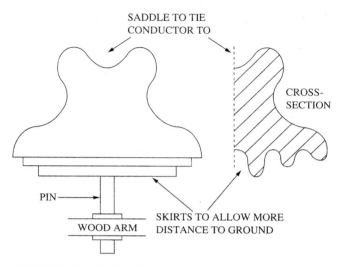

SADDLE TO TIE
CONDUCTOR TO

CROSS-
SECTION

PIN

WOOD ARM

SKIRTS TO ALLOW MORE
DISTANCE TO GROUND

**FIGURE 7.10** Pin insulator

**FIGURE 7.11** Distribution dead end insulators Clevis (lower) and ball and socket (upper)

changed out hot by experienced crews with hot sticks and proper maintenance tools. For this reason ball and socket insulators are used by nearly all electric utility companies. The mechanical loading must be supported by an auxiliary structure before the ball can be slipped from the socket, of course. Insulators are rated for both maximum withstand voltage and mechanical strength.

## 7.3

### DISTRIBUTION OVERHEAD CONDUCTORS

The only two metals that are cost effective and low enough in resistivity for electrical transmission and distribution systems are copper and aluminum. Copper is so much heavier than aluminum that it is used primarily in insulated wires and cables. Aluminum is suitable for transmission and distribution and allows the use of much lighter and more economical support structures.

The tensile strength of pure aluminum is not high enough for most applications so aluminum alloys or steel reinforced aluminum alloys are used. The primary cables used in distribution are ACSR, AAC, and AAAC, in that order.

### 7.3.1 ACSR (aluminum conductor, steel reinforced)

This is the most common transmission and subtransmission conductor, and it is the most commonly used distribution conductor for long spans and large, high capacity lines. The length of the span and the allowable sag determine the conductor characteristics required. ACSR maximizes span and minimizes sag.

ACSR consists of one or more layers of hard drawn alloy #1350 aluminum wire wrapped concentrically around a high strength galvanized steel core, which may consist of a single steel wire or several strands of steel wire as shown in Figure 7.12a. A single steel strand, seven steel strands, and for very large conductors 19 steel strands, are most common. The conductors are listed as aluminum conductors/steel conductors with 6/1 (six aluminum, one steel strand), 26/7, and 54/7 being common sizes. Many other sizes are available.

### 7.3.2 AAC (all aluminum conductor)

All aluminum conductor is available in 7, 19, 37, 61, and 91 strands of #1350 aluminum wire. It is shown in Figure 7.12b. AAC has slightly better conductivity at low voltages than ACSR, but it has less strength and more sag per span length. It is used for lines with short spans. AAC costs about the same as ACSR.

Distribution spans are generally short because the lines are built on public roads, streets, and easements. It is not permissible to serve a building with service drops across streets or across the property of another land owner. This means two lot widths is the general span for residential, and light commercial and business districts. This is typically between 120 and 200 feet.

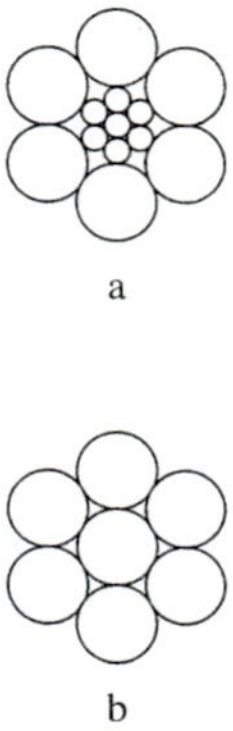

a

b

**FIGURE 7.12** Conductor cross sections (a) ASCR and (b) AAC

## 7.3.3 AAAC (all aluminum alloy conductor)

AAAC is constructed like AAC, shown in Figure 7.12b, but uses aluminum alloy #6201-T81. It has the strength of ACSR but is lighter in weight. It is also much more expensive. AAAC is used for very long spans where the strength of ACSR is needed but light weight is essential. This means its distribution use is limited to such applications as river crossings. For transmission and subtransmission there are applications in which the support structures must be very light so they can be transported by helicopter. Extremely rough mountains and swamps comprise most of these situations. AAAC is often used in these applications and in river crossings.

Most utilities select a few standard size conductors that meet their requirements and stock only those conductor sizes and accessories (such as splice sleeves and dead end shoes). They then use these sizes for all of their construction.

## 7.4

# UNDERGROUND DISTRIBUTION

Underground distribution systems consist of buried ducts (conduit systems for cables), vaults, and cables. An underground distribution system is very expensive compared to an equivalent overhead system, and has a shorter life span. Underground systems suffer fewer outages during their life spans because they are protected from storms, lightning, and vehicle accidents. When an outage does occur, such as when a buried cable is cut during construction excavation or a pad mount lightning strike that burns up a cable, it is longer and more expensive to repair than an overhead outage.

The NESC also governs underground distribution setting such standards as cable depth below grade, separation from other buried utility service such as gas and water pipes, insulation class, and access limitations. The two extremes of underground distribution are residential direct buried cable and downtown vault, utility access hole, and cable duct systems for networks.

## 7.4.1 Underground Cables

Underground cables normally use stranded copper instead of aluminum for conductors. The lower resistivity of copper helps keep losses lower since reactive losses are higher for underground cables. Some underground cable is shown in Figure 7.13.

Practical residential underground distribution began when direct buried underground cable was developed that used a relatively inexpensive plastic insulation, typically cross-linked polyethylene, and, as mentioned in Chapter 3, practical elbow connectors. The cable consists of an insulated cable of stranded conductors with a bare concentric neutral.

A variety of types and sizes of plastic insulated underground cable is available for industrial and commercial use, including three-phase and single-phase cable, shielded, bare neutral, and insulated neutral cable. Cable with operating voltages to over 25 kV is available for direct burial service. Guaranteed lifetimes are about 15 years, and seem accurate.

For many years the favored underground cable was stranded copper with an oil impregnated paper insulation wrapped in a lead sheath. It is still used in cable ducts, although plastic insulated cable is being used more and more. This cable is expensive to buy, install, and splice, but once in service it has a long life. Some are 50 years old and still in service in duct vaults. Oil impregnated paper insulated cable is available in single-and three-phase cable and in voltages up to 46 kV.

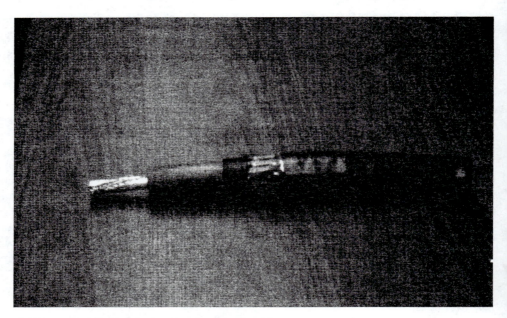

**FIGURE 7.13**   Plastic insulated underground cable

Oil-filled cables, in which liquid oil fills a void around the paper insulation, can withstand voltages between 23 and 69 kV for three-phase cables, and 69 to 230 kV for single-phase cables. The oil space is provided by a helical spring running the length of the cable. Periodic oil reservoirs are required for oil-filled cables. All cable can handle any voltage below the maximum operating voltage. Another type of cable that uses dry nitrogen at 10 to 15 psi (instead of oil) is available up to 46 kV.

Underground transmission, like underground distribution, is avoided when overhead can be used. Wide river crossings and service to downtown areas make both unavoidable. Underground transmission cables are more complex than underground distribution cables, and more expensive, because of the higher voltages involved. Underground transmission cables have three insulated single-phase cables that are usually contained inside pipes filled with high pressure (about 200 psi) oil or nitrogen. Cables of this type can handle up to 550 kV. Recently the use of $SF_6$ for underground transmission line pipe fill has become common. The $SF_6$ transmission line has the conductor supported in the center of a larger outside pipe by insulating disks, which have holes to let the insulating gas pass through. The $SF_6$ storage, pumping, pressure monitoring, and distribution equipment for $SF_6$ transmission systems is expensive and complex.

## 7.4.2 Residential Direct Burial

Underground residential distribution (URD) uses direct buried cable exclusively, allowing installation costs to be reasonable. The heavy three-phase feeders are overhead along the periphery of a development, and the laterals to the pad-mount transformers are buried about 40 inches deep. The lateral cable conductors come up from underground into the high voltage compartment of the transformer. The secondary service lines then run to the individual dwellings at a depth of about 24 inches, and come up into the dwelling meter through a conduit. The service conductors run along easements and do not cross adjacent property lines.

## 7.4.3 Commercial Direct Burial

To avoid the substantial cost of excavation and backfill rural telephone cable, highway lighting, and perimeter wiring in some industrial plants is installed with a cable plow. A cable plow is a small shoe-like plow connected to the draw bar by a long knife-shaped blade structure. The structure has a tube through which cable can be fed from the top down to the plow. The cable is then fed out behind the plow, usually at a depth of 48 inches. A D-8 Caterpiller tractor is usually needed to pull the cable plow, with a second needed to carry the cable reel. The tractor with the cable reel goes in front of the plow-pulling tractor and feeds cable back over it to the plow. Cable plows require a lot of room to bury cable. Many specifications prohibit the use of cable plows because the condition of the cable as it goes into the ground cannot be seen.

Most direct buried cable for commercial and industrial applications is placed in machine dug trenches. The trenches allow special backfill and cable handling requirements to be met. Some specifications require 3 inches of sand in the bottom of the trench under the cable, and another 3 inches over the cable to prevent rocks from damaging the cable insulation. Many specifications also require a 2- by 8-inch oak plank to be placed 8 inches below grade during backfill to deter accidental breaking of the cable during subsequent excavations. Bright yellow plastic is sometimes used instead of the oak plank to warn of high voltage.

### 7.4.4  Underground Duct Banks

Underground duct bank, with access at vaults and utility maintenance holes, is the most sophisticated, reliable, and costly method of underground distribution. The high reliability is why it is used for large industry such as chemical plants and refineries, and for downtown networks.

A duct bank is a set of parallel conduits made of steel, PVC covered steel, heavy walled PVC, or thin walled PVC in reinforced concrete. Large, long duct bank conduits are of thin walled PVC, evenly spaced and set in reinforced concrete as shown in Figure 7.14a. The concrete is often dyed red to warn of high voltage in the event an excavation should uncover the duct bank. The duct banks terminate in utility access holes or vaults, as shown in Figure 7.14b. Duct bank runs from a utility access hole at one intersection to another. At points between utility access holes smaller duct banks or rigid galvanized steel conduits leave the main bank to take power to a building. This is illustrated in Figure 7.15. Smaller duct banks leading from the main duct bank to buildings must be of thicker material than thin wall PVC if they are not enclosed in concrete. Transformers and network protectors are usually housed in vaults near the utility access hole, often under a building.

Duct banks are seldom filled to over 40% capacity (40% of the holes filled) upon installation to allow room for expansion of service capacity and fault repair. A serious cable fault in a duct frequently damages the duct. It is less expensive and quicker to run a new cable in another duct conduit, cut the old cable free, and splice to the new cable, than to remove the old cable first. The old cable can be removed later for fault analysis and scrap value.

The handling, splicing, and fireproofing of cables in vaults is costly. Additionally, the cost of equipment suitable for installation in vaults is high. The vault, cable duct bank, and utility access hole underground distribution system is the most reliable and safe system available, but it is very expensive.

## 7.5

### RESIDENTIAL DISTRIBUTION LAYOUT

The following exercise illustrates the design of an overhead distribution system for a small subdivision. The system will allow for future expansion and voltage conversion. The present feed is from a 12.47 kV, three-phase, 600 MCM feeder on Main Street. The area is to be converted to 34.5 kV in the future, as the load builds

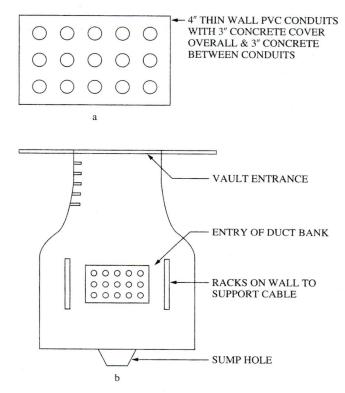

a

VAULT ENTRANCE

ENTRY OF DUCT BANK

RACKS ON WALL TO SUPPORT CABLE

SUMP HOLE

b

4" THIN WALL PVC CONDUITS WITH 3" CONCRETE COVER OVERALL & 3" CONCRETE BETWEEN CONDUITS

**FIGURE 7.14**   Duct bank (a) Cross-section and (b) location in maintenance access vault

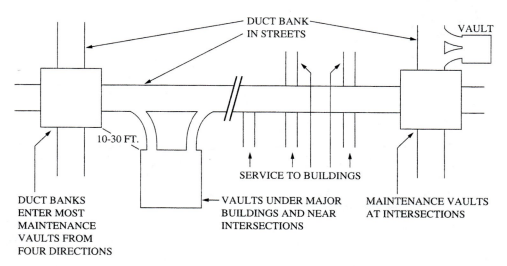

DUCT BANK IN STREETS

VAULT

10-30 FT.

SERVICE TO BUILDINGS

DUCT BANKS ENTER MOST MAINTENANCE VAULTS FROM FOUR DIRECTIONS

VAULTS UNDER MAJOR BUILDINGS AND NEAR INTERSECTIONS

MAINTENANCE VAULTS AT INTERSECTIONS

**FIGURE 7.15**   Underground distribution facilities in downtown area

up, to avoid excessive distribution loss. The symbols used for distribution design are shown in Figure 7.16, and the neighborhood is shown in Figure 7.17.

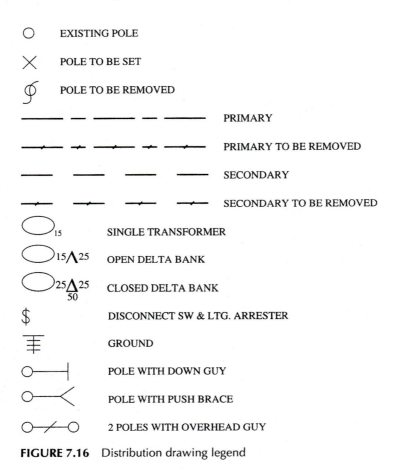

○    EXISTING POLE

✕    POLE TO BE SET

∅    POLE TO BE REMOVED

———  —  ———  —  ———    PRIMARY

———•—  ••—  ———•—  ••—  ——•—    PRIMARY TO BE REMOVED

———    ———    ———    ———    SECONDARY

——•—    ——•—    ——•—    ——•—    SECONDARY TO BE REMOVED

⬭₁₅    SINGLE TRANSFORMER

⬭15∧25    OPEN DELTA BANK

⬭25△25 / 50    CLOSED DELTA BANK

$    DISCONNECT SW & LTG. ARRESTER

⏚    GROUND

○—|    POLE WITH DOWN GUY

○—<    POLE WITH PUSH BRACE

○—✗—○    2 POLES WITH OVERHEAD GUY

**FIGURE 7.16**    Distribution drawing legend

## 7.5.1 Easement Requirements

The subdivision is to be fed from an existing 12.47 kV feeder. The transformers, lightning arresters, and usually the fuse cutouts, will all be 12.47 KV. Since future conversion to 34.5 kV is planned the framing will be insulated to 34.5 kV. The higher insulation capacity on the framing and insulators will make the voltage conversion faster and simpler when it occurs. The easement obtained must also be for voltages to 35 kV. To do the voltage conversion with minimum outage time, the subdivision must be fed from two points, marked E and W in Figure 7.17.

The 35 kV perimeter easement and interior easements required are shown in Figure 7.17. The interior easements must reach each lot not reached by a perimeter easement. The perimeter easement is necessary on three sides of the development.

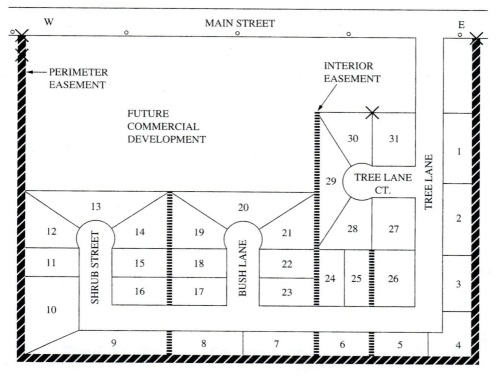

**FIGURE 7.17** Subdivision with easements

Overhead 35 kV easement vertical sections are shown in Figure 7.18a and b for perimeter and interior easements respectively.

Eventually a three-phase line must be built from point W on Main Street along the West perimeter easement to the Northwest corner of lot 12, and then along an interior easement bordering lots 13 and 20, to provide three-phase service to the commercial area. If the commercial load is 150 kVA or less then any two phases can be used in open Y primary configuration with the low voltage secondary windings connected in open delta. If the commercial load exceeds 150 kVA then normally all three phases are used. Until the commercial area is developed only one phase of the Main Street feeder need be tapped. Two taps are needed because of the planned voltage changeover. The East and West perimeter easements are the most logical places to tap for the lateral lines.

## 7.5.2 Pole Location

The pole locations are shown in Figure 7.19. They are set in the following order. First the poles in the existing Main Street line are set at the intersection of the line and the East and West laterals. These poles must be at least the length of the existing poles to insure proper ground clearance, but preferably they are about 5 feet

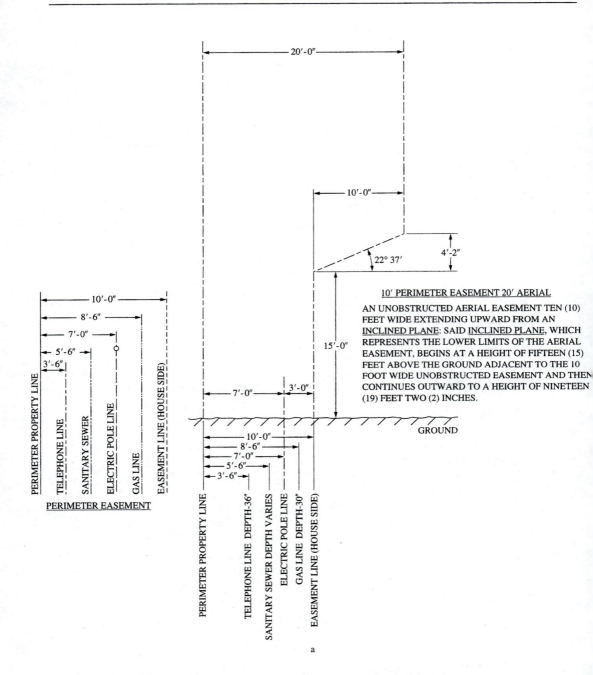

**FIGURE 7.18**    35 kV easement, vertical views (a) Perimeter easement

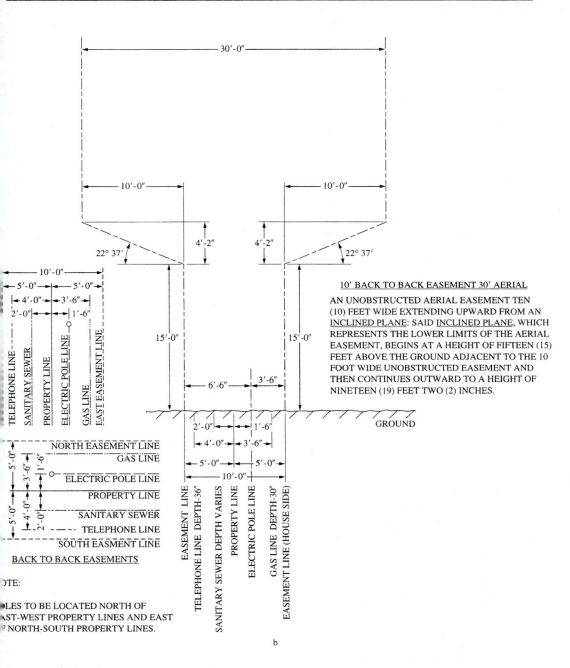

10' BACK TO BACK EASEMENT 30' AERIAL

AN UNOBSTRUCTED AERIAL EASEMENT TEN (10) FEET WIDE EXTENDING UPWARD FROM AN INCLINED PLANE: SAID INCLINED PLANE, WHICH REPRESENTS THE LOWER LIMITS OF THE AERIAL EASEMENT, BEGINS AT A HEIGHT OF FIFTEEN (15) FEET ABOVE THE GROUND ADJACENT TO THE 10 FOOT WIDE UNOBSTRUCTED EASEMENT AND THEN CONTINUES OUTWARD TO A HEIGHT OF NINETEEN (19) FEET TWO (2) INCHES.

BACK TO BACK EASEMENTS

NOTE:

LES TO BE LOCATED NORTH OF ST-WEST PROPERTY LINES AND EAST NORTH-SOUTH PROPERTY LINES.

(b) interior easement

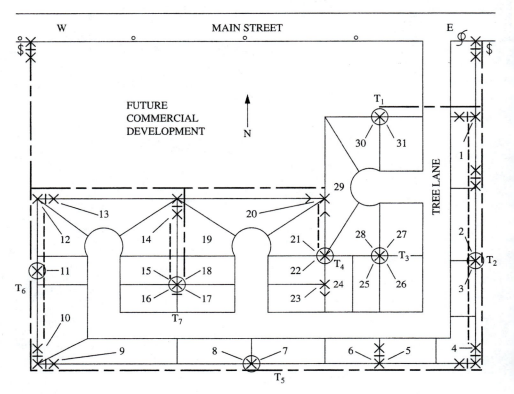

**FIGURE 7.19** Overhead distribution layout for subdivision of Figure 7.17

longer so that there is plenty of room for the lateral taps under the phase conductors. Recall that if the tap conductor is 30% smaller than the feeder conductor a fuse cutout and fuse are necessary to protect the lateral conductor. This project will need the fuses at the tap. Some poles along the Main Street line may have to be relocated to even up the span lengths so that a very long and a very short span are not adjacent.

Poles are now placed at each corner (lot 4, the intersection of lots 9 and 10, and the intersection of lots 20, 21, and 29). Next poles are placed at each tap point on the lateral: the intersection of lots 12 and 13; the intersection of lots 13, 14, 19, and 20; the intersection of lots 5 and 6; and the Northeast corner of lot 1. Finally the remaining poles are placed so that service drops to the remaining lots can be connected without crossing adjacent property (the intersections of: lots 7 and 8; lots 15, 16, 17, and 18; lots 30 and 31; lots 25, 26, 27, and 28; lots 21, 22, and 24; lots 22, 23, and 24; lots 2 and 3; and at lot 11).

Many corners must be used in our subdivision layout. The wire at a 90° connection or at a corner will tend to pull over an unreinforced pole. Recall that a guy wire and anchor are used to balance the wire tension on poles used for corners or dead ends. The owner of the subdivision does not own the property outside the perimeter easement and cannot give a guy easement at the outside subdivision corners or at the Main Street taps. The owners of the property adjacent to the sub-division have no incentive to donate their land for the corner guys for our subdivision overhead distribution system. A reinforcement scheme that requires no land outside the easement is needed. Two such schemes are commonly used: the slack span, in which a short mechanically unloaded, or slack, span is run to a nearby pole that has a down guy and anchor on the side opposite the tension span; and the push brace, in which a pole is lodged against another pole in the direction of the unbalanced tension. The slack span and push brace are shown in Figure 7.20.

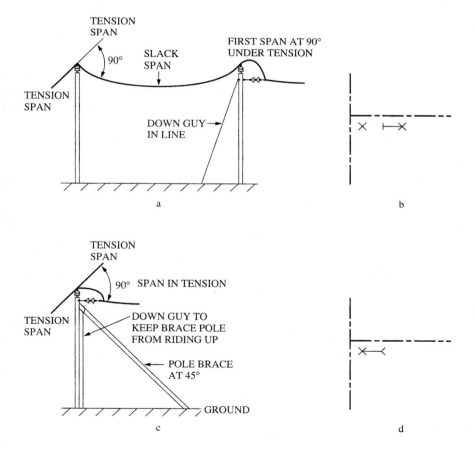

**FIGURE 7.20** Methods of in line pole reinforcement (single conductor shown) include: (a) Slack span (b) slack span symbol (c) push brace and (d) push brace symbol.

### 7.5.3 Transformer Selection

Estimating residential transformer capacity is not simply a matter of adding the individual residence loads. Lot sizes differ, with larger lots usually having larger houses that have more heating and cooling needs, which may or may not be electric. Larger houses do not necessarily have more electric appliance loads than smaller houses. Additionally it would be very rare for a house to have all of its electric appliances in use at once and even rarer for several houses to have most of their electric appliances on at once. The diversity of residential loads is great, so the estimates used for residential loads are the results of many historical studies that compile actual usage for residences of various types over a long period. These studies are updated from time to time. Most electric utilities use their own data to establish their own residential load estimates.

For apartment type dwellings the National Electric Code (NEC), which provides nationally recognized minimum safety standards for the installation of electrical wiring and equipment in buildings, provides rules for establishing the loading and wire size. The NEC is published by the Fire Underwriters Protection Association for the protection of life and property. It has the same force for buildings as the NESC has for electrical power utility installations. A similar set of rules could be established for single family dwellings, but the electric utilities are not subject to the NEC, so each utility makes up its own loading chart as described above. Many companies consider the average residential load to be 25 kVA.

A factor that is considered in making the loading charts (for all installations, not only residential) is the overload capacity of transformers. The transformer nameplate rating is the amount of power (kVA) the transformer can carry continuously without overheating. Oil filled distribution and power transformers can withstand considerable short-time overloading without materially reducing their reliability or lifetime. A transformer can carry 300% of its normal rating for a short while without damage, and 150% of its rating for several hours without overheating. Thus the load from the rare occasions that everyone turns on all of their electric appliances can be handled by the transformer overload capability.

If one house has a total electrical load of 25 kVA it can be served with a transformer smaller than 25 kVA because of the load diversity. A 15 kVA transformer will normally be adequate, but a 20 kVA transformer would be better. For two houses with a 25 kVA load each a 37.5 kVA transformer will be good. Four 25 kVA houses can be adequately served with a 50 kVA transformer, and six can be adequately served with a 67.5 kVA transformer, but since 67.5 kVA is not a standard size a 75 kVA transformer is normally used. Standard pole top residential distribution transformer sizes are 15, 25, 37.5, 50, 75, and 100 kVA. The normal transformer to connected load ratio is between 50 and 60% for most utilities. Table 7.3 summarizes the loading practices for distribution transformers at 50 and 60% transformer capacity to connected load ratios. The distance between houses normally prevents connecting more than six to eight houses to the same transformer.

In the rare instances when a customer reports low voltage, or the transformer is getting too hot, the utility company changes out the transformer to a larger capacity to handle the load. The 25 kVA assumed load with no more than about 50% of

◼◼◼ TABLE 7.3    Residential transformer capacities for normal 25 kVA connected load per residence

| Number of houses served by transformer | Capacity required for given connected load | | Transformer normally used |
|---|---|---|---|
| | **Percent Transformer capacity of load** | | |
| | **50%** | **60%** | |
| 1 | 12.5 kVA | 15 kVA | 15 kVA is the smallest distribution transformer built. Many utility companies use nothing smaller than 25 kVA. |
| 2 | 25 kVA | 30 kVA | a 37.5 kVA X-former is an acceptable choice. |
| 4 | 50 kVA | 60 kVA | a 50 kVA X-former is an acceptable choice because of the diversity of 4 houses. |
| 6 | 75 kVA | 90 kVA | 75 kVA is acceptable because of load diversity |
| 8 | 100 kVA | 120 kVA | 100 kVA is acceptable because of load diversity. 87.5 kVA would be acceptable for the same reason if that size is available. |

connected load turned on at a time is normally a good assumption for most residential areas. Situations in which the assumption breaks down are so rare as to be considered abnormal loading. Thus the utility can afford to upgrade the transformer capability when the abnormal loading occurs, and still provide economical service to the rest of the neighborhood.

Referring to Figure 7.19, we will now choose the transformer capacities for our subdivision. $T_1$ serves two houses so it is chosen to be 37.5 kVA, a conservative rating. $T_2$, $T_3$, and $T_6$ each serve four houses so they are chosen to be 50 kVA transformers. $T_5$ serves five houses and is chosen to be the next larger capacity available than that used for four houses, or 75 kVA. $T_4$ and $T_7$ both serve six houses and are chosen to be 75 kVA from Table 7.3. The total estimated connected load is (31 houses $\times$ 25 kVA) = 775 kVA. The individual transformer capacities add up to 412.5 kVA, or 53.23% of the estimated connected load. This ratio normally allows good service without dangerous transformer overheating because of the load diversity, because not everything is on at once.

## 7.5.4  Secondary Connections

The transformer secondary service drops must be made such that no service drop crosses adjacent property lines outside the easement. This means some pole to pole spans of secondary conductor are needed for our subdivision electrical distribution system. From Figure 7.19 we see that: $T_2$ needs two spans of secondary to serve four houses, as do $T_4$, $T_5$, and $T_6$; $T_7$ needs only one span; and $T_1$ and $T_3$ have only direct service drops, no secondary spans.

To choose the proper secondary conductor we must calculate the worst case voltage drop along a secondary span. The worst case span is the two lot span running East from $T_5$ with two houses connected at the end of the span. This is a span of about 240 feet with 50 kVA connected at the end. Number 2 all aluminum conductor (AAC) is a common secondary conductor with a resistance of 0.159 Ω /M ft. Thus the resistance for 240 feet is (120 ft./1000 ft.)(0.159 Ω /M ft.) = 0.3816 Ω. The current through this resistance is (50 kVA/240 V) = 208 A. Therefore the voltage drop across the 240 ft. span is (208 A)(0.03816 Ω) = 7.95 V, or 3.31%. This drop is at maximum loading that will rarely, if ever, occur, so #2 AAC conductor is a good choice for the secondary conductor. A larger conductor would be a waste of money because a lower voltage drop is not necessary, but a smaller conductor would be marginal in performance on the longest secondary span.

Most utilities stock only two or three standard secondary conductor sizes to reduce their inventory of wire sizes. This saves them money because of reduced warehousing costs, even if occasionally a conductor is oversized for the application.

## 7.5.5 Primary Conductor Selection

The primary conductor must be chosen. It too will be one of a few standard sizes stocked by the utility. The total transformer capacity connected is 412.5 kVA at 7.2 kV (12.47 kV/$\sqrt{3}$), so the primary current is 57.29 A at full transformer loading, no overloading included. ACSR conductor in the #2 size is a common primary wire size for small developments, and is the minimum size stocked by most utilities. It has a resistance of 0.319 Ω/M ft. The maximum distance for the primary from a Main Street tap (only one tap will be used at a time) is roughly 2500 feet. The total primary resistance is 2.5 M ft. × 0.319 Ω/M ft. = 0.798 Ω, making the voltage drop 45.68 V, which is only 0.63% of 7.2 kV. Number 2 ACSR will serve nicely as a primary conductor.

## 7.5.6 Primary Fusing

Primary fusing is usually set to 150% of full load and then raised to the next higher available continuous current rating. Though raising the fuse continuous current rating to the next higher available rating is allowed by the NESC, it is not always recommended. The fuse size for the West and East taps is calculated as follows:

(412.5 kVA)(1.5)/(7.2 kV) = 85.9 A

The next available size is 100 A, which is also the maximum current allowed for #2 ACSR. The transformer primary fuses are found in the same manner so for the transformers used in our subdivision the primary fuse sizes are: 37.5 kVA, 7.8 A, with a 7.5 A fuse recommended; 50 kVA, 10.4 A, with 10 A recommended; 75 kVA, 15.6 A, with 15 A recommended; and 100 kVA, 20.8 A, with 20 A recommended.

## 7.5.7 Voltage Changeover

### Subdivision Conversion

The subdivision will be converted to 34.5 kV in the following manner:

1. One lateral will be left on the 12.47 kV line and the other will be connected to the higher voltage. Figure 7.21 shows the West lateral left on the 12.47 kV line, and the East lateral connected to the 34.5 kV line on Main Street, but not energized yet. The cutout at the East lateral tap will be fused for 34.5 kV. The cutout will be changed if it is rated for 12.47 kV.

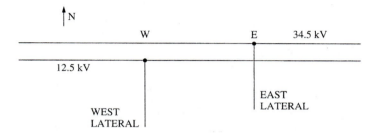

**FIGURE 7.21** 12.47 kV to 34.5 kV change over

2. The primary line will be opened at a jumper at the corner behind lot 4. If the primary insulators were not already 34.5 kV they would be changed as the process progresses. Transformers $T_1$ and $T_2$ will be replaced with 34.5 kV transformers, as will their disconnects and lightning arresters. The change out will normally require about $1\frac{1}{2}$ to 2 hours during which the houses on lots 1 through 4 and 30 and 31 will be without power. After the change out, the section of line to the corner of lot 4 will be energized so that those customers can resume their normal life.
3. Next the primary circuit will be broken at the intersection of lots 9 and 10. The transformers ($T_3$ and $T_4$), lightning arresters, and disconnects will be replaced with like items rated at 34.5 kV. This will require about $1\frac{1}{2}$ hours, after which this section of line will be energized at 34.5 kV.
4. Next the line disconnect at the West lateral tap will be opened and all the transformers, lightning arresters, and disconnects connected to the primary line between the intersection of lots 9 and 10, and the West lateral tap will be replaced with 34.5 kV rated ones. The disconnect and fuse cutout at the West lateral tap will also be replaced with 34.5 kV rated devices. The remaining line will then be energized at 34.5 kV from the intersection of lots 9 and 10.

Both taps will not be in use at once unless there is a break in the primary line. In this case both the East and West lateral taps may be energized while an isolated section of primary line is being repaired. Notice that the job is broken down into smaller tasks that can be accomplished in a relatively short time to reduce the time that any customer is without electricity. In our example no one would have been without power for more than 2 hours.

### Main Street Feeder Conversion

The voltage change over for the Main Street feeder is done as follows:

1. Temporary framing for 12.47 kV is installed below the present line as shown in Figure 7.22. Generally a 12 to 14 foot cross arm is used for 12.47 kV lines. This can be done without killing power.

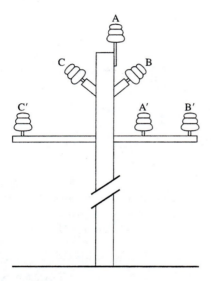

**FIGURE 7.22**    Pole configuration for feeder voltage change over

2. The 12.47 kV conductors are moved to the temporary framing with hot line tools. First A is moved to A', then B to B', and finally C to C'. All 12.47 kV loads are connected to these conductors.
3. The old framing and insulators are changed, if needed, and new conductor is installed for the higher voltage. The new conductor is then energized at 34.5 kV.

All laterals tapping off this feeder are changed out in the manner described for our subdivision. In this manner a major voltage change can be accomplished with no extended outage experienced by any customer on the Main Street line.

## 7.5.8 Underground Layout for Residential Neighborhood

If the subdivision of Figure 7.19 were to be laid out as an underground system instead of an overhead some changes would need to be made. First, the same easements are needed plus one across the North end of lots 1 and 31 (and across the street), but they are 10 feet wide at, above, and below ground level.

The portion that will be overhead, if any, has to be chosen. It is likely that the commercial section will be served with overhead because of the potential load size. If so, an overhead line will be built from the West lateral tap, along the West easement to the intersection of lots 12 and 13, and then over to intersection of lots 20 and 29. These may be underground but unless both are owned by the same developer, the residential developer will not pay to put the eventual commercial service underground. We will assume the lines along the commercial area are overhead.

The terminal poles (the location at which the service goes underground), with their cable terminators, fused disconnect switches, and lightning arresters, will be located at the West easement intersection of lots 12 and 13, at the intersection of lots 13 and 20, at the intersection of lots 20 and 29, and at the East easement at the Northeast corner of lot 1. Splice cabinets, to provide a protected environment for primary cable splices, will be needed at the intersection of lots 5 and 6, and at the Northeast corner of lot 1. Figure 7.17 illustrates this requirement. The primary line would be 34.5 kV cable if a voltage changeover is contemplated, to avoid the necessity of digging up and burying new cable in a completed neighborhood. The 34.5 kV cable is considerably more expensive than 12.5 kV cable.

The transformers will be pad mount types, mounted on concrete pads. The transformer capacities and locations can be the same as those chosen for the previous overhead illustration. The primary and secondary cables can be routed the same as the overhead lines shown in Figure 7.19. The poles, of course, will not be needed, except for the terminal poles. The lines will be energized from either the West or the East lateral tap for normal operation. The underground system will be safe, out of sight, and expensive.

## 7.6

### INDUSTRIAL DISTRIBUTION LOAD

Industrial loads tend to be much more predictable than residential and commercial loads. These loads must be known to select wire sizes and protective devices. Where large machines are used the choice of machine (such as synchronous instead of induction motor) can make a considerable difference in current. As we noted before, a high power factor is helpful in keeping total electricity costs down in an industrial plant. The outdoor distribution may be overhead or underground, and may be served by its own substation. The construction methods will be the same as those already discussed for outdoor distribution. The indoor, within a building, distribution system is planned using the methods studied in an electrical system design course. A practice problem to calculate currents using what is sometimes called the "P&Q" method will be helpful.

**Example 7.1:**

Calculate the steady state currents $I_1$, $I_2$, and $I_3$ for the circuit of Figure 7.23.

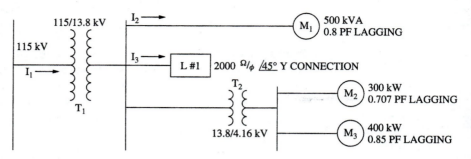

**FIGURE 7.23**   Circuit for Example 7.1

**Solution:**

Starting with $M_1$, which is 500 kVA at 0.8 PF lagging.

$\text{Cos}\theta = 0.8$ lagging so $\theta = -36.87°$

$S = 500 \text{ kVA} \angle -36.87° = 400 \text{ kW} - j300 \text{ kVAR}$

three-phase power $P = (\sqrt{3})IV\text{cos}\theta$ so

$$I_2 = \frac{P}{(\sqrt{3})V\text{cos}\theta} = \frac{400 \text{ kW}}{(\sqrt{3})13.8 \text{ kV} (0.8)} = 20.92 \text{ A} \angle -36.87$$

Now for load L#1 = 2000 $\Omega$/phase $\angle 45°$, Y connected.

L#1 = (1.414 + j 1.414) k$\Omega$ in rectangular coordinates.

The phase voltage is

13.8 kV/$\sqrt{3}$ = 7.97 kV

thus

$$I_3 = \frac{V_P}{Z} = \frac{7.97 \text{ kV}}{2000 \text{ } \Omega \angle 45°} = 3.985 \text{ A} \angle -45°$$

We will need $P$ and $Q$ later so

$P_P = I^2R = (3.985)^2(1.414\text{k } \Omega) = 22.4 \text{ kW}$

$P_{3P} = 3 \times P_P = 67.2 \text{ kW}$

By inspection we see that since the absolute values of R and X are equal

$Q_{3P} = 67.2 \text{ kVAR}$

To find the final current we must know the $P$ and $Q$ for $M_2$ and $M_3$. $P$ is given for each so we need only find Q. Recall $Q_{M2} = S\text{sin}\theta = (P/\text{cos}\theta)\text{sin}\theta = (300 \text{ kW}/0.707)\text{sin}(-45°) = -300\text{kVAR}$ similarly $Q_{M3} = 247.9 \text{ kVAR}$

Except for the power losses within the transformers (which we will assume negligible here) the power and reactive volt amperes remain constant as they flow through the transformers. The changing voltage levels result in the power being produced by different current levels on the high and low voltage sides of a transformer. To find $I_1$ we must only add the total power and VARs supplied at that point in the circuit, and calculate the current.

$$P_T = P_{M1} + P_{L\#1} + P_{M2} + P_{M3}$$
$$= (400 + 67.2 + 300 + 400)kW = 1167.2 \text{ kW}$$

$$Q_T = Q_{M1} + Q_{L\#1} + Q_{M2} + Q_{M3}$$
$$= (^-300 - 67.2 - 300 - 247.9) \text{ kVAR} = ^-915.1 \text{ kVAR}$$

now $S = P + Q$ so

$$S_T = 1167.2 - j915.1 = 1483.16 \text{ kVA}\angle^-38.1°$$

$$PF = \cos\theta = P/S = 0.7870$$

and from the three-phase power equation $P = (\sqrt{3})VI\cos\theta$

$$I_1 = \frac{P}{(\sqrt{3})V\cos\theta} = \frac{1167.2 \text{ kW}}{(\sqrt{3})\ 115 \text{ kV } (0.787)} = 7.45 \text{ A}\angle 21.5°$$

With the currents quickly calculated in this manner, the resistivity of the conductors, and the distance they must run, the proper size conductors can be chosen for an acceptable voltage drop. Usually a worst case voltage drop of 5% is acceptable. Once the proper conductors for steady state operation are chosen the calculation must be repeated using the motor locked rotor values to assure that the voltage dip from the motor starting current is not excessive. The motors may never all be started simultaneously so several variations of the problem may have to be worked. This problem could be worked more precisely with pu quantities, but this quicker method is accurate enough for this purpose.

## 7.7

## SUMMARY

The most common distribution voltages are 12.47 kV line to line (7.2 kV line to ground), and 34.5 kV line to line (20 kV line to ground). Public easements are provided for utility services to commercial, industrial, and residential loads. The safety standards for distribution line construction are published the National Electric Safety Code (NESC).

The common supports for overhead distribution lines are wood poles and steel poles. Wood poles are normally Southern Yellow Pine (about 70%) or Douglas Fir (about 25%). Wood poles are simple, inexpensive, and reliable. Steel poles are more expensive than wood poles, but become competitive in lengths greater than

90 feet. Steel poles are considered more attractive than wood poles by many people. Wood poles are normally set by direct imbedment in firm soils. Concrete stabilized imbedment and guy wire anchoring are used in less stable soils and for dead end poles. Steel poles can be set by imbedment, but reinforced concrete foundations are often used for steel pole foundations.

Distribution insulators are normally made of glass or porcelain but fiberglass insulators with silicone rubber skirts are becoming more popular. The most common overhead conductors used for distribution are, most common to least: ASCR (aluminum conductor, steel reinforced), which is strong because of the steel reinforcement; all aluminum conductor (AAC), which has slightly better conductivity than ACSR but has less strength; and AAAC (all aluminum alloy conductor), which is stronger than AAC but more expensive.

Underground distribution is less subject to weather outages and is less visible than overhead distribution. Direct buried underground cable has markedly improved in recent years with the reliability increasing and voltages to over 34.5 kV now available. Oil impregnated cable is still used in many networks.

Factors critical in distribution system layout are the easement requirements, which are set by the voltage, the geometry of the service area, the types and density of the loads, and the expected load growth. An example distribution layout is shown in section 7.5.

## 7.8
■■■■■■

### QUESTIONS

1. State the function of the electrical power distribution system.
2. List two advantages of a grounded wye distribution system.
3. What is the NESC, and what organization is responsible for its publication?
4. How often are revisions to the NESC published?
5. What is the procedure for making changes in the NESC?
6. List the four parts of the NESC.
7. List the clearances required by the NESC for a 12,470 V line: a. over a railroad track; b. from a building; c. from any area accessible by a pedestrian.
8. Why does the required line to line clearance increase as sag increases?
9. What trees are most commonly used for wood poles?
10. How is wood pole failure under load defined?
11. Calculate the tangent load a 70 foot wood pole can carry at its top, assuming the bottom is securely anchored, if it is of the following class: H1, 1, 2, 3.
12. Why are ft-k values listed in the tables for the bottom of the wood poles? Look carefully at Figure 7.2 before you answer.
13. Name the two worst enemies of wood poles and the defense(s) against them.
14. State the major advantage of steel poles.
15. Under what circumstances do steel poles become competitive with wood poles?

16. What is meant by direct imbedment?
17. What is concrete stabilized sand?
18. How much of the length of a 90 foot pole must be imbedded?
19. How are directly imbedded poles reinforced at turns and dead ends?
20. Define framing.
21. State the two characteristics of insulators that must be specified.
22. What is the advantage of ball and socket insulators?
23. State the major advantage of each of the following conductor types: ACSR, AAC, AAAC.
24. List three types of commonly used underground distribution cable.
25. What is meant by direct buried cable?
26. Describe a duct bank.
27. Why are only 40% of the duct bank conduits filled at the time of the system installation?
28. What is an easement?
29. For the neighborhood of Figure 7.24 lay out an overhead distribution system including easements needed, pole and transformer placement, transformer

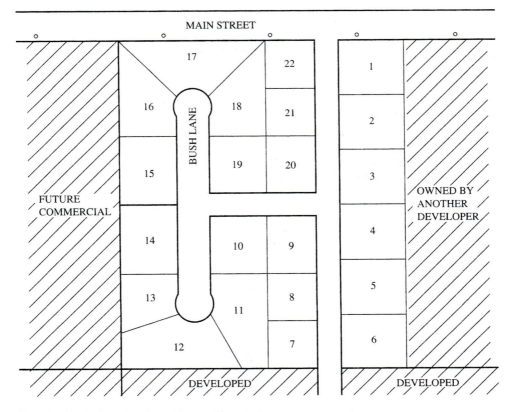

**FIGURE 7.24** Neighborhood for Problem 7.29

sizing assuming that the maximum connected load for each house is 25 kVA and the (capacity/connected load) ratio used by our company is 60%, and primary fuse requirements. The distribution primary voltage is 34.5 kV and no change is contemplated, and the secondary voltage is 240 V. Show the primary and secondary conductor paths, and calculate the worst case voltage drop for each if the primary conductor is #2 ACSR and the secondary is 2/0 AAC.

30. What changes will be needed if the system of problem 30 is to be underground instead of overhead?

31. Describe the process by which one distribution voltage level is changed to a higher voltage.

32. For the circuit of Figure 7.25 calculate the total $S$, $P$, $Q$, and currents $I_1$ through $I_4$.

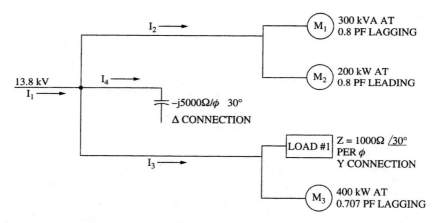

**FIGURE 7.25**   Circuit for Problem 7.32

# Electrical Power Transmission

# CHAPTER 8

# Transmission System Overview

## 8.1

### SYSTEM OPERATION

The job of the electrical power transmission system is to transport electrical power from the generating stations to the points of distribution, the distribution substations, or the interties with other utilities. The electrical power system should serve all its customers and interconnect partners economically and reliably. The system voltage and frequency must be maintained within acceptable limits, and accurate records should be kept of all power transferred to both customers and other utility companies.

## 8.1.1  Operation Components

The major components of operation are planning, control, and accounting. Digital computers are now heavily used in all three components of operation.

Planning is the process of choosing among alternate courses of action in the utility operation. Planning has both long-term and short-term aspects. Load forecasting is an example of long-term planning. Load forecasting is necessary to assure that needed state permits, financing, right of ways, and finally generating equipment, transmission lines, distribution lines, and substations are constructed by the time the customer needs the power. Short-term planning includes such things as load diversity forecasting for efficient machine utilization, generating unit commitment, spinning reserve commitment, and interconnect power transfers. Planning also includes choosing emergency procedures for various malfunctions such as emergency load shedding procedures for when a critical generator or line is lost. Maintenance scheduling, fuel selection and acquisition, personnel requirements, and financial needs are also planned ahead of time if the company is to perform its function efficiently.

Control in this context refers to operation control, the minute by minute management of the utility system. This includes the myriad calculations and decisions required in such things as automatic generation control (AGC), all of the checks and control needed to assure proper generating station operation, automatic VAR and power dispatch, allocation of generation to various generating units for maximum efficiency and minimum fuel use, and keeping net interchange of power with neighboring utilities at desired values. The large number of decisions to be made constantly throughout a system would be impossible without digital computers to provide timely preplanned responses within the system to varying load conditions.

Accounting is more than billing and financial record keeping, although these are very important jobs. Accounting also refers to the collection of operating and production statistics, maintenance information, and records of abnormal conditions. The records allow evaluation of performance from a factual base.

## 8.1.2  Control Function

If a generator is operating with a given load and the load increases, the increased electrical load causes more mechanical counterforce in the rotor, and the rotor counterforce causes the prime mover to attempt to slow down. A drop in prime mover speed causes the generator frequency to drop, as shown in Figure 8.1a. Most governors allow a generator to drop about 5-6% from full load to no load. Such a drop in frequency is not acceptable for a power system that contains frequency sensitive loads such as electric motors. The generator control system responds to the load increase by increasing prime mover power so that the increased load can be met with only a very small drop in the generator frequency, as shown in Figure 8.1b. The frequency still drops, but less. A drooping characteristic of a generator, where frequency drops as load increases, is important if more than one generator is to be used because it enables proportionate sharing of load by generators.

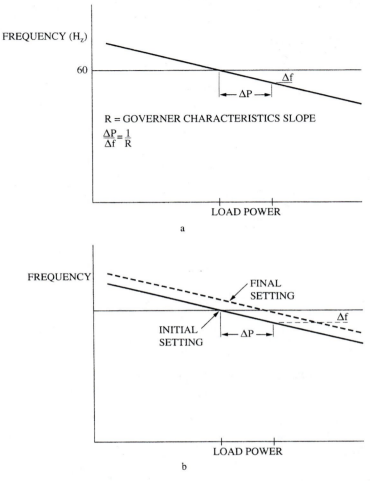

**FIGURE 8.1** Generator regulation (a) Frequency drop with load (b) Corrective action of governor and turbine control

Figure 8.2 shows the regulation curve of two generators, each handling a portion of the total load. If the load is increased each generator picks up an equal amount of the load increase, and the system frequency drops. If one of the generators has a rising characteristic, frequency increasing with increasing load, it would pick up all of the load, possibly overload and fail. The generators whose regulation curve is shown in Figure 8.2 would pick up proportional amounts of the load increase if the slopes of the drooping regulation slopes were set to be inversely proportional to generator capacity instead of identical. In an actual system the control system of the generators acts to drive the frequency back to almost what it was before the load increase. A large frequency drop, nearing a Hz, indicates an emergency condition in the system in which load exceeds generating capacity.

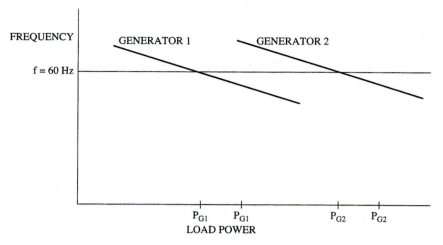

**FIGURE 8.2**  Generators with same governor regulation slope divide increased load proportionately.

Frequency sensitive relays are used to monitor the power system frequency and automatically shed noncritical load if the system frequency drops too much. Many books on electrical machines and generation deal at length with load sharing between generators.

The coordination of many generators in a system is a very complex problem. Interties between systems complicate the problem further. Each intertied system must operate so as to maintain a fairly uniform instantaneous frequency, the correct integrated frequency (over time), divide loads between systems as needed, and assure the generating stations within the system are sharing the load in such a way that the system is operating with maximum economy for the load condition. This is no small task. Computers are essential to solve the large equations associated with economical operation and to keep up with all of the details of system operation.

## 8.1.3 Economic Generation Allocation

Economic allocation of generation means taking into account individual generating unit efficiency and location in the network to decide how much each unit should contribute to most economically serve a given total system load. Figure 8.3a shows a plot of fuel input versus power output for two generators. Figure 8.3b shows the plots of the incremental fuel cost (change in fuel input/change in power output) for the same two generators. It has been shown mathematically that the optimum operating economy occurs when the incremental cost of operating the units is equal. This is true as long as line losses, which depend on the location of the unit within the system, are neglected. The problem of economic allocation (also referred to as economic dispatch) is more complicated when line losses are included. It is still approximately true that maximum generation economy occurs when the total load is divided among the generating units so that their

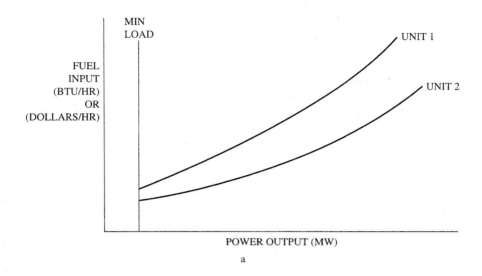

a

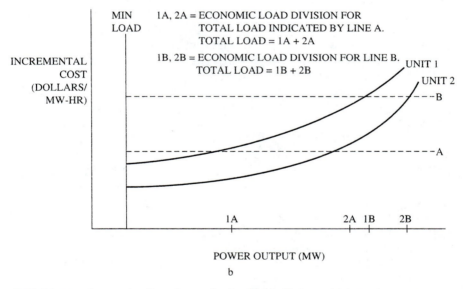

b

**FIGURE 8.3** Economic allocation on basis of fuel efficiency (a) Input-output curves and (b) Incremental cost curves

incremental costs are the same. The availability of fuels may mean that the equal incremental cost criteria may have to be violated at times. Also a new unit may be so much more efficient than older units that it is operated at maximum capacity all of the time, with the load the new unit cannot handle being allocated among the older generating units on an equal incremental cost basis.

## 8.1.4 Control Methods

Originally control systems were analog, in which all of the control signals and circuitry are continuously variable. A mechanical governor for speed control is an example of an analog control, as is the field current control of many older generators. The field current is manually set (perhaps remotely) to provide the desired output voltage and the circuitry holds the voltage constant as the generator load varies. The problem with analog control is that it is hard to send analog signals long distances accurately, making remote control from a central station, or coordinating several remotely controlled stations, difficult with analog control.

The next step in control was digitally directed analog control. In this type of control a digital signal is sent to the controlled machine from a central computer to indicate the desired machine output (this signal is called a set point). At the machine site a digital to analog converter (D/A) converts the digital signal to an analog signal, which is used to set the analog control output to the proper value. Digitally directed analog control for a generating unit is illustrated in Figure 8.4a. The computer set output is held by the analog control except when the value is modified by the area control error (ACE). The ACE is a signal that indicates the amount of power needed by an area when the power demand is changing because of increased or decreased loading. The ACE signal is derived from frequency and power measuring equipment (recall that system frequency changes as the power demand varies). The ACE signal is converted to the form required, digital or analog, at the control. The ACE signal is in digital form at the computer input in Figure 8.4b. The desired generator output is converted to field current and governor settings by the control signal generator (CSG). The output voltage, current, and power of the generator is sensed by monitoring equipment to indicate to the control when the generator has reached the desired output. The digital computer used to provide the generator set point is usually a large one that takes into account many system factors in arriving at the generator set points.

As computers became less costly and capable it became economical to use a smaller computer at the generating site to perform direct digital control. The system information needed to arrive at the generating unit set point is often sent from a larger central computer to the unit computer over communication lines. Originally, and still in some locations, direct digital control was backed up with an analog control that held the generator output at the last set point, or a preset set point, in the event of computer failure. Normally the unit control computer is large enough to handle several computing and status monitoring jobs at the generating site on a time-share basis. The transducers for many measurements, such as voltage, and control actuators, such as field current supplies, are analog. The analog transducer information must be converted to digital for the computer and the digital control signals must be changed to analog for actuator control. Direct digital control with analog back-up is illustrated in Figure 8.4b.

The next step in control is to use a small computer for back-up control instead of an analog controller. The reduced cost and increased capability of microprocessor-based microcomputers, which outperform large minicomputers of only a few years ago, have made possible this form of control, illustrated in Figure 8.4c.

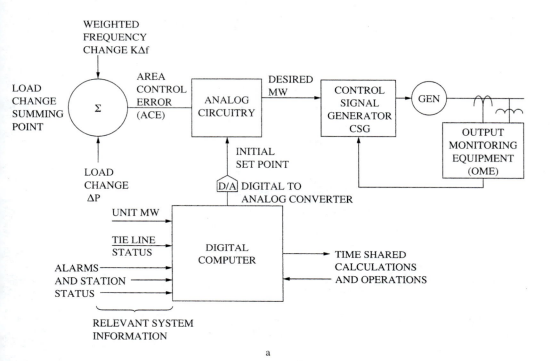

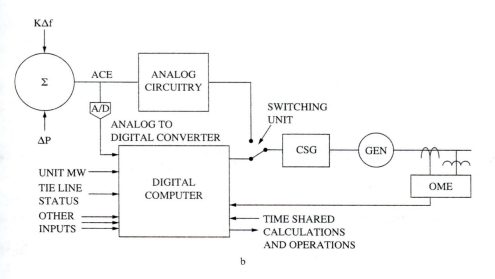

**FIGURE 8.4** Generator control (a) Digitally-directed analog control (b) Direct digital control—analog backup

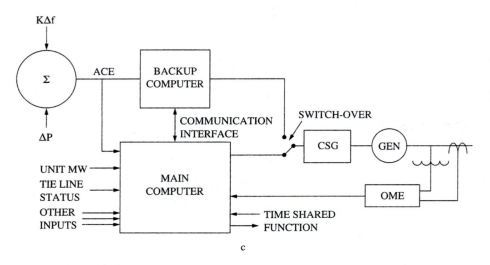

**FIGURE 8.4**  Generator control (c) Direct digital control—digital backup

Direct digital computer control is now the norm in the power industry. It is used for generating station control, substation control, economic dispatch of power, and interchanges of power over interties. In addition to the use of computers by individual utilities to control their own activities, groups of utilities use them for pool dispatch. A pool is the total power available from a group of power companies that have agreed to cooperate in serving the power needs of an area larger than that served by any one of the cooperating utilities. Pool dispatch is the routing of power from and to individual power companies to most economically serve the area load. Figure 8.5 shows the relationship between the pool and one generating unit of one utility.

The communication system between the computers is the critical link in the operation of direct digital control systems. Supervisory control and data acquisition (SCADA) systems, described in the next section, supply the communication system. Not all communication is between machines. Voice communication must be available for communication between people at the central control station, generating stations, and maintenance crews at manned and unmanned stations.

## 8.1.5 Supervisory Control and Data Acquisition

SCADA is the equipment and procedures for controlling one or more remote stations from a master control station. It includes the digital control equipment, sensing and telemetry equipment, and two-way communications to and from the master station and the remotely controlled stations.

The SCADA digital control equipment includes the control computers and terminals for data display and entry. The sensing and telemetry equipment includes the sensors, digital to analog and analog to digital converters, actuators, and relays used at the remote station to sense operating and alarm conditions and to

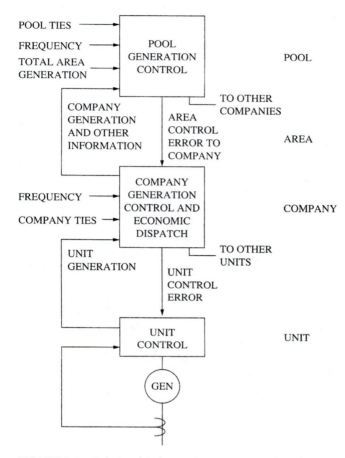

**FIGURE 8.5** Relationship between company and pool

remotely activate equipment such as breakers. The communications equipment includes the MODEMs (modulator-demodulator) for transmitting the digital data, and the communications links (radio, phone line, microwave link, or wire line). The procedures are the man-machine interfaces and the programs needed to match the control to the system needs. Figure 8.6 shows a block diagram of a SCADA system.

A SCADA system performs at least one, but usually more, of the following functions:

1. Alarm sensing for such things as fire or the performance of a non-commanded function.
2. Control and indication of the position of a two or three position device such as a circuit breaker or motor driven switch respectively.
3. State indication without control such as transformer fans on or off.
4. Control without indication such as capacitors switched in or out.

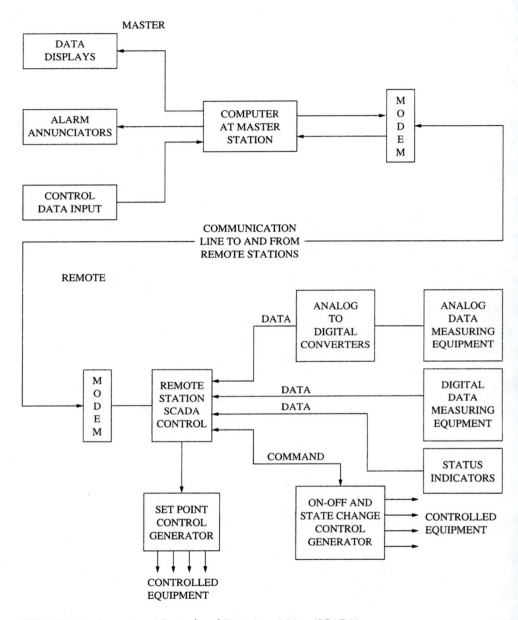

**FIGURE 8.6**  Supervisory Control and Data Acquisition (SCADA)

5. Set point control of remote control station such as nominal voltage for an automatic tap changer.
6. Data acquisition from metering equipment, usually via an A/D converter and digital communication link. Examples of this are the power transmitted over a particular tie line or the power used by a particular customer.

7. Initiation and recognition of sequences of events such as sectionalizing a bus with a fault on it or routing power around a bad transformer by opening and closing circuit breakers.
8. Allow operators to initiate operations at remote stations from a central control station.

SCADA systems have enough capabilities that it has been many years since there has been a manned substation in the United States. Almost all routine substation functions are remotely controlled. A complete SCADA system, with the appropriate relays and auxiliary equipment, can perform the following substation control functions: synchronism check, automatic reclosing after a fault, protection of equipment in a station, automatic bus sectionalizing, voltage and VAR control, fault reporting, transformer load balancing, equipment condition monitoring, data acquisition, status monitoring, and data logging.

SCADA systems provide two-way data and voice communication between the master and the remote stations. Digital codes are used for information exchange with various error detection schemes to assure that all data are received correctly. MODEMs at the sending and receiving ends modulate (put information on the carrier frequency) and demodulate (remove information from the carrier) respectively. The remote terminal unit (RTU) contains circuitry to properly code remote station information into the proper digital form for the MODEM to transmit, and to convert the signals received from the master into the proper form for each piece of remote equipment.

During operation a SCADA system scans all routine alarm and monitoring functions periodically by sending the proper digital code to interrogate, or poll, each device. Upon being polled the device sends its status or data to the master station. The total scan time for a substation might be 30 seconds to several minutes depending on the speed of the SCADA system and the substation size. An alarm condition interrupts a normal scan. Upon an alarm the computer polls the equipment at the station that initiated the alarm. An alarm may trigger a computer-initiated sequence of events such as breaker action to sectionalize a faulted bus. Each activated device has a code to activate it, in other words to make it listen, and another to cause the controlled action to occur. Some alarm conditions may sound an alarm at the control station that indicates action is required by an operator. The operator then initiates the action via a keyboard or a CRT.

The computers used in SCADA systems must have considerable memory to store all the data, codes for the controlled devices, and the programs for automatic response to abnormal events. SCADA is computer control, and it is the communication link for direct digital control.

## 8.1.6  System Security

A secure system is one that is operating normally. Automatic equipment, such as direct digital control and SCADA, help system security by automatic monitoring and detecting abnormal conditions early, so that damage can be minimized, and by automatically implementing corrective strategies. Automation cannot replace good system design and construction with quality components.

The system states are:

*Normal*: No emergencies exist on the system. All customer loads are being met and no lines or equipment is overloaded.

*Alert*: An insecure state in which one or more contingencies could cause an emergency condition. Examples are: an approaching hurricane or winter storm, concurrent events such as an intertie problem at the same time a generating unit is down.

*Emergency*: A condition in which critical operating constraints are being violated but the system is still carrying the load. Examples are: an overloaded transmission line based on capacity or system stability limits, voltage limits exceeded, or a transient that causes an oscillatory response that could cause a portion of the system to loose synchronization. An emergency state requires immediate action.

*Extremis*: This state is one in which system integrity has been lost by load shedding or separation of the system into islands. Extremis states usually result in some portion of the system being blacked out.

*Restorative*: This state is the one in which repairs are being made after an extremis state.

## 8.2

### SYSTEM STABILITY

A stable system is one in which all opposing forces are in equilibrium, or will be returned to equilibrium by a restoring force in the event of a system disturbance. Electrical power system stability is defined in the same way. Often power system stability refers to the synchronism of all synchronous machines, basically generators, in the system. A synchronous machine is said to be operating stably when the forces tending to accelerate and decelerate the rotor are balanced. A generator is stable when the mechanical power in from the turbine is equal to the electrical power out plus the generator mechanical and electrical losses.

If a generator, operating in equilibrium, is disturbed by a load increase, the mechanical counter-torque on the rotor caused by the load current is increased. The counter-torque causes the rotor to begin to decelerate. The generator control system and governor increase the turbine accelerating torque to prevent a decrease in the generator frequency. The decelerating torque from the load increases and the accelerating torque from the generator control system in response to the load increase must be temporary or the rotor will begin oscillations of increasing amplitude that continue until the rotor loses synchronism with the rotating stator magnetic field. Damping within the generator, which absorbs energy from the rotor when it rotates at a rate different than the stator magnetic field, provides a restoring force to prevent loss of synchronism by the machine.

The maximum power that can be transferred between machines or groups of machines with no loss in synchronism following a disturbance is the stability limit of the system. There are two categories of instabilities: steady state and transient. Steady state instability occurs when the system is slowly put into a condition for which no equilibrium exists. For example, steady state instability occurs when the power limits of the system are gradually exceeded. Transient instability occurs when a sudden system disturbance causes a loss of equilibrium. A large fault can induce transient instability.

## 8.2.1 Power Transfer

Figure 8.7 shows a three-phase synchronous generator supplying power to a three-phase motor through connecting line impedance Z. The line impedance is predominantly reactive so a change in $V_1 - V_2$ results in mostly reactive current. If $X/R = 10$ then the per phase current is

$$I = \frac{V_1 - V_2}{Z} = \frac{V_1 - V_2}{R + j10R} = \frac{V_1 - V_2}{10.05 \, |R| \, \angle 84.29°} = I \angle \text{-}84.29°$$

which is mostly reactive. The representation of a line by a reactor is thus justified for approximate calculations. The reactive current exchange between the synchronous machines provides synchronizing force on the machine rotors.

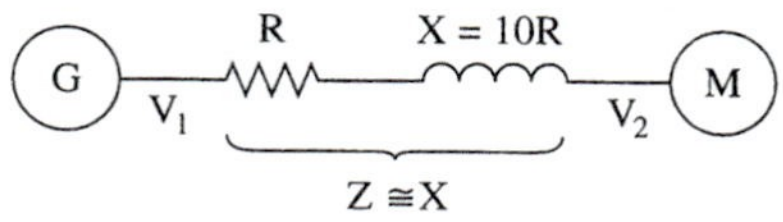

**FIGURE 8.7** Line mostly reactive

Recall that a load change on a synchronous machine causes the angle of the rotor to change with respect to the rotating magnetic field of the stator. The angle tends to advance more in a generator, and lag more in a motor. This is illustrated in Figure 8.8. The angles are designated $\delta_1$ and $\delta_2$ for the generator and motor respectively. The total angle is the sum of the two

$$\delta = \delta_1 + \delta_2$$

and the power exchanged depends on the total angle.

We will now obtain an expression for power transfer from a source to a load. Refer to Figure 8.9. The generator is operating into a receiving bus large in capacity compared to the sending bus. We will use the receiving bus as our reference and the angle, $\delta$, will all be from the sending bus, which may be a single generator or an entire system. The connecting line impedance includes the generator and receiving bus impedance, for simplicity we are lumping all of these into one. By Kirchoff's law we write

$$V_S - IZ - V_R = 0$$

Solving for $I \angle \Phi$ we find

$$I \angle \Phi = \frac{V_S \angle \delta}{Z \angle \theta} - \frac{V_R}{Z \angle \theta}$$

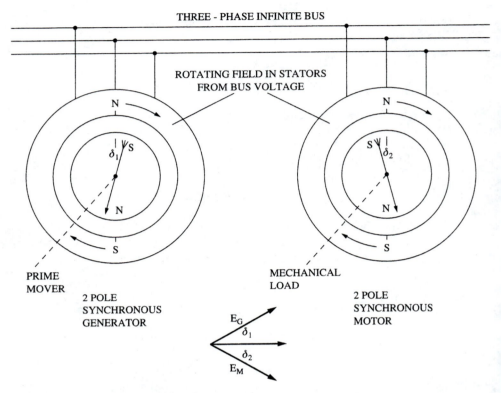

**FIGURE 8.8** Synchronous machine rotor with respect to rotating magnetic field

$$= \frac{V_S \angle (\delta - \theta)}{Z} - \frac{V_R \angle -\theta}{Z} \tag{8.1}$$

Recall that apparent received power, $S_R = V_R I^*$, and that the apparent power sent is, $S_S = V_S I^*$. The complex conjugate of the current, where $I^* = I \angle -\theta$ instead of $I \angle \theta$, is used to make the power flow sign convention consistent with the sign convention chosen for reactance. Thus we can write

$$I^* = \frac{V_S}{Z} \angle (\theta - \delta) - \frac{V_R}{Z} \angle \theta \tag{8.2}$$

now

$$S_R = V_R I^* = \frac{V_R V_S}{Z} \angle (\theta - \delta) - \frac{V_R^{\,2}}{Z} \angle \theta \tag{8.3}$$

from which

$$P_R = \frac{V_S V_R}{Z} \cos(\theta - \delta) - \frac{V_R^{\,2}}{Z} \cos\theta \tag{8.4}$$

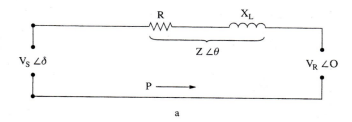

a

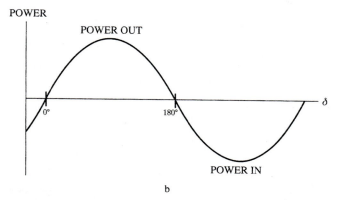

b

**FIGURE 8.9** Power transfer (a) Sending and receiving voltage linked by line impedance (b) Power angle vs sending power flow (sending machine on infinite bus)

and of course

$$Q_R = S_R \sin\angle(V,I^*) \tag{8.5}$$

If we recall from trigonometry that

$$\cos(a \pm b) = \cos a \cos b + \sin a \sin b$$

equation 8.4 reduces to

$$P_R = \frac{V_R V_S}{Z} \cos\theta\cos\delta + \frac{V_R V_S}{Z} \sin\theta\sin\delta - \frac{V_R^2}{Z} \cos\theta \tag{8.6}$$

If we now consider that $R << X$, a very good approximation, so that $\theta = 90°$ equation 8.6 reduces to

$$P_R = \frac{V_R V_S}{X} \sin\delta \tag{8.7}$$

The angle $\delta$ is the total angle between the receiving and sending voltages. In our development the receiving voltage angle is zero because the receiving voltage is our reference, but this is not the general case. A similar analysis would reveal that

$$Q_R = \frac{V_R V_S}{X} \cos\delta \tag{8.8}$$

The power sent, by a similar analysis is

$$S_S = V_S I^* = \frac{V_S^2}{Z} \angle\theta - \frac{V_S V_R}{Z} \angle(\theta + \delta)$$

$$P_S = \frac{V_S^2}{Z} \cos\theta - \frac{V_S V_R}{Z} \cos(\theta + \delta)$$

Using the trigonometric identity again

$$P_S = \frac{V_S^2}{Z} \cos\theta - \frac{V_S V_R}{Z} \cos\theta\cos\delta + \frac{V_S V_R}{Z} \sin\theta\sin\delta \qquad (8.9)$$

and if $R = 0$ so $\theta = 90°$

$$P_S = \frac{V_R V_S}{X} \sin\delta \qquad (8.10)$$

and

$$Q_S = \frac{V_S V_R}{X} \cos\delta \qquad (8.11)$$

Note that the maximum occurs at 90° in the approximate equations above. This is approximately true for nonsalient pole generator, but for salient pole machines (poles stick out instead of being smooth) the maximum occurs before 90° because the higher reluctance magnetic path between the poles causes the magnetic flux to be more concentrated at the poles than in a nonsalient pole machine.

A generator operates with a power angle of 20° to 25°, and any angle beyond 90° is unstable because the restoring force on the rotor is lost. Any power angle beyond 180° results in a reversal of power flow, as shown in Figure 8.9b. A reversal of power flow in a generating station results in extreme mechanical force being applied to the generator rotor and the turbine. This severely damages many station components as well as the generator and turbine. Frequency relays are used to protect the generator by detecting the frequency drop caused by an overload and removing the generator from the line before damage occurs. A reverse power relay is used to back up the frequency relay. The power angle across a long transmission line seldom reaches 10°, and seldom exceeds 40° across a whole system.

The loss components of equation 8.6 reduces the peak power transferred by $(V_R^2/Z)\cos\theta$, and shifts the curve of Figure 8.9b to the left by $90° - \theta$.

From the preceding equations we can see that VAR flow on a line is controlled by adjusting the voltage $V_S - V_R$. The power flow between two points in a system must be varied by changing the power angle, $\delta$, between the two points. Phase shifting transformers, both fixed and variable, are available to control power flow by adjusting $\delta$. Note that the angle varies at different points in the system and that the power flow is from the point with the most leading angle to the point with the most lagging angle. VAR flow is from the higher voltage on a line to the lower voltage. Power and VAR flow need not be in the same direction.

**Example 8.1:**

Calculate the power flow between the buses in Figure 8.10.

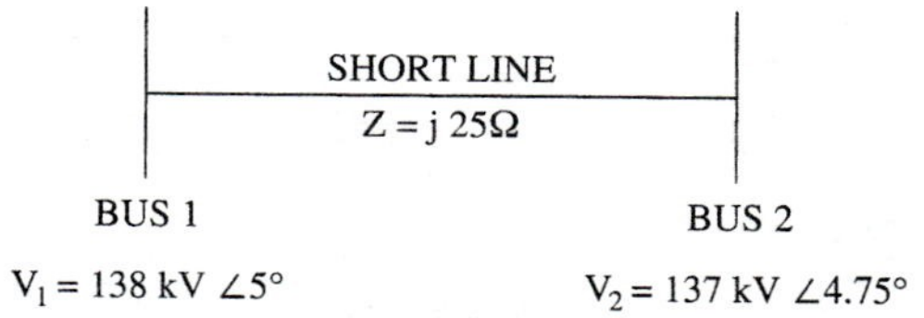

| | SHORT LINE | |
| --- | --- | --- |
| | $Z = j\ 25\Omega$ | |

BUS 1                          BUS 2

$V_1 = 138\ \text{kV} \angle 5°$          $V_2 = 137\ \text{kV} \angle 4.75°$

**FIGURE 8.10**    Figure for Example 8.1

**Solution:**

Using equation 8.7 and converting to phase voltages we find

$$P = 3\ \frac{V_{P1} V_{P2}}{Z} \sin(\delta_1 - \delta_2)$$

$$= 3\ \frac{79.67\ \text{kV}\ 79.1\ \text{kV}}{25\ \Omega} \sin(0.25°)$$

$$= 3.3\ \text{MW}$$

Calculating the power transferred using per unit (pu) and a base of 10 MVA and 138 kV we find

$$V_1 = 1.0\ \text{pu}$$

$$V_2 = 0.99275\ \text{pu}$$

$$\text{base}\ Z = \frac{(\text{base kV})^2}{\text{base MVA}}$$

$$= \frac{(138\ \text{kV})^2}{10\ \text{MVA}}$$

$$= 1.904\ \text{k}\ \Omega$$

$$\text{pu}\ Z = \frac{Z}{\text{base}\ Z} = \frac{25\ \Omega}{1.904\ \text{k}\ \Omega} = 0.01313\ \text{pu}$$

$$\text{pu}\ P = \frac{1.0\ (0.99275)}{0.01313} \sin 0.25° = 0.32997\ \text{pu}$$

$$P = \text{base MVA}\ (0.32997) = 3.2997\ \text{MW}$$

The difference in the two values of transferred power is from round off error.

Note that if the line voltage doubles the pu impedance is a quarter of its lower voltage value.

$$\text{base } Z = \frac{(2 \times 138 \text{ kV})^2}{10 \text{ MVA}} = 7.618 \text{ k}\Omega$$

$$\text{per unit } Z = \frac{25 \text{ }\Omega}{7.618 \text{ k }\Omega} = 0.00328 \text{ pu}$$

## 8.2.2 Power Flow Division

If two parallel transmission lines are used to transfer power the lowest impedance line transfers the most power, all other factors being equal. Recall from ac circuits that for two impedances in parallel

$$\text{path 1 current} = \frac{(\text{path 2 impedance})}{(\text{total impedance})} \text{ total current}$$

Multiplying both sides by base voltage we obtain

$$\text{path 1 power flow} = \frac{(\text{path 2 impedance})}{(\text{total impedance})} \text{ total power flow} \qquad (8.12)$$

Similarly more apparent power flows in the lower impedance line. If the lines are not at the same voltage the line with the lower pu impedance transmits more power. The following example demonstrates this.

**Example 8.2:**

Calculate the power flow for each path of Figure 8.11. All pu impedances are to the same base.

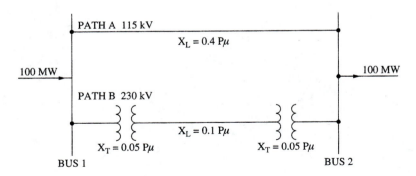

**FIGURE 8.11**   Power flow division

**Solution:**

The power flowing in line a is found by equation 8.12 after the path impedances are found.

Z path A = j0.4 pu

$$Z \text{ path B} = j(0.05 + 0.1 + 0.05) \text{ pu} = 0.2 \text{ pu}$$

$$\text{path A power flow} = \frac{(0.2 \text{ pu})}{(0.2 \text{ pu} + 0.4 \text{ pu})} \; 100 \text{ MW} = 33 \text{ MW}$$

and by subtraction path B power flow is 67 MW.

Given equal line impedances the higher voltage line carries the most power because its pu impedance is lower. In general the losses are lower on the higher voltage line because the current is lower for a given power transfer.

The division of the power in Example 8.2 can be altered by inserting a phase shift transformer in one line, changing δ for that line. Making the phase shift larger from one end of a line to the other increases the power flow on that line. If possible the voltage transformation and phase shifting functions are combined in one transformer.

If one line opens then the remaining line must carry all of the load, which usually results in an overload of the remaining line until corrective action is taken. More than two paths for power flow helps but does not guarantee that no remaining lines will be overloaded. Referring to Figure 8.12 we see three lines are transferring power from bus A to bus C. The power flow on the lines has divided in inverse proportion to the pu impedance of the lines. The 230 kV lines carry more power than the 115 kV line. The two 230 kV lines have a pu Z of 0.1 each and the 115 kV line has a pu Z of 0.4 including transformer $T_3$. To keep the calculations direct and illustrative assume that the impedance of $T_4$ is negligible, and that the power angle is the same across all three lines. Then, with the 230 kV line parallel impedance equal to 0.05 pu

$$230 \text{ kV line power} = (0.40/0.45)(600 \text{ MW}) = 533.3 \text{ MW}$$

and each 230 kV line carries half the power. The remaining 67.7 MW is carried by the 115 kV line. If one 230 kV line is removed from service the power split will be 120 MW on the 115 kV line and 480 MW on the remaining 230 kV line. This is an increase of about 80% on each line and $T_3$. If the MVA rating (thermal limit because of $I^2R$ heating) of either line or the transformer is exceeded action must be taken by the control station dispatcher. The most likely action is to increase the power delivered to bus B or bus C with reserve generating capacity, or possibly intertied power, and reduce the power to bus A until the lines are operating at an acceptable level of power transfer. The power transfer level of the two remaining lines may be higher than normal until the other 230 kV line is restored to operation. The last resort is to shed load on bus B or C to avoid overloading the lines. A line is not removed from service unless it is tripped out on a fault, or removed for maintenance. If the line is removed from service for maintenance, it is at a time when the load is at its lowest.

*Steady state stability* refers to a system's ability to withstand small disturbances from equilibrium without losing synchronism between two or more synchronous machines on the system. Two subdivisions of steady state are defined: classical and dynamic.

Classical steady state stability is that in which the generator field excitation is not changed for small deviations from equilibrium. Early steady state stability studies

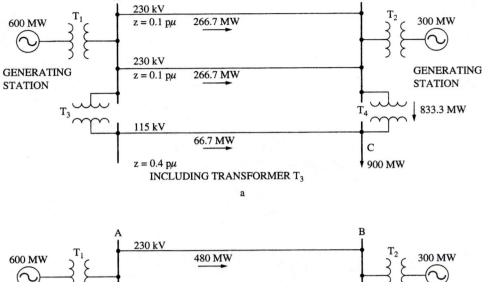

**FIGURE 8.12**   Effect of opening a line (a) Three lines in operation and (b) one line removed

were made with this assumption. The early stations had voltage regulators with *dead bands,* a span of output changes allowed with no control action, to prevent control action for small variations from the set point. The early voltage regulators were too slow and insensitive to effectively adjust for very small rapid changes in the output voltage. If such response was allowed it often resulted in a regulator induced oscillation that was worse than the original change. Thus the regulators were set to respond only to changes that represented a significant deviation from the set point. Small changes in the regulator setting were made manually.

Dynamic steady state stability is that in which the system generators are equipped with automatic field controls that are sensitive enough and fast enough to respond to small deviations in the output voltage from the set point, even if they occur rapidly. Such controllers resulted from the use of electronic circuitry instead of electromechanical systems for field control. The automatic voltage regulators allow higher power transfers with stability than did the older regulators. Thus the only steady state stability studies done today are dynamic.

Steady state stability limits are exceeded under the following conditions: (1) When the slow increase in load in some section is such that a generator is pulled out of step. The ability to transfer load from one place to another within the system reduces the possibility of this type of stability loss. (2) When a small load change causes an oscillation in the system to start, and the oscillation continues to grow until a machine is pulled out of synchronization. Good system damping reduces the possibility of this occurring.

Steady state stability can be enhanced by a number of actions that increase the ability of the system to transfer power.

*Multiple transmission circuits:* Two circuits carry twice the power of one. Additionally, one circuit can provide partial service while one is down for repair. Double circuit transmission lines are very common.

*Higher line voltage:* As we calculated earlier, increasing line voltage decreases pu impedance and increases line capacity, if all other factors remain constant. Additionally, resistive losses are lower.

*Reducing series line reactance:* This can be done with series capacitance, as discussed in Chapter 4, and by bundling conductors (which also reduces corona).

*Shunt reactors:* As we will calculate in Chapter 9, long transmission lines have a lot of shunt capacitance, both from line to line, and from line to ground. The capacitance can cause the line charging current to be excessive when the line is energized, and cause the line voltage to rise above acceptable limits, as discussed in Chapter 4. Shunt inductive reactance at the beginning and end of the line (and possibly at points in between) cancel some of the capacitive reactance. The inductors at the generator end provide lagging VARs that cause the field current to be higher for a given output voltage. This is because lagging load VARs cancel some of the rotor flux, thus a higher field current is needed to maintain the generator output voltage.

This causes the rotor to be harder to pull out of synchronization, which improves stability. The inductors at the receiving end help hold the line voltage at acceptable levels by cancelling the capacitive reactance.

*Damping:* Damping in a generator is provided by the production of a counter torque on the rotor when it tries to change speed. The generator rotor damper windings provide direct counter torque to the rotor proportional to any instantaneous speed variation. The generator rotor inertia and the prime mover inertia tend to prevent deviations in speed. Anything that removes power from the generator also provides damping, as resistance does in a resonant circuit. The friction and wind losses of the generator also provide damping. High capacity lines that can move a lot of power away from a generator in the event of a disturbance also provide damping, as long as the power demand by the lines does not exceed the generator capacity.

*Voltage regulators:* Voltage regulators that adjust the terminal voltage very quickly to the set point value in the event of a load change increase the ability of the generator to supply power because the effect of the generator impedance is not seen by the line, as shown in Figure 8.13. The increased capacity provides a stabilizing effect on the system.

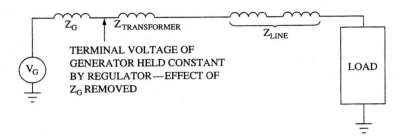

**FIGURE 8.13** Generator regulation to improve stability

*Computer power dispatching:* The network of a modern electric power utility can consist of several generators, several transmission lines, a vast distribution network, and many tie points. The complexity of the system means that a computer must be used to provide the coordination of all the parts of the system in a timely manner. The computer can solve the complicated power flow equations quickly, providing prompt action in the event a generator, or a part of the system, is approaching a potentially unstable condition. The computer's capability to quickly route power where it is needed improves the stability of the system.

### 8.2.4 Transient Stability Definition

Transient stability refers to the ability of a system to withstand sudden and severe changes in system conditions without loosing synchronization. These sudden and severe changes are usually the result of switching or fault conditions, but sudden large load changes can also introduce transient disturbances.

### 8.2.5 Transfer Impedance

A fault on a circuit that connects two system sections reduces the ability of the circuit to transfer power between the two sections. A three-phase fault with zero fault impedance reduces the power transfer capability to zero. A single phase fault reduces power transfer to a lesser degree. The higher the fault impedance, the less power transfer capability is reduced. All of the faults must be removed from service by opening a circuit breaker, which reduces the power transfer capability to zero. Figure 8.14a illustrates a line with a fault.

The transfer impedance is the impedance of the faulted section prior to opening a breaker. Figure 8.14b is the reactance diagram for the faulted line of Figure 8.14a. The transfer impedance is defined as

$$Z_{TR} = \frac{E_A}{I_B} \tag{8.13}$$

If we let $R = 0$ then

$$X_{TR} = \frac{E_A}{I_B}\bigg|V_B = 0 \tag{8.14}$$

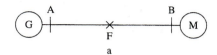

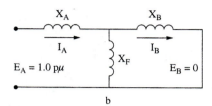

**FIGURE 8.14** Transfer impedance (a) One-line representation (b) reactance diagram

If we let $E_A = 1.0$ pu, the equation reduces to

$$X_{TR} = 1/I_B \tag{8.15}$$

with $V_B = 0$. Now let the components in Figure 8.14b be defined as:

$X_A$ = reactance to the fault

$X_F$ = fault reactance

$X_B$ = reactance from fault to receiving end of line

Then

$$I_A = \frac{1}{X_A + \dfrac{X_B X_F}{(X_B + X_F)}}$$

and

$$I_B = \frac{X_F}{X_B + X_F}(I_A)$$

$$= \frac{X_F}{X_B + X_F} \times \frac{1}{X_A + \dfrac{X_B X_F}{X_B + X_F}}$$

$$= \frac{X_F}{X_A(X_B + X_F) + X_B X_F}$$

rearranging

$$I_B = \frac{X_F}{(X_A + X_B)X_F + X_A X_B}$$

$$X_{TR} = \frac{1}{I_B} = X_A + X_B + \frac{X_A X_B}{X_F} \tag{8.16}$$

From equation 8.16 we see that:

1. If $X_F = \infty$ there is no fault and $X_{TR}$ equals the through impedance, $X_A + X_B$.
2. If $X_F = 0$, $X_{TR} = \infty$, and no power is transferred to the end of the line.

In most faults the fault impedance is low enough that the power transfer of the line is seriously impaired.

Most faults involve arcing, which is resistive. The system impedance is largely reactive so the arc resistance is usually neglected in all studies not done on a computer. The fault impedance, and therefore the transfer impedance vary with the type of fault. It is least in a three-phase fault, but this is the rarest type of fault. Fault impedances are calculated in Chapter 9.

In practice faulted lines are disconnected from the system as quickly as the relays can detect them and the circuit breakers can open. A common practice, as noted earlier, is to use a double circuit line to connect two parts of a system. Before a fault the total power transfer capability is described in Equation 8.10, where the impedance (or reactance, if resistance is assumed zero) is the parallel impedance of the two lines. If a fault occurs on one line the transfer impedance is larger than the line through impedance, and the power transfer capability of the line is reduced. After the protective relay detects the fault condition and isolates the faulted line by opening the circuit breakers connecting the line to the system, the transfer impedance is now the through impedance of one circuit of the double circuit line. This impedance is higher than that of the two circuits in parallel so the power transfer capability of the line is reduced, usually to about half of the double circuit capability. The more tie lines that two system sections have between each other, the less that the loss of one line reduces the power transfer capability.

## 8.2.6 Inertia

The rotors of large generators and turbines have considerable inertia because they are massive. (Recall from physics that inertia, $In$, of a body is the sum of each mass element, $m$, over its radius, $R$, squared; $In = \Sigma\, mR^2$.) The inertia tends to keep a body in motion once the motion has been established. In other words inertia opposes changes in motion of a mechanical body in much the same way inductance opposes changes in current in an electrical circuit. Before a disturbance a motion of a generator rotor is a constant speed rotation that is the result of a balance of all the torques applied to the rotor. These include the prime mover torque, the counter torque from the electrical output, and the damping torques such as friction and windage.

In the event of a transient disturbance in which the mechanical torque, which is proportional to power, the electrical counter torque, which is proportional to electrical power out, and the damping torques, which are proportional to the loss factors such as windage and friction, are not in balance the power angle changes because of the acceleration or deceleration from the imbalance. *Note:* the damping effect of the damping windings occurs only during a disturbance because they provide damping only in response to a change in the relative motion between the rotor and the rotating stator magnetic fields. If the transient disturbance results in the mechanical power in ($P_M$) exceeding the electrical power out ($P_E$) there is an accelerating torque (or power) on the rotor that causes the power angle ($\delta$) to advance. The opposite occurs if $P_E > P_M$.

The mechanical power into the rotor can be expressed as

$$P_M = \omega\, T_{IN}$$

where $\omega = 2\,\pi\, f$, and $T$ = torque. If $\omega$ is written as 1 pu, then

$$P_M = T_{IN}$$

The electrical power out is equal to

$$P_E = P_{E\,MAX} \sin\delta$$

for round rotor machines. Salient pole machines have a more complex equation because of the concentration of magnetic flux at the poles. $P_{E\,MAX}$ during the transient is expressed as

$$P_{E\,MAX} = \frac{e_1 e_2}{X'} \tag{8.17}$$

where:

$e_1$ and $e_2$ are the constant voltages supplying the transient reactances during the transient. Most faults are lagging, which causes the generator flux to weaken. The fast acting generator voltage regulators as well as the action of the damping windings force the gap flux to remain almost constant during the first 0.5 seconds or so of a transient disturbance. Thus the generator voltage does not change appreciably during this period of time. The inertia of any receiving machines holds their voltage almost constant during this time. Referring to equation 8.10, $E_1 = e_1$, and $E_2 = e_2$.

$X'$ is the sum of the transient reactance of the generator ($X_d'$), the connecting system reactance ($X_{TR}$), and the transient reactance of the receiving system machines ($X_d'$). The transient reactance in a machine results from damping windings, and in a generator the voltage regulator, trying to hold the gap flux constant.

$$X' = X_{d\,G'} + X_{Tr} + X_{d\,M'}$$

$\delta$ is the angle between $e_1$ and $e_2$

The accelerating torque on the generator rotor is then the difference in the mechanical input torque and the electrical output torque, all of which are proportional to their associated power. Expressed as power in per unit terms

$$P_A = P_M - P_E = \frac{H}{180\,f} \cdot \frac{d^2\delta}{dt^2} \tag{8.18}$$

where

$f$ = frequency

$\delta$ = power angle in electrical degrees

$H$ = per unit inertia constant in kW-seconds/kVA

The inertia constant can be found from the following equation

$$H = \frac{(0.231)(WR^2)\,(\text{rpm})^2 \times 10^{-6}}{\text{base kVA}}$$

The resulting rotor acceleration ($a$) from an accelerating power

input $P_A$ is

$$a = \frac{180\,f\,P_A}{(H)\,(\text{kVA})} \tag{8.19}$$

where

$H \times$ kVA is the rotor stored energy.

In summary, a transient condition results in an imbalance between the mechanical and electrical torque acting on the rotor because the power into and out of the rotor are different. The torque imbalance causes the rotor to accelerate or decelerate. The inertia resists the change in rotational speed from the imbalanced torque, but a change in the power angle still occurs.

## 8.2.7 Equal Area Criterion

Figure 8.15a illustrates two electrical power system segments tied together by a double circuit tie line. S is the sending circuit segment and R is the receiving circuit segment (S could be a single generator and R a single synchronous motor). S is at power angle $\delta_S$ and R is at $\delta_R$. One line has faulted at F. Transferred power during the fault is given by Equation 8.18

$$P = \frac{e_S e_{sR}}{X'}\,\sin(\delta_S + \delta_R)$$

The power transferred before and after the fault is given by equation 8.10.

The power transfer equation is plotted for three conditions in Figure 8.15b. Curve a is the pre-fault condition. The impedance is the through impedance of the lines in parallel plus the normal terminal impedance of the two system segments. The peak power is not the maximum power that can be transferred between the two systems, but the maximum synchronizing power (proportional to torque) capability of the two system areas. The lines will often lack the capacity to transfer this peak power. The prime mover power is represented as 1 pu on the diagram. The power angle of the system is at the intersection of the power curve and the prime mover power line. Before the fault this is at the prime mover power line and

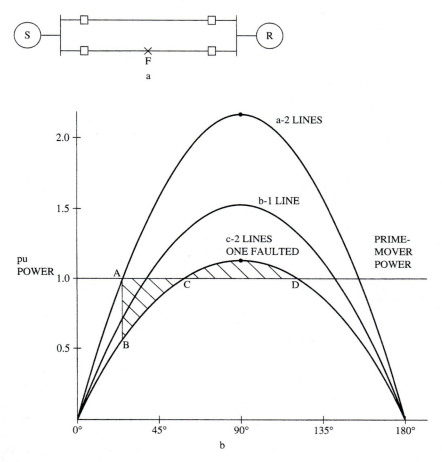

**FIGURE 8.15** Equal area criterion (a) Double circuit connection between system areas (b) Transient power (or torque) angle curves for double circuit line

curve a. Curve b is the power curve with two lines, one of which is faulted. The fault causes the transfer impedance between the system segments to rise (recall this is not the fault impedance to ground), reducing the power that can be transferred between them. Curve c is the P-δ curve after the faulted line has been isolated by opening circuit breakers at either end. The power transfer capability has been improved in comparison to the faulted condition, but it is not as high as the original double circuit line.

We will now trace the conditions that occur during the fault relative to the power curves. At the instant of the fault the operating point of the system moves from point A in Figure 8.15b, the intersection of the unfaulted power curve and the prime mover power line, to point B, the same power angle on the faulted power curve. The prime mover power exceeds the transferred power so the rotor is accelerated along curve c, the operating curve, toward point C, at which the power

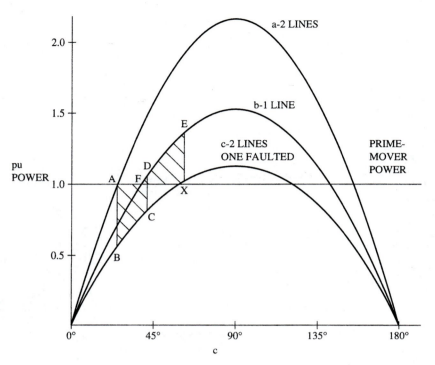

**FIGURE 8.15** Equal area criterion (c) Faulted line circuit breaker tripped

angle and the new power curve intersect. The inertia of the rotor causes the rotor to overshoot the power angle represented by point C. After the rotor passes point C the rotor is decelerated. If the rotor reaches point D the synchronism of the system is lost. If the shaded area ABC of Figure 8.15b, which is equivalent to the accelerating energy acting on the generator rotor, is less than area CD, the decelerating energy, the system tends to stay in synchronism because the decelerating force causes the operating point to move back toward point C. This is called the equal area criterion. If the area ABC is greater than CD the accelerating torque drives the system out of synchronism as the operating point moves past D. If area ABC < CD the operating point (the power angle) oscillates around point C along the power curve with decreasing amplitude excursions until the damping absorbs the oscillatory energy. The relays should open the circuit breakers on the faulted line long before the oscillations are damped.

Removing a faulted line from the system quickly improves the transient stability of the line. To see why, refer to Figure 8.15c. As in the previous explanation, the system is operating at the power angle shown at point A, the intersection of the unfaulted power curve, a, and the prime mover line when a fault occurs. The fault moves the operating point to B and the operating point moves along curve c as the power angle advances. The power angle is advancing because $P_M > P_E$. At point C the faulted line is isolated from the system. The operating point moves to point D on the single line power curve, b. Inertia causes the power angle to advance, forcing the operating point along curve b toward point E. The force on the rotor is now decelerating because

$P_E > P_M$. At point E the area ABCD = DEX so the power angle stops advancing and moves back to point F. The operating point now oscillates around point F with excursions that decrease in amplitude with time as the rotor (or system) damping absorbs the energy of the oscillations. The reason the fast isolation of the faulted line improves stability is because the power curve of the single unfaulted line has a greater range of power angle over which $P_E > P_M$, the power curve is above the prime mover line, thus providing more decelerating torque than the power curve with a faulted line.

During a typical day a power system experiences many transient disturbances, caused by sudden load changes or switching (faults cause only a few of the transient disturbances), that cause oscillations on a portion of the system. These oscillations, which may have periods of up to several seconds, are damped by the system generator damping, line capacity, losses, and inertia. Their amplitude is normally low enough that they are not noticed by the utility customers.

Transient and steady state stability calculation for a system is a massive job for a system of any size. Thus, digital computers are used to do the calculations. The system impedances, machine characteristics, and network configuration must all be known. A possible disturbance must be chosen and the resulting system response calculated for small increments of time after the disturbance, from the minimum time to the maximum time of interest. The step by step method of calculation must be used because the machine power equations are non-linear.

### 8.2.8 Swing Curves

One calculation of great interest is the advance of rotor angle versus time for each generator involved in a fault. The curve, called the swing curve, then indicates the stability, or lack of it, of a generator for a given length of disturbance. Figure 8.16 shows a set of swing curves. The power angle-time curves vary depending on when the fault is removed, so the swing curves must be plotted for each fault clearing time of interest. The generators of Figure 8.16 appear to be stable for a fault cleared in 0.23 seconds because the curves have begun to swing back toward the original power angles. The first swing may not indicate stability because the dynamic stability of a machine may be poor, allowing the oscillations to build up after the disturbance until synchronism is lost. A plot of several seconds may be necessary to spot the tendency toward post-fault isolation synchronism loss. Figure 8.17 shows the swing curves for one generator for several fault clearing times. Note that the quicker the fault is cleared the better the post-fault stability.

## 8.3

## TRANSMISSION VOLTAGE LEVELS

The transmission system is made up of the high voltage lines and bulk power substations that connect the generating stations with the distribution substations. The transmission system can be further broken down into the transmission and subtransmission systems. The high voltage transmission lines must often be terminated in a bulk power substation some distance from the loads because the loads are dispersed and the very high voltage stations are not allowed in populated areas. Lines with voltage levels and capacities between those used for transmission and those used for distribution, called subtransmission lines, are then used to carry power from the bulk power substations to the distribution substations and large individual customers. The right of way width is smaller for subtransmission lines than for transmission lines.

Distribution voltages are normally between 4 kV and 34.5 kV. Subtransmission voltages are normally between 69 kV and 138 kV, with 69 kV, 115 kV, and 138 kV being common. Transmission voltages are usually in the EHV range, greater than 230 kV, and UHV range, over 800 kV. Common ac transmission voltages are 230 kV, 345 kV, 500 kV, 550 kV, and 765 kV. 1100 kV and 1500 kV lines are planned. Voltages for dc transmission range from about 100 kV to 1200 kV with ± 100 kV (200 kV), ± 250 kV, ± 400 kV, ± 500 kV, and ± 550 kV being common. Higher dc transmission voltages are being planned.

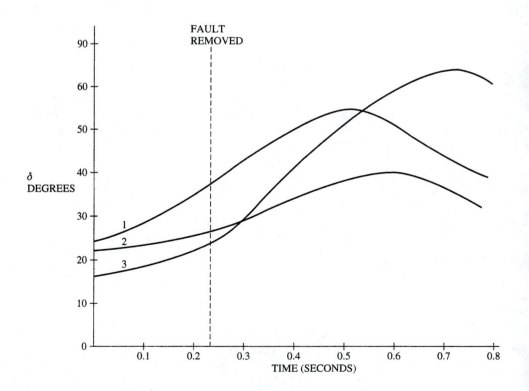

**FIGURE 8.16**  Swing curves

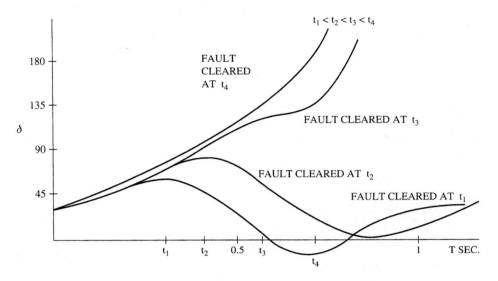

**FIGURE 8.17** Swing curve vs time to clear fault

8.4

## TRANSMISSION LINE CONSIDERATIONS

Electrical power transmission lines can be ac, dc, underground, or overhead. Overhead ac is the most used method of electrical power transmission. Transmission lines represent from 10 to 20% of the investment of a typical electric utility company. This significant amount must be carefully spent so as to assure reliable, efficient, and economical electric power transmission.

### 8.4.1 Overhead Line Considerations

Overhead transmission line construction is much less expensive than underground because it is simpler. Bare wires can be used in overhead transmission line construction with insulation used only at the points that the wire is suspended, as shown in Figure 8.18. The use of bare conductors alone cuts both the cost and losses of an overhead line compared to an underground line. Wood or galvanized steel towers are used to support the conductors. The insulators at the conductor support points are usually ball and socket porcelain or fiberglass rods covered with skirts made of a compound similar to silicon rubber.

Overhead transmission lines must be very reliable because an outage on one leaves many homes and businesses without electricity. Thus transmission lines are well protected against lightning with lightning arresters placed periodically along the line, and shield wires along the entire line. The conductors and conductor

a

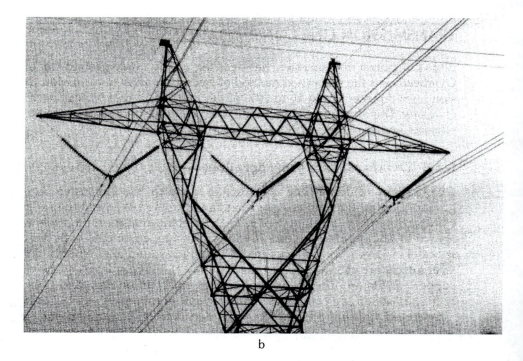

b

**FIGURE 8.18**    Overhead transmission line (a) 345 kV (Courtesy of Houston Lighting and Power) (b) 500 kV (Courtesy of Gulf States Utilities)

c

**FIGURE 8.18** Overhead transmission line (c) double circuit 138 kV H structure with one circuit installed on the left, with a distribution underbuild on the right (Courtesy of Gulf States Utilities)

supports must be strong enough to withstand severe weather, which can subject the lines to considerable wind, snow, and ice loading. Bundled conductors, which are used almost universally at 230 kV and above, to reduce series reactance and corona, experience more weather related loading than single conductors. Thus bundled conductors must be supported better, at a higher cost, than single conductor lines.

There is a trade off between the losses on a transmission line and the cost of constructing the line. Lowering $I^2R$ losses usually means using larger conductors or more conductors to increase the conductor area. Larger conductor area means increased conductor weight, which means the strength of the supporting towers and insulators must be greater. The increased strength required means that the manufacturing cost, transportation cost, and construction cost of every line component is higher. Transmission line costs are also effected by the terrain that the line must pass through. The rougher the terrain the higher the cost. Terrain, such as swamp land, that is not suitable for normal foundations or temporary access roads also increases the cost of construction because helicopter pole placement and special support guys become necessary.

The magnitude of the power that a transmission line carries makes the reduction of loss a major consideration. For example, a 2% loss on a 1000 kVA line, at 6 cents/kW-hr, costs only $1.20/hr. A 50% reduction in loss amounts to a savings of only 60 cents an hour. The same loss on a 1000 MVA line at 6 cents/kW-hr costs $1,200/hr. A 50% reduction in this loss is a savings of $600/hr, or $600/hr × 24hr/day × 365 days/yr = 5.26 million dollars a year. Over the expected 40-year lifetime of a transmission line this is a savings of 210.4 million dollars. Thus the reduction in line loss must be carefully weighed against increased construction costs to plan a transmission line that is optimally economical.

Obtaining right of way is not a large problem for distribution system construction. The areas that do not grant right of ways do not get electric service. Very few distribution right of way requests are denied. Transmission lines do not effect people so immediately, are considered unsightly by some people, require larger right of ways than distribution lines, and are considered a health hazard by a few people, so people are not as willing to grant transmission right of ways. The courts tend to agree and support the reluctance of people to grant right of ways for transmission lines. As a result the right of way cost is a very large portion of the cost of constructing a transmission line, especially when litigation is necessary.

In rural areas the utility pays a landowner a fee for each transmission structure on that owner's property, for which the owner allows the structure, or its replacement, to be on the property as long as it is needed. The landowner must also allow access to each structure on the right of way via an easement. The utility maintains the right of way by clearing bush, weeding, and mowing grass. The landowner can graze or farm the right of way, but cannot build on it. The fee is paid one time, as in a purchase, or annually. In and around cities the usual practice is to purchase right of ways. The purchased right of ways are called fee strips.

## 8.4.2 Underground Line Considerations

Less than 1% of the total transmission line miles in the United States are underground. The most extensive underground system in the United States is in the New York City area. Underground transmission lines are most commonly used to feed urban substations in high load density areas, such as downtown. Industrial plants use underground transmission and/or distribution lines in areas where overhead lines are not practical because of the clearance required. The highest underground transmission line voltage is 525 kV. This installation used nine oil type cables (three circuits), running about 6500 feet to connect the generation plant to the switch yard at the Grand Coulee Dam. Since the cost of an underground transmission line is 9 to 15 times the cost of an overhead line, they are installed only where they offer a clear advantage, or there is no alternative.

Underground lines are more expensive than overhead for a variety of reasons. Underground cables must be insulated, and EHV insulation is expensive. It must be installed in pipe, which is expensive, and cooled with a oil circulation system, which is expensive both to install and maintain. Should a failure occur in an underground system, it is expensive to repair because of the difficulty in access.

The pipes used for underground transmission cable conduit are usually steel or bronze, 6 to 8 inches in diameter. Bronze, while more expensive, is not magnetic so it has less effect on cable reactance. The pipes are insulated on the outside to prevent cathodic currents (essentially from battery action in the soil involving the pipe as one electrode), and buried from 4 to 8 feet deep, enveloped within reinforced concrete. The pipe comes in 30 and 40 foot lengths. It must be welded securely enough to hold a vacuum and the joint must be insulated on the outside. After the run of pipe is in place a cup driven by compressed air, called a rabbit, is used to pull a light line through it. Then a mandril, called a pig, is pulled through the pipe to remove any welding burrs. The pipe is then put under a vacuum to test for leaks. The cables are then installed. The pipe is then put under a vacuum (to exclude moisture and air) with the cables installed and filled with oil. Figure 8.19 shows a cross section of a pipe with cables installed.

The cables are constructed much like the oil impregnated paper insulated underground distribution cables, except that the transmission cables have more oil impregnated paper insulation in more layers. The cable has a spiral (or double spiral) metal wire wound around the outside of the insulation to prevent damage to the cable when it is pulled through the pipe, or from movement caused by expansion or contraction caused by heating or cooling as load current changes. The entire cable is covered with a thin lead sheath that keeps the oil impregnation in the insulation and prevents handling damage during shipping and installation. During cable pulling the lead sheath is removed just as the cable enters the pipe. The cable must enter the pipe, and cable splices must be made, in a clean, humidity controlled environment so an air conditioned building is constructed where the cable is fed into the pipe.

After the cable is installed the pipe is filled with oil. The oil is pumped through the pipe to cool as well as insulate the cable. Pumping cool oil through the pipe can greatly increase the capacity of the cable. The oil circulation system includes pumps, filters, and oil tanks. Sulphur hexaflouride ($SF_6$) is sometimes used to insulate underground transmission cables, but is not as popular as oil because it does not provide an increase in capacity as does cooled oil. The popularity of $SF_6$ is increasing because of its relative simplicity.

We see that all of the complexity associated with underground transmission lines cause them to be extremely expensive in comparison to overhead transmission lines. However, properly installed underground transmission lines are very reliable.

## 8.4.3 DC or AC Considerations

DC transmission systems are costly because of the converters at the line taps. DC lines have lower losses than ac lines because the skin effect is negligible, allowing the entire conductor cross-section to carry current, and the line reactance is zero for steady state loads (line reactance almost disappears after the line is charged). The fast acting controllers used in the dc to ac, and ac to dc, converters react very quickly to load change, so a dc line can improve the stability of a power system. DC lines also require less right of way than an ac line of the same voltage. At present dc lines are used for long transmission lines to lower line loss, and for short tie lines between systems that are not in synchronism.

**FIGURE 8.19**   Fayette Power Project, Units I and II (courtesy of Lower Colorado River Authority).

## 8.5

## HVDC TRANSMISSION

High voltage dc (HVDC) transmission, as noted in Chapter 1, originated in Germany in WWII. The first commercial line was installed in Sweden in 1954. The use of HVDC transmission lines has grown steadily ever since. The growth was accelerated when high voltage, high current (to 5 kV and 2000 A) silicon controlled rectifiers (SCR) were developed to replace the bulkier, less efficient mercury rectifiers that had been used until that time. The invention of effective dc air blast and $SF_6$ circuit breakers and lightning arresters also did much to encourage the use of HVDC transmission lines. Today there are over 40 HVDC lines throughout the world serving as high capacity long haul lines or assynchronous ties between systems.

Walter Coffer, one of the authors of this book, was in charge of constructing the Inga-Shaba extra high voltage direct current (EHVDC) converter stations and intertie line between Inga and Kolwezi in Zaire, Africa. The 1700 kM ± 500 kV line serves to transport up to 1120 MW from the hydroelectric generators in the Inga area to the mining district of Shaba. The project was completed in 1976.

### 8.5.1  Thyristor Valves and Bridges

An SCR is a four-layer device of the thyristor family. It blocks voltage both ways until it receives a gate signal of sufficient magnitude to turn it on while it is forward biased (anode positive with respect to the cathode). A reverse biased SCR cannot be turned on. Once an SCR is on it then remains on until its anode current is held at a very low value for sufficient time for the device to turn off. In phase control, used by dc transmission converters, the negative line voltage (with respect to the cathode) is used to turn the SCR off. The time from the beginning of the positive, or forward biasing, cycle until the SCR is fired, expressed in degrees, is the retard angle, $\alpha$. These items are illustrated in Figure 8.20.

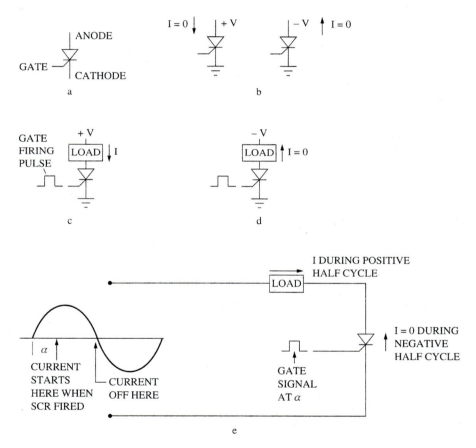

**FIGURE 8.20** Silicon controlled rectifier (SCR) (a) Symbol (b) SCRs block current both ways when off (c) Current flows when forward biased SCR fired (d) Reverse biased SCR cannot fire (e) Negative line voltage turns SCR off

HVDC converters use three-phase full controlled bridges to convert ac to dc and vice versa. Conceptually they work like the single-phase full controlled bridge shown in Figure 8.21a. Essentially the bridge is a full wave rectifier that does not rectify until the SCRs are fired. They are fired in pairs, SCRs 1 and 3 control the line half cycle that is positive with respect to the reference line, and SCRs 2 and 4

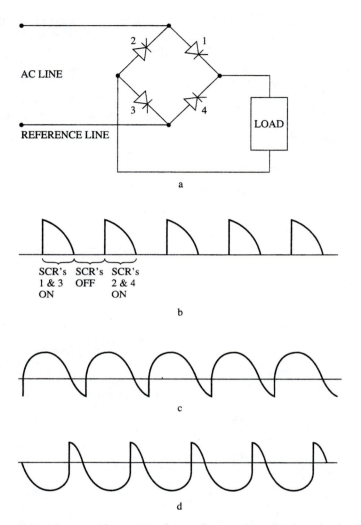

**FIGURE 8.21**   Phase control (a) Full control single-phase bridge (b) Load current (and voltage) waveform, resistive load, $\alpha = 90°$ (c) Load voltage waveform, highly inductive load, $\alpha = 45°$, dc current flow constant. Net load voltage is positive, power flow into load. (d) Load voltage waveform, highly inductive load, $\alpha = 135°$, dc current flow constant. Net load voltage negative, power flow to ac line from load.

control the ac half cycle that is negative with respect to the reference line. The output waveform for $\alpha = 90°$ and a resistive load is shown in Figure 8.21b. Notice that the voltage waveform is discontinuous because the load current stops flowing twice a cycle as the SCRs turn off. Two current pulses are delivered to the load each ac input cycle, so the circuit is called a two-pulse converter.

When a full controlled converter (all SCRs, no diodes) has a highly inductive load, the negative half cycle of the input ac does not turn the SCRs off because the inductive load discharge from collapsing magnetic flux keeps the cathode more negative than the anode. This keeps the SCRs forward biased. Only firing the positively biased SCRs turns the SCRs with negative anodes off, by diverting their current to the supply. During all of this time the dc load current never changes direction. In other words, if SCRs 1 and 3 are on during the positive ac input half cycle they do not turn off as the input ac becomes negative with a highly inductive load. To turn SCRs 1 and 3 off we must turn SCRs 2 and 4 on so that the inductor current is diverted to the positive source, away from SCRs 1 and 3. Then SCRs 1 and 3 can turn off. The load voltage for this situation is shown in Figure 8.21c for an $\alpha = 45°$. Note that the net load voltage is positive. This is called the rectifying mode of operation. If the retard angle is advanced to 135° (the SCRs are fired later in the cycle) the net load voltage is negative because the inductive load spends more time discharging from contracting flux than charging (flux expanding) from being connected to the positive line. This is shown in Figure 8.21d. The negative voltage with no change in current direction results in negative power, which signifies power transfer from the dc to the ac side of the bridge. This is called the inverting mode of operation. Of course an inductive load soon discharges its energy, but if a dc supply is connected to the bridge in place of the load, with the negative dc connected at the cathode junction of SCRs 1 and 4, energy can be transferred to the ac line as long as the SCR retard angle is between 90° and 180°. Where can we find such a dc source? How about the dc side of a similar bridge operated so that the retard angle is more than 0° and less than 90°. Note that the same bridge can be used to turn ac to dc and dc to ac by simply varying the retard angle of the SCRs and the polarity of the dc.

A single SCR can block only a small portion of the voltage used for dc transmission. To block such high voltages a large number of SCRs are connected in series forming modules called valves. An SCR of the style used in dc converters and two heat sinked SCRs are shown in Figure 8.22. The series SCRs must have resistor-capacitor voltage dividers across them, as shown in Figure 8-23a, to assure that they each block an equal share of the applied voltage. The resistors assure that the voltage across each SCR is within acceptable limits, and the capacitors assure that transient voltages and voltage changes are shared evenly. If the converter current is higher than a single series SCR stack can handle (about 2000 A) then the SCRs must be paralleled. A current sharing reactor forces the current to divide equally between the SCRs, as shown in Figure 8.23b. If the current in the right hand SCR tries to increase, the voltage drop across the right hand winding of the reactor gets larger with the dot becoming more negative with respect to the common point. Transformer coupling causes the voltage on the anode of the left hand SCR to become more positive, causing the current in that SCR to increase. This process keeps the current shared evenly by the SCRs.

The bridges used in dc transmission system converters are six pulse, as shown in Figure 8.24a. These have three-phase input ac and deliver six pulses of dc current to the load every input cycle. The three-phase rectification delivers smoother dc than single phase. The firing sequence of the bridge SCRs is shown in Figure 8.24b. Each SCR is fired only during the portion of the ac cycle in which it is

**FIGURE 8.22** Converter valve: SCR (bottom) and two SCRs in a water cooled heat sink (top).

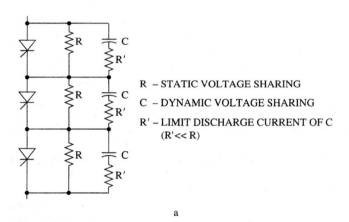

R – STATIC VOLTAGE SHARING

C – DYNAMIC VOLTAGE SHARING

R' – LIMIT DISCHARGE CURRENT OF C
  (R' ≪ R)

a

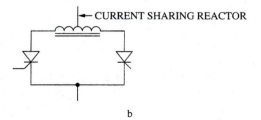

← CURRENT SHARING REACTOR

b

**FIGURE 8.23** SCR series and parallel operation (a) Series, forced voltage sharing (b) Parallel, forced current sharing

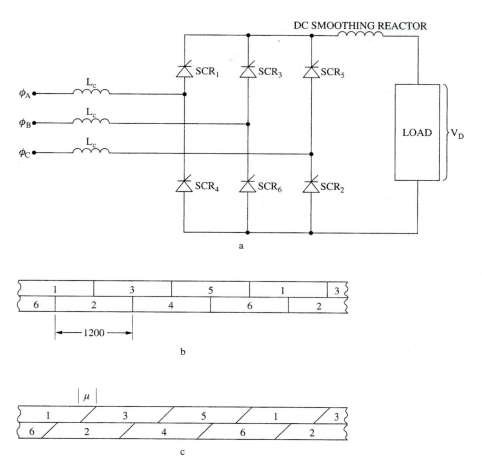

**FIGURE 8.24** Six-pulse full control SCR bridge (a) Circuit (b) Firing order and (c) Overlap angle, μ

forward biased. The overlap of the three phase voltages means that any one SCR is forward biased between 30° and 150° of a half cycle. If the bridge retard angle is between 30° and 90° the bridge is rectifying and power is delivered from the ac to the dc side of the bridge. If the retard angle is between 90° and 150° and the voltage is reversed, the bridge is inverting, and power is delivered from the dc to the ac side of the bridge.

When SCRs are turned on, the current does not switch from the one SCR set to another instantaneously. That means that for a short time both the SCRs turning on and the SCRs turning off are conducting current while switching. This time is called the overlap angle, μ, and it is illustrated in Figure 8.24c. Inductors are used to prevent excessive current change during the overlap angle. These inductors are called commutating (when used with SCRs this term means turn off) inductors. They are inductors $L_C$ in Figure 8.24a.

Better, smoother dc with fewer harmonics can be obtained from a 12-pulse converter. The switching of the SCRs in a six-pulse bridge generates a lot of fifth, seventh, eleventh, and thirteenth harmonics. The fifth and seventh harmonics cancel in the 12-pulse converter. A 12-pulse converter is constructed by connecting a six-pulse converter fed by a delta transformer secondary in series with a six-pulse converter fed by a wye transformer secondary, as shown in Figure 8.25. The wye

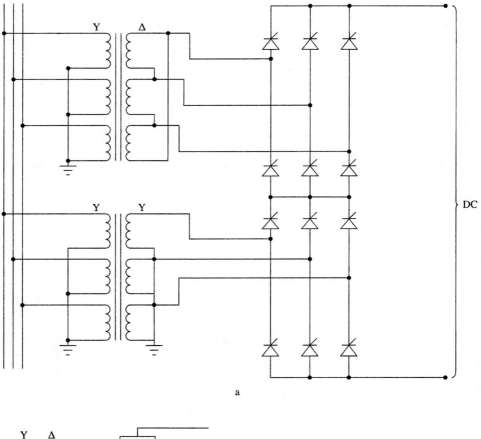

a

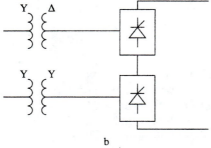

b

**FIGURE 8.25**   Twelve-pulse converter (a) Circuit and (b) One-line symbol

and delta secondaries are 30° out of phase causing 12 pulses of current to be delivered to the load each cycle. The transformers are usually wound so that each six pulse bridge has the same input line voltage. The converter output voltage is varied by adjusting the retard angle on the SCR firing control for small changes, and by transformer tap changers for large changes.

The dc converters require a lot of reactive power. Reactive power of about 40 to 60% of the power transferred is needed to provide energy for the harmonics from which the nonsinusoidal converter waveforms are made. Since the ac to dc converters can only send real power (dc contains only real power) the VAR support must come from the input. Shunt capacitor banks and the harmonic filters provide much of the VAR support, because it is usually not desirable to obtain all of the necessary VARs from the generators. In some stations the VARs are generated with synchronous condensers (synchronous motors run over excited with no mechanical load). The dc to ac converters cannot receive VARs from the dc line, so they get them from the ac system they supply. Again, capacitors along with the harmonic filters supply much of the VAR support. Synchronous condensers are often used to provide VAR support in systems for which the dc line is the sole source of electrical power. A receiving system that does not have the short circuit capacity to absorb three to five times the power delivered by the dc converter is unstable. VAR support must be added until this condition is met. In other words, the receiving system must be able to provide an ac pilot signal for the bridges in the converter to start. An initial oscillatory response or a synchronous condenser can provide this. Figure 8.26 shows a block diagram of a sending and receiving station showing the placement of the VAR support and the dc polarity relationship of the two stations.

Thyristors dissipate about 30 to 40 W/cm$^2$ of thyristor area. The heat must be removed or the valves will fail. The valves have finned metal heat sinks over which clean, dry, cool air is blown by fans. The warm air is then forced through a heat exchanger to cool it. SCRs can be water cooled but air is simpler for high voltage applications because of the insulation requirements.

All of the series and paralleled SCRs in a valve must be fired in unison to avoid over voltage and/or overcurrent conditions in some of the SCRs during turn on. Because of the high voltage involved the pulse shaping circuit for the SCR gate drive is powered from the voltage the SCR is blocking. The firing command signal is carried by light pipes to a light activated semiconductor on each firing circuit board. Older systems use Xenon flash tubes for the light firing signals, but more recent systems use cesium-mercury flash lamps, which send UV light to small light activated SCRs that then fire the larger SCRs.

Light guides in the form of fiber optic cables are used to carry valve status information back to the central monitoring and control circuitry. The fiber optic cables provide the isolation from high voltage for the control system.

The earth return shown in Figure 8.26 allows the system to be operated at reduced capacity when one line is down or a converter is malfunctioning.

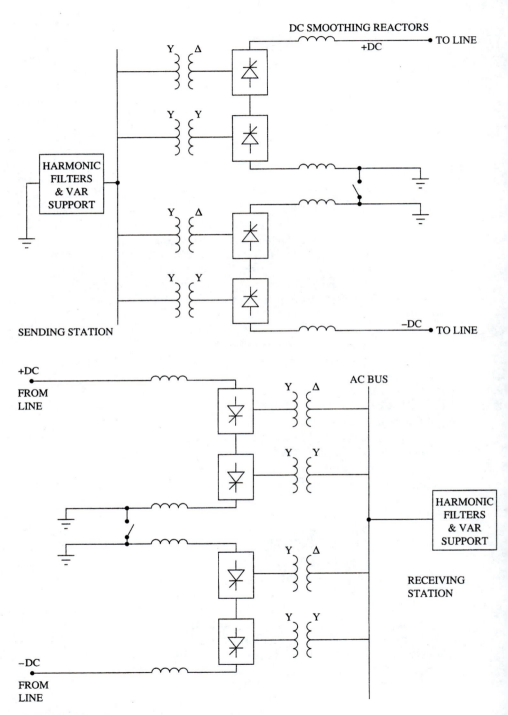

**FIGURE 8.26**   Converter stations

## 8.5.2 Converter Operation

Figure 8.26 shows the ac to dc converter, the dc line, and the dc to ac converter. The two converters look alike, only the operating modes and polarity of dc connection differ. If the receiving station has excess ac power from generators and the sending station needs power, the stations can reverse their function, under proper coordination and control, and the power flow on the dc line is reversed. This is a very useful capability for lines used as system ties. Long haul dc lines normally transmit power only one way.

Recall, the time expressed in degrees from the beginning of a half cycle that an SCR is on until it is fired is called the retard or firing angle ($\alpha$). In a three-phase system the half cycle may be considered to begin 30° after the voltage zero crossing because the SCR is not forward biased until then. The time expressed in degrees from the SCR firing until the end of the cycle is called the conduction angle. If the converter is operating as an inverter, the conduction angle is called the extinction angle ($\gamma$). In normal operation the inverter retard angle is about 130°, and the rectifier retard angle is about 10° to 20°. The power transferred is

$$P_{DC} = \frac{(V_{D1} - V_{D2})}{R} V_{D2} \tag{8.20}$$

where, referring to Figure 8.24a

$$V_{D1} = \frac{3\sqrt{2}}{\pi} V_1 \cos\alpha - \frac{3\omega}{\pi} (L_{C1} I_D - I_D R_1 - 2V_T) N \tag{8.21}$$

$$V_{D2} = \frac{3\sqrt{2}}{\pi} V_2 \cos\gamma - \frac{3\omega}{\pi} (L_{C2} I_D - I_D R_2 - 2V_T) N \tag{8.22}$$

where

$V_1$ = converter 1 rms ac line voltage

$V_2$ = converter 2 rms ac line voltage

$\alpha$ = retard angle

$\gamma$ = extinction angle

$I_D$ = dc line current

$V_D$ = dc converter voltage

$R$ = dc resistance of converter circuit

$L_C$ = commutation inductor reactance

$N$ = number of six pulse bridges

The sending voltage can be changed using the retard angle or tap changing, as can the receiving station output voltage. The retard angle is used for small, fast adjustments, and the taps are used for larger voltage changes.

The normal operating modes are illustrated in Figure 8.27. The constant current mode shown in Figure 8.27a holds the current to that demanded by the receiving station by adjusting the sending converter (rectifier) voltage. Frequency relays capable of sensing minute fractions of a Hz change are often used to indicate the need for more receiving station power. As the receiving station ac frequency

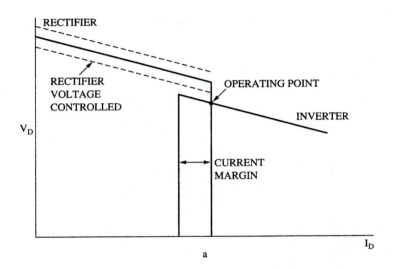

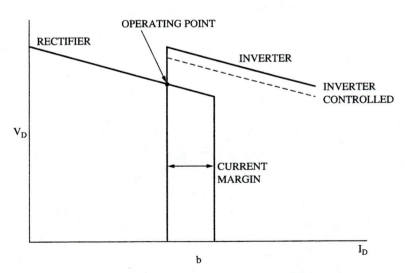

**FIGURE 8.27**   DC control (a) Constant current and (b) Constant power (current margin)

drops, more power is required. A communication system signals the sending station to send more power, and the sending converter raises the rectifier voltage in response. Notice the inverter and rectifier curves have some overlap to allow for rapid adjustment of power flow in response to changing loads.

Figure 8.27b illustrates the constant power mode, also called current margin control. In this mode the rectifier is held constant except for very large load condition changes, and the inverter changes in voltage and accepted current from the available margin to provide the power its ac load needs. Frequency relays are usually used to indicate power needs. An advantage of this mode of control is that it can continue with no communication to the sending station. For this reason it is used as a back-up mode for constant current control in the event communication between the sending and receiving stations is lost.

### 8.5.3 DC Converter Stations

The area requirements for a dc converter station are approximately 15 to 20 $m^2$/MW. A major portion of the station area requirements is for the harmonic filters and reactive power support. The valve hall usually requires only 1 to 2 $m^2$/MW. Where space is a major factor the use of $SF_6$ for insulation wherever possible has made it possible to build a station in as little as about 8 $m^2$/MW, at a higher cost.

Since the thyristor bridges generate harmonics the grounded metal clad buildings in which they are housed double as electromagnetic shielding. Proper spacing sets limits on the size of each part of a station. Insulating the dc side is more difficult than the ac side, and requires more space. Figure 8.28 shows a HVDC electrical layout, and Figure 8.29 shows a corresponding physical layout without synchronous condensers.

## 8.6

## SUPERCONDUCTORS

Superconductivity was discovered in 1911 by Dutch scientist Heike Onnes. He discovered that mercury lost its resistance when cooled to 4°K (‾269°C). Superconducting materials lose all resistance and exclude external magnetic fields when they become superconducting. The temperature at which they become superconducting is the transition (or critical) temperature $T_C$. At temperatures above $T_C$ the resistance of the material is not zero. In like manner the magnetic field strength at which a superconducting material loses its superconductivity is the critical field. A useful superconductor has a high critical temperature and field. In the early 1960s it was discovered that niobium-titanium and some other compounds were superconducting at temperatures as high as 18°K, and a niobium-germanium alloy as high as 23.2°K. In 1973 alloys of yttrium, barium, and copper oxide exhibited superconductivity at temperatures as high as 28°K. All these must be cooled

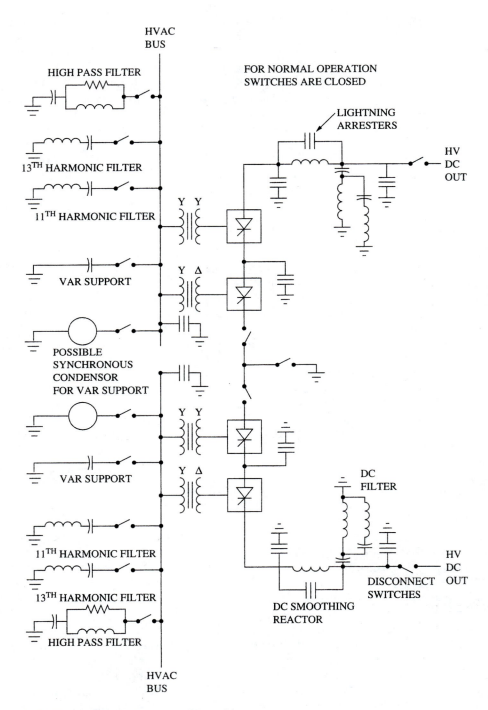

**FIGURE 8.28**    HVDC station electrical layout

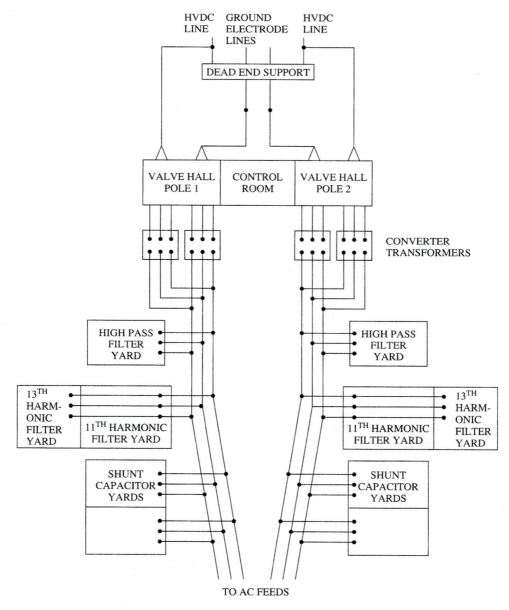

**FIGURE 8.29**  HVDC station physical layout

by liquid helium, which boils at 4.2°K. In 1986 workers at IBM Zurich discovered that certain ceramic compounds of lanthium, barium, and copper oxide would superconduct at 30°K under pressure. A race began to find compounds whose fit into the crystal lattice would replace (act like) pressure. In January of 1987 the superconductivity transition temperature of a compound of yttrium, barium, and copper oxide was found to be about 90°K. This is above the temperature of liquid nitrogen, which boils at 77°K. Higher transition temperatures have been reported, but have not been reliably reproduced yet. The new superconducting compounds are brittle, but have been drawn into fine wire. The mechanical properties and current densities of the new materials show promise for electronic devices. Much work remains before they can be used for superconducting generators, motors, and transmission lines.

Superconducting magnets for high energy particle research have been around since the 1960s. The magnets are used to guide and focus the particle beams. Thin superconducting films have been used to measure minute voltage, current, and magnetic field levels. They have also been used to make very fast digital logic circuit elements that show promise for computer applications. The higher $T_C$ of the new superconducting material should encourage research in all these areas.

Short experimental superconducting lines have been built and operated, but have not yet been shown economical. Superconducting generators have been built and operated. It has been estimated that generators with superconducting fields can be made 40% smaller, 1% more efficient, and up to 30% less expensive than conventional generators. The cost, size, and efficiency all come from the superconducting rotor. The magnetic field from the superconducting rotor windings is so strong that no magnetic core is needed, although a metallic shield is needed to help contain the coolant and keep ac electric and magnetic fields from the stator from reaching the rotor. Several superconducting generators in the 20 to 50 MVA range have been built by concerns in the United States and abroad. Generally they have worked well. The largest superconducting generator, a 300 MVA unit, to be built by Westinghouse and funded by the EPRI, was cancelled. Economic reasons were cited. Large generating units have recently become less attractive because of economic uncertainty.

At this time superconducting electrical power components are in the very early development stage. The rate at which the development of practical power components made from superconducting materials will develop is unpredictable at this time.

## 8.7
<hr>

### HIGH PHASE ORDER LINES (SIX-PHASE)*

High phase order lines are those designed to transmit more than three phases of electrical power, normally six, but nine- and twelve-phase transmission are possible. As shown in Figure 8.30 three-phase ac consists of three phasors offset by 120° and six-phase ac consists of six phasors offset by 60°. Similarly, nine-phase ac

*The high order phase information contained in this section was made available by New York State Electric & Gas Corporation. Credit and thanks is extended to all involved in the project, the Empire State Electric Energy Research Corp. (ESEERCO), the New York State Energy Research Development Authority (NYSERDA), the NY State Energy Office (NYSEO), the US Department of Energy (DOE), and the Electric Power Research Institute (EPRI).

consists of nine phasors offset by 40° and 12-phase ac consists of 12 phasors offset by 30°. Three-phase transmission was accepted as the standard for ac transmission for two reasons. The first is that three phases are the least number required for power flow that is constant with time, and the second is that electrical machine power does not increase as phases are increased beyond three. Electrical power transmission with more than three phases has some advantages, so a 93 kV demonstration line was built by New York State Electric & Gas Corporation between Goudey and Oakdale in New York to demonstrate the commercial feasibility of the concept.

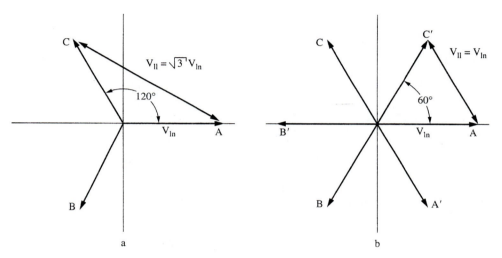

**FIGURE 8.30** Polyphase voltage phasors (a) Three-phase phasor and (b) Six-phase phasor

Notice from Figure 8.30 that the ratio between line and phase voltage for three phases is $\sqrt{3}$ but for six phases it is one. The phase and line voltages are the same for six-phase systems. Thus for a phase to ground voltage of 79.6 kV the line to line voltage is 138 kV for three-phase systems and 79.6 kV for six-phase systems. In like manner, for a phase to ground voltage of 132.8 kV the line voltage is 230 kV for three-phase and 132.8 kV for six-phase systems, and for 199.2 kV the line voltages are 345 for three-phase ac and 199.2 kV for six-phase ac. The maximum power, assuming unity power factor, that a six phase line can handle is

$$P = 6\,I_{line}V_{\text{phase to neutral}}$$

and the maximum power that a double circuit three-phase line of the same line to neutral voltage is

$$P = 2(3\,I_{line}V_{\text{phase-to-neutral}})$$
$$= 6\,I_{line}V_{\text{phase to neutral}}$$

which is the same as a six-phase line. The lower line to line voltage of the six-phase line allows the support structures to be smaller than for a double circuit three-phase line of the same phase to neutral voltage, as illustrated in Figure 8.31. Thus smaller six-phase right of ways can be used to transmit the same power as a double circuit three-phase line with the same phase to neutral voltage, assuming the same conductor current capacity.

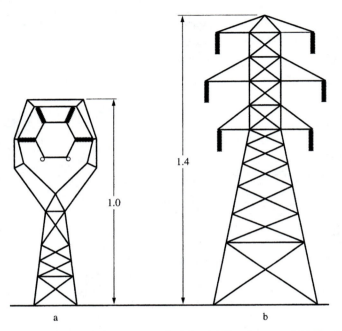

**FIGURE 8.31**   Relative tower heights (a) Six phase and (b) Double circuit three phase

The power capacity of a six-phase line with the same line voltage as a double circuit three-phase line is

$$P = 6\,I_{\text{line}}\sqrt{3}\;V_{\text{phase to neutral}}$$

Thus, the capacity of a six-phase line is 173% that of a double circuit three-phase system with the same line to line voltage, and similar space requirements.

An additional advantage of six-phase transmission is that the magnetic fields are three to four times lower than a double circuit three-phase line at similar distances from the center line of the right of way. The corona loss should be lower on six-phase lines because the electric field gradients on the conductors are lower.

Six-phase transformer connections are more complex than three-phase, which drives the installation cost up. If six-phase transmission becomes common transformers will be designed and built for six-phase substation applications. Two conventional three-phase transformers can be connected to obtain six-phases. One transformer is connected delta-wye and the other delta-inverted wye or one wye-wye and the other wye-inverted wye. The secondary connections are then interspaced to obtain six-phase ac as shown in Figure 8.32.

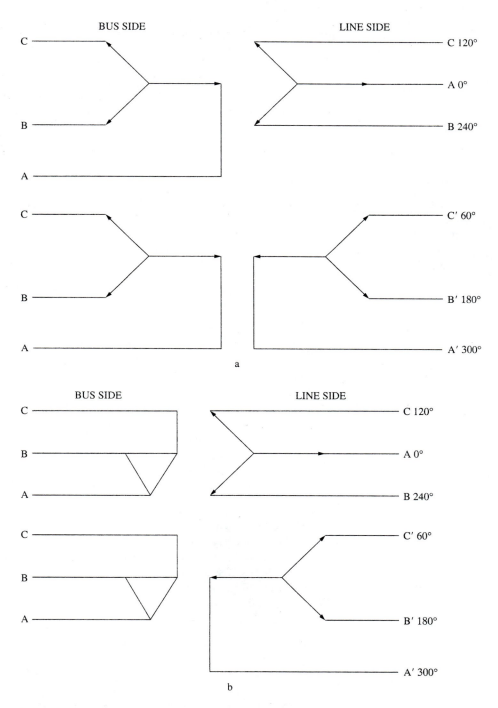

**FIGURE 8.32** Six-phase connection from two three-phase transformers (a) Wye-wye wye-inverted wye (b) Delta-wye delta-inverted wye

Protection of a six-phase line is more complex than for a double circuit three-phase line. There are more combinations of line to line and line to neutral faults that can occur and the tools for calculating fault current are not as well developed for six-phase as for three-phase lines.

Three types of protection are used for the New York State Electric & Gas Corporation Goudey-Oakdale line:

1. Phase distance (impedance) relays with directional ground current relays.
2. Current differential relaying using pilot signal over a fiber optic based communication channel.
3. Segregated phase comparison relays for back up. These are relays that compare the phase at the ends of each conductor, via a fiber optic communications system, to determine if an internal fault has occurred.

The relays are microprocessor based for speed and flexibility.

Three trip schemes are reasonably possible. All can, and normally would, be reclosed after a delay for temporary faults, and then locked out if a second trip occurs. *Tripping all six phases* on a fault is the easiest to implement but provides the lowest line availability. *Alternate phase trip,* in which only the breaker for the three-phase segment that is faulted is tripped, either phases 1-3-5 or 2-4-6. Both phase group breakers are tripped if a fault occurs on a conductor of each phase group. *Single pole tripping* is a scheme in which the appropriate breaker pole trips for a single phase to ground fault. A single reclose is done after a suitable time and the breaker is tripped and locked out if the line is still faulted. For line to line faults on adjacent lines the faulted poles open, and a reclose then lock out sequence is initiated. If faults occur on lines 120° or 180° apart single pole tripping is not attempted. Instead the breaker on the faulted phase group is opened, or both breakers if both phase groups are involved in the fault. Single pole tripping provides the best line availability, but is the most complex and expensive scheme.

The relay arrangements possible for reliability are:

1. The *one out of two arrangement,* in which two independent relay systems are used, either of which can initiate a trip. This is a highly dependable arrangement in that few faults will not cause a trip, but not very secure in that transient events may cause spurious trips.
2. The *two out of two arrangement,* in which the operation of relays in two independent systems is necessary to cause a trip. This system is very secure, but is not as dependable as the one out of two.
3. The *two out of three arrangement,* in which any two out of three independent relay systems can cause a trip is both secure and dependable.

The New York State Electric & Gas Corporation Goudey-Oakdale line used one out of two with an independent third system for a back up.

Six-phase transmission may gain in popularity if difficulty in obtaining right of way and concern about electromagnetic fields from electric transmission lines increases. The Goudey-Oakdale line demonstrated that an existing double circuit three-phase line could be upgraded to a six-phase line that integrates into a three-phase system.

## STATIC VAR COMPENSATION

Transmission lines are very costly so it is important to load them to as near maximum capacity as stable operation allows. Loading can be closer to the design capability of the line if the line can be damped quickly in response to transient events and the line losses can be minimized. Static VAR compensation using thyristor switching helps to provide VARs in a matter of milliseconds in response to transient events, thus providing damping, and helps hold terminal voltages constant by minimizing line losses (by improving load power factors) and reactive line drops for a given load/power factor combination.

In *switched VAR compensation* one or more reactors (capacitive or inductive depending on the need) are switched onto the line, as shown in Figure 8.33. The reactors can be switched in series or shunt with the line depending on the line type and needs, although shunt switching is shown in Figure 8.33. Stepping down the voltage to the reactors results in less expensive lower voltage switching and reactive elements. The harmonics of switched VAR compensation are low because the thyristors, connected in inverse parallel, can be switched when the line and reactor voltage is equal and the thyristors turn off as line current passes through zero.

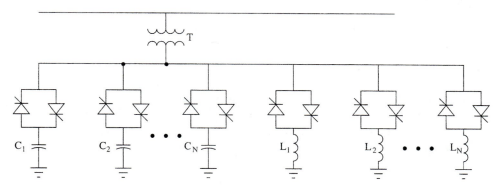

**FIGURE 8.33**   Multistep switched reactor

*Thyristor controlled shunt reactance (TCR)* is shown in Figure 8.34. The retard angle of the thyristors is controlled between 90° and 180° for maximum to minimum reactance resulting in a continuously varying reactance. The thyristor switching causes discontinuous current resulting in harmonics, but most of the energy is in the fundamental. Filtering is required to make TCR practical. A system similar to that in Figure 8.33, controlled by an automatic voltage regulator, can be used for TCR. The reactors are often connected in delta configuration on the secondary to eliminate third harmonics. Each delta branch has a reactor and inverse parallel connected thyristors. Gapped core inductors are commonly used to increase the range of linear control.

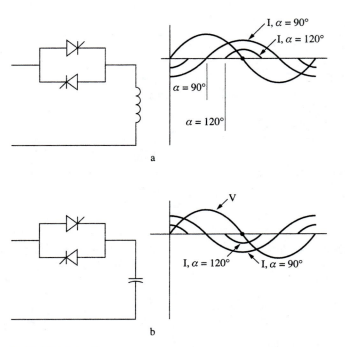

**FIGURE 8.34**   Thyristor controlled reactance (a) Inductive and (b) Capacitive

A common TCR scheme is to use a fixed capacitor with an inductive TCR branch in shunt with it (one set per secondary phase). The TCR adjusts the total reactance from capacitive to inductive as needed.

*Thyristor-controlled high-impedance transformer control (TCT)* consists of a transformer with a high secondary leakage reactance (almost 100%) controlled by inverse parallel connected thyristors, as shown in Figure 8.35. The controlled delta

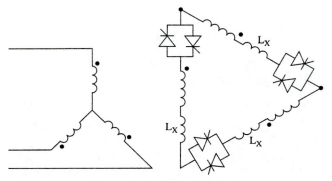

$L_X$ = ADDITIONAL REACTANCE AS REQUIRED

**FIGURE 8.35**   Thyristor controlled high impedance transformer

secondary eliminates third harmonic. Ungrounded wye secondary connected transformers also eliminate third harmonics but have the highest fifth harmonic amplitude. Transformers with controlled grounded wye secondaries and delta tertiary windings have the highest third but the lowest fifth and seventh harmonic amplitude. Additional reactance can be added to the secondaries if required. TCR and TCT can be used together for a wider range of control.

## 8.9

## SUMMARY

The job of the electrical power transmission system is to transport large amounts of electrical power safely, economically, and reliably. The major components of power system operation are planning, control, and accounting. Digital computers are now used to control most real time power flow and routing. They are used in accounting and planning also.

Planning consists of load forecasting and all the steps necessary to secure needed financing, right of way, permits, and all equipment and construction required to meet the forecast load. Operation is the real time control of power flow and the ancillary functions such as maintenance, equipment utilization, and scheduling that go with the day-to-day operation of a system. Accounting includes not only billing customers for power usage and financial record keeping, but also keeping records of the performance factors that help in efficient and effective planning and operation.

Supervisory control and data acquisition (SCADA) systems are needed to oversee the system operation and gather the operating and performance data needed for power system operation. SCADA systems are complex communication systems coupled to data acquisition and remote control facilities throughout the power system. SCADA systems often use a variety of communication and control technologies and media to perform the critical function of tying together system function and status information. They can be envisioned as the nervous system of the electrical power system.

System stability refers to the ability of the system to provide power without oscillations in the system or unplanned shut downs. The ability to quickly dispatch power and VARs as load needs change is important to stability. High capacity lines that can handle the dispatched power also contribute to stability.

Normal ac transmission voltage levels are: subtransmission—69 kV and 138 kV; transmission—230 kV, 345 kV, 500 kV, and 765 kV. Common dc transmission voltages are $\pm$ 200 kV, $\pm$ 400 kV, and $\pm$ 500 to $\pm$ 550 kV.

Overhead transmission lines are the most common and the least costly to construct and maintain. Obtaining right of way for transmission lines is becoming more difficult. Underground transmission lines are much more expensive than overhead lines and are used only for relatively short line segments or in places that overhead lines cannot be built. Underground lines are constructed with insulated lines inside a steel or bronze conduit. Circulating oil and increasingly $SF_6$ are used to insulate underground lines.

HVDC have become increasingly attractive for long lines because dc lines have lower loss than ac and no synchronization is needed between the power source and user. Additionally, the conversion equipment for HVDC has become more efficient and reliable.

Six-phase transmission is being investigated because less space is required to transmit the same power as a double circuit three-phase line. If the line to line voltages are the same the six-phase line can transmit more power than a double circuit three-phase line. Also, the magnetic field of a six-phase line is lower than an equivalent double circuit three-phase line.

Superconducting transmission lines, while feasible, are not economically attractive at the present time. The use of buried superconducting rings to store large amounts of power that can be switched onto a line in milliseconds promise to add the ability to stabilize lines to the extent that they can be loaded almost to capacity because the loading margin for stable operation is reduced.

## 8.10

## QUESTIONS

1. State the function of the electrical power transmission system.
2. Explain briefly the three major components of utility operation.
3. Define ac generator regulation.
4. What is the function of a turbine governor?
5. Discounting line loss, under what power sharing condition do generating units operate together most economically?
6. Define ACE.
7. What is a power pool?
8. What is the most common configuration of generator control system in use today?
9. What is a SCADA and what does it do?
10. List and briefly explain the system security states.
11. Define steady state and transient stability.
12. Referring to Figure 8.10, let $V_1 = 345$ kV $\angle \delta_1$ and $V_2 = 343$ kV $\angle \delta_2$. If $Z = j90$ $\Omega$ calculate $\delta_1 - \delta_2$ for power transfers of 100 MW and 500 MW.
13. Repeat problem 12 for $V_1 = 765$ kV $\angle \delta_1$ and $V_2 = 760$ kV $\angle \delta_2$.
14. In Figure 8.11 path A is a 345 kV line with pu $Z = j0.1$ and path B is a 765 kV line with pu $Z = j0.02$. The transformers each have pu $Z = j0.05$. 400 MW is being transferred from bus 1 to bus 2. Calculate the power transferred on each path if $\delta$ path A is the same as path B.
15. Explain a reasonable emergency operating strategy to avoid line overload if one line of Figure 8.12a are down.
16. List five steps that can be taken to improve steady state stability.

17. Define transfer impedance.
18. State the effect of generator rotational inertia in the event of a transient disturbance.
19. Why is it important to isolate faulted lines from a system as quickly as possible?
20. State the equal area criterion.
21. What is a swing curve?
22. What are subtransmission lines?
23. Discuss the economic trade off between minimum transmission line loss and construction costs.
24. What factors make underground transmission lines so much more expensive than overhead lines?
25. Why are underground transmission lines used in light of their great expense?
26. State the two major advantages of HVDC transmission over HVAC.
27. State the difference between the rectifying and inverting modes of a six pulse full controlled SCR bridge.
28. State the two major advantages of a twelve pulse converter over a six pulse.
29. List the two modes of control used for dc lines.
30. List four possible uses for superconductors in the electrical power industry.

# CHAPTER 9

# Transmission Line Parameters

Transmission lines consist of resistance, inductance, and capacitance. Thus a transmission line is an RLC circuit with the values of $R$, $L$, and $C$ constrained by the function and geometry of the line. The job of the transmission line is to transport large amounts of electrical power from one location to another with the lowest loss that can be economically attained. The materials available, cost restraints set by available capital, terrain the line must pass through, right of way availability and cost, and other factors prevent transmission lines from being optimized for electrical characteristics alone. The designers of transmission lines must obtain the lowest loss line that can be built to carry the needed load over the desired path at a reasonable cost.

This chapter covers transmission line resistance, inductance, capacitance, and equivalent circuits.

## 9.1

### TRANSMISSION LINE RESISTANCE

Recall that the resistance of a conductor can be expressed in equation form (see equation 2.3) as

$$R = \frac{\rho\, l}{A}$$

where

$\rho$ = resistivity

$l$ = conductor length

$A$ = conductor area

Also recall that the direct current (dc) resistance is not equal to the alternating current (ac) resistance because of the skin effect.

The ac and dc resistance of most commonly used transmission line conductors have been calculated and are available in tables such as Table 9.1. Table 9.1 is for ACSR conductors, which are the most commonly used transmission line conductors. Similar tables are available from conductor manufacturers for other types of conductors and for cables. In Table 9.1 the dc and ac resistance is given at 20°C, which is considered the light load temperature of a transmission line. Because power losses along the line cause self heating within the line 50°C is considered the standard loaded temperature. The standard temperature for short circuit calculations is 80°C. By standard temperature we mean the temperature used when the actual operating temperature is not known. Computer programs are now available that estimate closely the actual operating temperature of the line if the weather conditions along the line and the load are known.

Notice that in Table 9.1 the conductors are referred to by code names that are the names of birds. The area of the conductors is given in circular mils (CM). A circular mil is the area of a circle with a diameter of one thousandth of an inch $(1 \times 10^{-3})$. One circular mil is equal to 0.7854 square mil. The circular mil and thousand circular mils (MCM) are popular units of conductor area in power work. For this reason tables often give resistivity in $\Omega$—CM/ft for convenient use. The aluminum area is given, excluding the steel, because virtually all of the current flows in the aluminum. The resistivity of steel is higher than that of aluminum, and the aluminum strands are effectively in parallel with the steel strands, so most of the current flows through the lowest resistance.

Table 9.1 also gives the stranding of the conductor in aluminum strands/steel strands. Recall the steel strands provide strength for the conductor and the aluminum strands conduct almost all of the current. The number of layers of aluminum strands is also given. Figure 9.1 illustrates the stranding of ACSR conductor. The conductor strands are twisted, which causes the length of all but the center strand to be greater than that of the composite conductor. Thus the stranded aluminum has a resistance that is 1 to 2% higher than that of a solid aluminum conductor of the same total area. The table also gives the outside diameter of the cable in inches.

The standard of conductivity $(1/\rho)$ is annealed copper (silver is more conductive but is not a practical metal for transmission line conductor fabrication). Hard drawn copper, used for most copper conductors, has 97.3% of the conductivity of annealed copper because the drawing process causes imperfections in the grain structure of the metal. ACSR conductors have a conductivity 61% that of annealed copper conductors. The resistivities of Al and annealed Cu at 20°C are 17 $\Omega$—CM/ft and 10.66 $\Omega$—CM/ft respectively.

■■■■ **TABLE 9.1   Electrical characteristics of bare aluminum conductors steel-reinforced (ACSR)[†]**

| Code Word | Aluminum area, cmil | Stranding Al/St | Layers of aluminum | Outside diameter, in | $D_C$, 20°C, Ω/1,000 ft | Resistance Ac, 60 Hz 20°C, Ω/mi | 50°C, Ω/mi | GMR $D_s$, ft | Reactance per conductor 1-ft spacing, 60 Hz Inductive $X_a$, Ω/mi | Capacitive $X'_a$, MΩ·mi |
|---|---|---|---|---|---|---|---|---|---|---|
| Waxwing | 266,800 | 18/1 | 2 | 0.609 | 0.0646 | 0.3488 | 0.3831 | 0.0198 | 0.476 | 0.1090 |
| Partridge | 266,800 | 26/7 | 2 | 0.642 | 0.0640 | 0.3452 | 0.3792 | 0.0217 | 0.465 | 0.1074 |
| Ostrich | 300,000 | 26/7 | 2 | 0.680 | 0.0569 | 0.3070 | 0.3372 | 0.0229 | 0.458 | 0.1057 |
| Merlin | 336,400 | 18/1 | 2 | 0.684 | 0.0512 | 0.2767 | 0.3037 | 0.0222 | 0.462 | 0.1055 |
| Linnet | 336,400 | 26/7 | 2 | 0.721 | 0.0507 | 0.2737 | 0.3006 | 0.0243 | 0.451 | 0.1040 |
| Oriole | 336,400 | 30/7 | 2 | 0.741 | 0.0504 | 0.2719 | 0.2987 | 0.0255 | 0.445 | 0.1032 |
| Chickadee | 397,500 | 18/1 | 2 | 0.743 | 0.0433 | 0.2342 | 0.2572 | 0.0241 | 0.452 | 0.1031 |
| Ibis | 387,500 | 26/7 | 2 | 0.783 | 0.0430 | 0.2323 | 0.2551 | 0.0264 | 0.441 | 0.1015 |
| Pelican | 477,000 | 18/1 | 2 | 0.814 | 0.0361 | 0.1957 | 0.2148 | 0.0264 | 0.441 | 0.1004 |
| Flicker | 477,000 | 24/7 | 2 | 0.846 | 0.0359 | 0.1943 | 0.2134 | 0.0284 | 0.432 | 0.0992 |
| Hawk | 477,000 | 26/7 | 2 | 0.858 | 0.0357 | 0.1931 | 0.2120 | 0.0289 | 0.430 | 0.0988 |
| Hen | 477,000 | 30/7 | 2 | 0.883 | 0.0355 | 0.1919 | 0.2107 | 0.0304 | 0.424 | 0.0980 |
| Osprey | 556,500 | 18/1 | 2 | 0.879 | 0.0309 | 0.1679 | 0.1843 | 0.0284 | 0.432 | 0.0981 |
| Parakeet | 556,500 | 24/7 | 2 | 0.914 | 0.0308 | 0.1669 | 0.1832 | 0.0306 | 0.423 | 0.0969 |
| Dove | 556,500 | 26/7 | 2 | 0.927 | 0.0307 | 0.1663 | 0.1826 | 0.0314 | 0.420 | 0.0965 |
| Rook | 636,000 | 24/7 | 2 | 0.977 | 0.0269 | 0.1461 | 0.1603 | 0.0327 | 0.415 | 0.0950 |
| Grosbeak | 636,000 | 26/7 | 2 | 0.990 | 0.0268 | 0.1454 | 0.1596 | 0.0335 | 0.412 | 0.0946 |
| Drake | 795,000 | 26/7 | 2 | 1.108 | 0.0215 | 0.1172 | 0.1284 | 0.0373 | 0.399 | 0.0912 |
| Tern | 795,000 | 45/7 | 3 | 1.063 | 0.0217 | 0.1188 | 0.1302 | 0.0352 | 0.406 | 0.0925 |
| Rail | 954,000 | 45/7 | 3 | 1.165 | 0.0181 | 0.0997 | 0.1092 | 0.0386 | 0.395 | 0.0897 |
| Cardinal | 954,000 | 54/7 | 3 | 1.196 | 0.0180 | 0.0988 | 0.1082 | 0.0402 | 0.390 | 0.0890 |
| Ortolan | 1,033,500 | 45/7 | 3 | 1.213 | 0.0167 | 0.0924 | 0.1011 | 0.0402 | 0.390 | 0.0885 |
| Bluejay | 1,113,000 | 45/7 | 3 | 1.259 | 0.0155 | 0.0861 | 0.0941 | 0.0415 | 0.386 | 0.0874 |
| Finch | 1,113,000 | 54/19 | 3 | 1.293 | 0.0155 | 0.0856 | 0.0937 | 0.0436 | 0.380 | 0.0866 |
| Bittern | 1,272,000 | 45/7 | 3 | 1.345 | 0.0136 | 0.0762 | 0.0832 | 0.0444 | 0.378 | 0.0855 |
| Pheasant | 1,272,000 | 54/19 | 3 | 1.382 | 0.0135 | 0.0751 | 0.0821 | 0.0466 | 0.372 | 0.0847 |
| Bobolink | 1,431,000 | 45/7 | 3 | 1.427 | 0.0121 | 0.0684 | 0.0746 | 0.0470 | 0.371 | 0.0837 |
| Plover | 1,431,000 | 54/19 | 3 | 1.465 | 0.0120 | 0.0673 | 0.0735 | 0.0494 | 0.365 | 0.0829 |
| Lapwing | 1,590,000 | 45/7 | 3 | 1.502 | 0.0109 | 0.0623 | 0.0678 | 0.0498 | 0.364 | 0.0822 |
| Falcon | 1,590,000 | 54/19 | 3 | 1.545 | 0.0108 | 0.0612 | 0.0667 | 0.0523 | 0.358 | 0.0814 |
| Bluebird | 2,156,000 | 84/19 | 4 | 1.762 | 0.0080 | 0.0476 | 0.0515 | 0.0586 | 0.344 | 0.0776 |

[†]Most used multilayer sizes.

[‡]Data, by permission, from Aluminum Association, Inc. "Aluminum Electrical Conductor Handbook," New York, September 1971.

The variation of resistance with temperature is nearly linear over the normal temperature range that transmission lines experience. At very low temperatures, near absolute zero, the change of resistance with temperature is no longer linear. By assuming that the change of resistance with temperature is linear and plotting

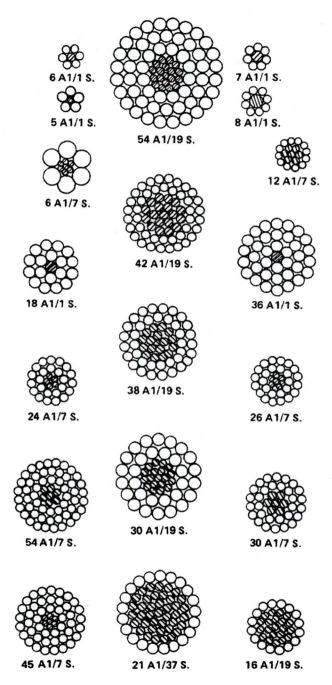

6 A1/1 S.

5 A1/1 S.

54 A1/19 S.

7 A1/1 S.

8 A1/1 S.

12 A1/7 S.

6 A1/7 S.

42 A1/19 S.

36 A1/1 S.

18 A1/1 S.

38 A1/19 S.

24 A1/7 S.

26 A1/7 S.

54 A1/7 S.

30 A1/19 S.

30 A1/7 S.

45 A1/7 S.

21 A1/37 S.

16 A1/19 S.

**FIGURE 9.1**    ACSR structure, 26/7 stranding, 2 Al layers (courtesy of the Aluminum Association's Aluminum Electrical Conductor Handbook [AECH]).

the mythical $R = 0$ temperature, as shown in Figure 9.2 provides a useful constant $(C_{R = 0})$ that can be used to calculate resistance at a second temperature if the resistance at one temperature and the constant are known. Equation 9.1 gives the relationship. $C_{R = 0} = 234.5$ for hard drawn Cu, and 236 for hard drawn Al.

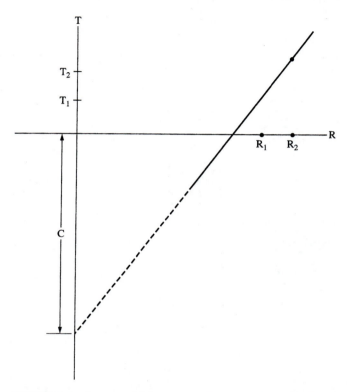

**FIGURE 9.2**   Source of the constant, $C_{R = 0}$, for Equation 9.1

$$\frac{R_2}{R_1} = \frac{C_{R = 0} + T_2}{C_{R = 0} + T_1}$$

(9.1)

Where $T_1$ and $T_2$ are temperature in the same units as the constant, usually degrees C. Equation 9.1 is good only for conductors of a single material such as AAC, but not ACSR. For example, 37 strand hard drawn Cu wire of 700 MCM area has a tested value of 0.0871 $\Omega$/mi at 25°C and 0.0947 $\Omega$/mi at 50°C. From Equation 9.1 we find

$$R_{ac(50°C)} = 0.0871 \ \Omega/\text{mi} \ \frac{(234.5 + 50)°C}{(234.5 + 25)°C} = 0.0955 \ \Omega/\text{mi}$$

This is very close to the tested value.

Drake is a very commonly used transmission conductor. From Table 9.1 we find for Drake: Al area = 795 MCM; stranding is 26/7 (Al/st) with 2 layers of Al strands; outside diameter (OD) = 1.108 in; $R_{DC}$ = 0.0215 $\Omega$/1000 ft (or 0.1135 $\Omega$/mi); $R_{ac}$ = 0.1172 $\Omega$/mi at 20°C; and $R_{ac}$ = 0.1284 $\Omega$/mi at 50°C. It is interesting to note that at 20°C $R_{ac}/R_{DC}$ = 1.0325, which means that the increase in resistance from the skin effect is 3.25%. The increase in resistance from skin effect gets larger as conductor diameter increases.

## 9.2

## TRANSMISSION LINE INDUCTIVE REACTANCE

To obtain the impedance of a line we need both the resistance and inductive reactance. The magnetic flux produced by the current in a conductor cuts the conductor itself as well as the adjacent conductors. Thus to calculate the series $X_L$ of a transmission line both the self and mutual (with other conductors) flux coupling must be considered.

When possible we obtain the inductive reactance of a transmission line conductor from the tables. However, there are situations in which the tables do not provide accurate results and it is instructive to see the equations used to obtain the tables.

### 9.2.1 Mutual Inductance

Recall from Chapter 2 that L is the proportionality constant between the voltage induced on a conductor and the current change that caused the flux change that induced the voltage.

$$e = L\frac{dI}{dt} \tag{2.4}$$

where

$e$ = induced voltage

$dI/dt$ = time varying current in amperes/second

$L$ = inductance in Henries (H)

Recall also that the voltage induced on a conductor in a magnetic field is

$$e = \frac{d\Phi}{dt} \tag{4.2}$$

where

$\Phi$ = magnetic flux in Webers (Wb)

$t$ = time in seconds

If the flux is given in Weber turns (Wbt) = $N\Phi$ then

$$e = N \frac{d\Phi}{dt} \tag{9.2}$$

Equating equations 9.2 and 2.4 results in

$$L \frac{dI}{dt} = N \frac{d\Phi}{dt}$$

so that

$$L = \left(\frac{d\Phi}{dt}\right) \left(\frac{dt}{dI}\right) N$$

where $L$ is in Henries, and if the permeability is held constant for all flux levels, as in air then

$$L = \frac{N\Phi}{I} \tag{9.3}$$

This results in an expression for flux linkages of

$$N\Phi = Li \qquad \text{Wbt} \tag{9.4}$$

where $i$ and $\Phi$ are instantaneous values.
If $I$ is rms current and $N\Phi = \psi$, phasor flux linkages in Wbt,

$$\psi = LI \qquad \text{wbt} \tag{9.5}$$

and

$$L = \psi/I \, \text{H} \tag{9.6}$$

The mutual inductance (M) between two conductors is the flux linkage in conductor 1 caused by the current in conductor 2 ($\psi_{12}$)

$$M_{12} = \psi_{12}/I \qquad \text{H} \tag{9.7}$$

The mutual reactance is $X_{M12} = j2\pi f \, M_{12}$, so the voltage drop on conductor 1 from the flux produced by conductor 2 is

$$V_1 = j \, \omega \, M_{12} \, I_2 \tag{9.8}$$

$$= j\omega_{12} \, \psi_{12} \tag{9.10}$$

The mutual inductance between power line conductors is a large part of the total line inductance.

## 9.2.2 Conductor Self Inductance

The self inductance of the conductor is caused by the flux generated by current within the conductor interacting with charges within the conductor. Figure 9.3a shows the flux distribution inside and outside of a conductor caused by current

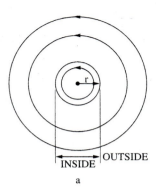

a

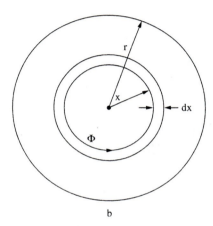

b

**FIGURE 9.3** Magnetic flux lines in a conductor (a) Distribution, I out of page (e⁻ into page) (b) Figure for developing Equation 9.11

within the conductor. We will now develop an equation for the self inductance per unit (pu) length of a conductor.

From basic magnetics we know that the magnetomotive force around any closed path is equal to the current enclosed by the path. Thus at a distance $x$ from the center of the conductor in Figure 9.3b the current enclosed ($I_x$) is

$$I_x = \left(\frac{\pi x^2}{\pi r^2}\right)I \tag{9.11}$$

where

$I$ = total conductor current

$r$ = radius of conductor

Additionally we know that the magnetomotive force ($H$) is equal to the magnetic field intensity times the distance over which it is applied. For the circle at x, if $H_x$ = $H$ at distance x

$$H_x(2\pi x) = I_x \tag{9.12}$$

Equating equations 9.11 and 9.12 we find

$$2\,\pi x H_x = \left(\frac{\pi\,x^2}{\pi\,r^2}\right)I$$

$$H_x = \left(\frac{x}{2\,\pi\,r^2}\right)I \qquad \text{in Ampereturns per meter}$$

The flux density $B = \mu H$ so

$$B_x = \mu H_x = \frac{\mu x I}{2\pi r^2}$$

In the tube $dx$ the flux $d\phi$ perpendicular to the flux lines is the area of the tube cross section times the flux density (recall $\phi = BA$, where A is area). For an infinitesimal dx the area is dx times the length of interest, in our case 1 m.

$$d\phi = \frac{\mu x I}{2\pi r^2}\,dx \qquad \text{Wb/m}$$

$d\psi$ is the linkages pu length caused by the flux in the tubular element on the area outside the tubular element. It is equal to the flux times the current ($i_x$) enclosed by the ring dx.

$$d\psi = d\phi i_x = \frac{\mu I x^3}{2\pi r^4}\,dx$$

Solving for $\psi_I$, the internal flux linkages, by integrating (summing) from $x = 0$ to $x = r$ yields

$$\psi_I = \int_0^r \frac{\mu I x^3}{2\pi r^4}\,dx$$

$$= \frac{\mu I x^4}{2\pi r^4}\bigg|_0^r$$

$$= \frac{\mu I}{8\pi} \tag{9.13}$$

which is the internal flux from the internal current.

The permeability ($\mu$) within Cu or Al (but not within the steel reinforcing strands of ACSR) is almost equal to the permeability of a vacuum, $4\pi \times 10^{-7}$ Wb/m. For AAC or Cu

$$\psi_I = \frac{I}{2} \times 10^{-7} \qquad \text{Wbt/m} \tag{9.14}$$

and

$$L = 0.5 \times 10^{-7} \quad \text{H/m} \tag{9.15}$$

if the current density is uniform. This is approximately true for ACSR also because almost all of the current flows in the Al.

### 9.2.3 Flux Linkage Outside a Conductor

We now need to find an expression for the inductance caused by flux linkages between two points external to a current carrying conductor, as shown in Figure 9.4. The two points $P_1$ and $P_2$ are at distances $D_1$ and $D_2$ from the conductor respectively. The flux paths are represented as concentric circles around the conductor. All of the flux between $P_1$ and $P_2$ lie within circles through $P_1$ and $P_2$ in Figure 9.4. In a manner almost identical to that of the last section we find that $H_X$ in a tubular section a distance x from the conductor is $2 \pi x H_X = I$, thus

$$H_x = \frac{I}{2\pi x}$$

and

$$B_x = \frac{\mu I}{2\pi x} \quad \text{Wb/m}^2$$

The flux, $d\phi$, in a small element dx at distance X is

$$\text{B dA} = \text{B dx } \pi \text{ X}$$

so

$$d\phi = \frac{\mu I}{2\pi x} dx \quad \text{Wb/m}$$

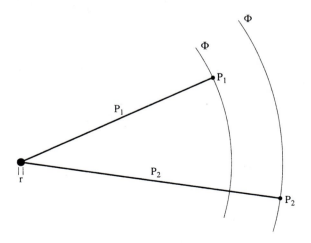

**FIGURE 9.4** Flux outside a conductor

Note that $d\phi = d\psi$ per meter since flux external to the conductor links all of the current in the conductor (that is what caused the flux to begin with). As in section 9.2.2 we calculate the flux between $P_1$ and $P_2$ by integrating $d\phi$ from $D_1$ to $D_2$.

$$\psi_{12} = \int_{D_1}^{D_2} \frac{\mu I}{2\pi x}\, dx$$

$$= \frac{\mu I}{2\pi} \ln x \Big|_{D_1}^{D_2}$$

$$= \frac{\mu I}{2\pi} (\ln D_2 - \ln D_1)$$

$$= \frac{\mu I}{2\pi} \ln\!\left(\frac{D_2}{D_1}\right) \tag{9.16}$$

which for $\mu_R = 1$, (recall $\mu = 4\pi \times 10^{-7}$    H/m)

$$\psi_{12} = (2 \times 10^{-7})\, I \ln\!\left(\frac{D_2}{D_1}\right) \qquad \text{Wb}\ \frac{t}{m} \tag{9.17}$$

Thus L from flux between $P_1$ and $P_2$ is

$$L_{12} = (2 \times 10^{-7}) \ln\!\left(\frac{D_2}{D_1}\right) \qquad \text{H/m} \tag{9.18}$$

### 9.2.4 Inductance Between Two Single-phase Conductors

Note in Figure 9.5 that a line of flux produced by current in conductor 1 that is at a distance greater than $D + r_2$ from conductor 2 cannot link conductor 2. Flux at a distance less than $D + r_2$ from conductor 1 will link with conductor 2. The fraction

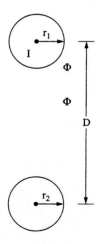

**FIGURE 9.5**   Two conductors, magnetic field from current from one conductor

of flux from conductor 1 that links with conductor 2 current varies from 1 to 0 between $D - r_2$ and $D + r_2$. If $D >> r$ the fractional flux linkage can be considered 1 from $D - r_2$ and 0 from $D$ to $D + r_2$ with very little error. If we let $D_2 = D$ and $D_1 = r$ for conductor 1 of Figure 9.5 and use Equation 9.18 we obtain

$$L_{1\,EXT} = (2 \times 10^{-7})\ln(D/r_1) \qquad H/m$$

and from Equation 9.15

$$L_{1\,INT} = (0.5 \times 10^{-7}) \qquad H/m$$

Thus $L_1$ total $= L_{1\,INT} + L_{1\,EXT}$ giving

$$L_1 = (0.5 + 2\ln(D/r_1) \times 10^{-7} \qquad H/m \tag{9.19}$$

A favored form of Equation 9.19 is found as follows

$$L_1 = (2 \times 10^{-7})[1/4 + \ln\left(\frac{d}{r_1}\right)]$$

$$L_1 = (2 \times 10^{-7})[\ln e^{1/4} + \ln\left(\frac{d}{r_1}\right)]$$

but $1/4 = \ln(e^{1/4})$, so

$$= (2 \times 10^{-7})\ln\left(\frac{D}{r_1 e^{-1/4}}\right)$$

$$= (2 \times 10^{-7})\ln\left(\frac{D}{r_1'}\right) \tag{9.20}$$

where $r_1' = r_1 e^{-1/4} = 0.7788 r_1$. Equation 9.20 has no self inductance term, but $r' = 0.7788 r$ compensates for the self inductance.

Now we will use Equation 9.20 for $L_2$

$$L_2 = (2 \times 10^{-7})\ln\left(\frac{D}{r_2'}\right)$$

The total inductance is found from $L_T = L_1 + L_2$ yielding

$$L_T = (2 \times 10^{-7}) [\ln\left(\frac{D}{r_1'}\right) + \ln\left(\frac{D}{r_2'}\right)]$$

$$= (4 \times 10^{-7}) \frac{1}{2} \ln\left(\frac{D^2}{r_1' r_2'}\right)$$

$$= (4 \times 10^{-7})\ln\left(\frac{D}{\sqrt{r_1' r_2'}}\right) \tag{9.21}$$

which for $r_1 = r_2$ reduces to

$$L_T = (4 \times 10^{-7})\ln\left(\frac{D}{r_1'}\right) \tag{9.22}$$

## 9.2.5 Geometric Mean Distance (GMD)

Two single phase multi–filament lines with equal current shares ($I = I_T/n$ and $I = I_T/m$ in conductors A and B respectively) are shown in Figure 9.6. The GMD, written as $D_m$ in equations, is the distance between the geometric centers of the two lines. The following equation defines the GMD.

$$D_m = \sqrt[mn]{(D_{aa'} D_{ab'} C_{ac'} \dots D_{am'})(D_{ba'} D_{bb'} C_{bc'} \dots D_{bm'}) \dots (D_{na'} D_{nb'} C_{nc'} \dots D_{nm'})} \tag{9.23}$$

where

$D_{aa'}$ = distance between filament a in conductor A and filament a' in conductor B.

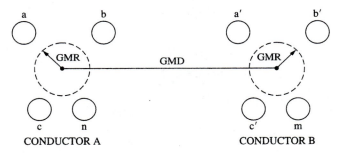

a        b                          a'        b'

GMR            GMD            GMR

c        n                          c'        m
CONDUCTOR A                  CONDUCTOR B

**FIGURE 9.6**   Two multifilament conductors used for single-phase line

Note that the multiplication is continued until the distances between all filaments are used in the product. Thus the GMD is a root-mean-power distance.

The geometric mean radius (GMR), written as $D_S$ in equations (the s is for self), is the effective or root-mean-power radius of a multi filament conductor. The equation for the GMR is

$$D_s = \sqrt[n^2]{(D_{aa}D_{ab}C_{ac} \dots D_{an})(D_{ba}D_{bb}C_{bc} \dots D_{bn}) \dots (D_{na}D_{nb}C_{nc} \dots D_{nn})} \tag{9.24}$$

where

$D_{aa} = r_a'$, $D_{bb} = r_b'$, $D_{cc} = r_c'$, and $D_{nn} = r_n'$

$D_{ab}$ = distance from filament a center to filament b center

Note that Equation 9.20 can be modified to calculate the inductance of one filament of one conductor. For example the inductance of filament a is

$$L_a = \frac{\psi}{I/n} = (2n \times 10^{-7}) \ln\left(\frac{D_m}{D_s}\right) \quad \text{H/m} \tag{9.25}$$

The inductance of each filament can be written in the same way. The total inductance of conductor A is the sum of the inductances of the filaments in conductor A.

The inductances may not all be the same, even with the same current in each filament, because the spacing of each filament with respect to the rest of the filaments may not be the same. The average inductance can be used to compensate for this.

The inductance of each filament is effectively in parallel with that of the other filaments in the conductor because each has the same source and destination. If all filaments in a conductor have the same inductance, or if the average inductance is used

$$L_{ave} = (L_a + L_b + \ldots + L_n)/n$$

Now $L_A$ can be written as

$$L_A = \frac{L_{ave}}{n} \tag{9.26}$$

so, in the form of Equation 9.21 and $L_T = L_A + L_B$ for two conductors A and B

$$L_T = (4 \times 10^{-7})\ln\left(\frac{D_m}{\sqrt{D_{sA}D_{sB}}}\right) \quad \text{H/m} \tag{9.27}$$

An example with very large filament spacing, to make the example both easy to see and easy to work, follows.

**Example 9.1:**

For the two conductors in Figure 9.7a calculate GMD, GMR, $L_A$, $L_B$, and $L_T$. Conductor A consists of three 0.4 cm diameter filaments and B of two 0.8 cm diameter filaments. The filaments share the conductor current evenly.

**Solution:**

First we calculate the m × n root of m × n distances to obtain the GMD.

$$D_m = \sqrt[6]{D_{aa'} D_{ab'} D_{ba'} D_{bb'} D_{ca'} D_{cb'}}$$

where

$$D_{aa'} = D_{bb'} = 12 \text{ m}$$
$$D_{ab'} = C_{ba'} = D_{cb'} = \sqrt{12^2 \text{ m} + 6^2 \text{ m}} = 13.416 \text{ m}$$
$$D_{ca'} = \sqrt{12^2 + 12^2} = 16.97 \text{ m}$$

yielding

$$D_m = \sqrt[6]{(12 \text{ m})(13.416 \text{ m})(13.416 \text{ m})(12 \text{ m})(16.97 \text{ m})(13.416 \text{ m})} = 13.443 \text{ m}$$

Now we calculate the GMR of conductor A, the 9th root of 9 distances, from Equation 9.24.

$$D_{sA} = \sqrt[9]{D_{aa}D_{ab}D_{ac}D_{ba}D_{bb}D_{bc}D_{ca}D_{cb}D_{cc}}$$

where

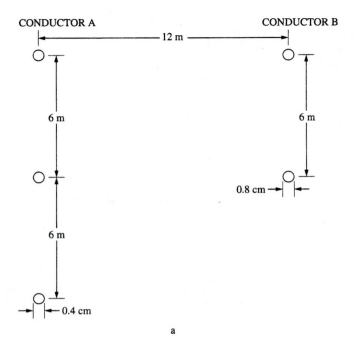

a

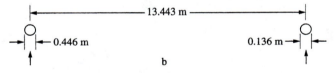

b

**FIGURE 9.7** Two single-phase conductors (a) Physical layout and (b) Equivalent geometry

$D_{aa} = D_{bb} = D_{cc} = 0.2 \text{ cm e}^{-0.25} = 0.2 \ (0.7788)$

$\qquad\qquad = 0.15576 \text{ cm} = 1.5576 \times 10^{-3} \text{ m}$

$D_{ab} = D_{ba} = D_{bc} = D_{cb} = 6 \text{ m}$

$D_{ac} = D_{ca} = 12 \text{ m}$

resulting in

$D_{sA} = \sqrt[9]{(1.5576 \times 10^{-3})^3 (6\text{m})^4 (12\text{m})^2} = 0.446\text{m}$

Similarly for conductor B

$D_{sb} = \sqrt[4]{D_{a'b'} D_{a'a'} D_{b'a'} D_{b'b'}}$

$\qquad = \sqrt[4]{(0.4 \times 0.7788 \times 10^{-2})^2 (6\text{m})^2}$

$\qquad = \sqrt[4]{349.36 \times 10^{-6}}$

$\qquad = 0.136\text{m}$

Figure 9.7b shows the equivalent geometry of the conductors in Figure 9.7a. We now use Equation 9.26 to find the inductance of each conductor.

$$L_A = (2 \times 10^{-7}) \ln \frac{D_m}{D_{sA}}$$

$$= (2 \times 10^{-7}) \ln \left( \frac{13.443m}{0.446m} \right)$$

$$= 0.681 \times 10^{-6} \quad H/m$$

$$L_B = (2 \times 10^{-7}) \ln \frac{D_m}{D_{sB}}$$

$$= (2 \times 10^{-7}) \ln \left( \frac{13.443m}{0.136m} \right)$$

$$= 0.918 \times 10^{-6} \, H/m$$

and from Equation 9.27

$$L_T = (4 \times 10^{-7}) \ln \left( \frac{D_m}{\sqrt{D_{sA} D_{sB}}} \right)$$

$$= (4 \times 10^{-7}) \ln \left( \frac{13.443m}{0.246m} \right)$$

$$= 1.599 \times 10^{-6} \quad H/m$$

## 9.2.6 Inductive Reactance From the Conductor Table

The GMR of each conductor is given in Table 9.1. $X_L$ at 60 hz is given in $\Omega/mi$, and is listed as $X_a$. $X_L$ is found by multiplying the inductance obtained from Equation 9.25, 9.26, or 9.27 by jω. Xa in the table is the inductive reactance of a line with a 1 foot spacing. The inductive reactance of the line can be expressed as the reactance for a 1 foot spacing plus a spacing factor reactance. To find the spacing factor first we multiply Equation 9.25 by $2\pi f$. The inductive reactance for one conductor is then

$$X_L = 2\pi f (2 \times 10^{-7}) \ln \frac{D_m}{D_s}$$

$$= (4\pi \times 10^{-7}) f \ln \frac{D_m}{D_s} \tag{9.28a}$$

Changing meters to miles (1609 m/mi), and separating

$$X_L = (2.022 \times 10^{-3}) f \ln \frac{1}{D_s} + (2.022 \times 10^{-3}) f \ln D_m$$

$$= X_a + X_d \tag{9.28b}$$

where $X_d$ is the spacing factor reactance. Now we see that

$$X_a = (2.022 \times 10^{-3})f \ln \frac{1}{D_s} \tag{9.29}$$

Which is in Table 9.1

$$X_d = (2.022 \times 10^{-3})f \ln D_m \tag{9.30}$$

Which is in Table 9.2.

$X_d$ calculated for distances other than 1 foot is found in Table 9.2. Note that $X_d$ for conductor separation less than a foot is negative.

**Example 9.2:**

Drake is a widely used ACSR conductor in transmission. Calculate the $X_L$ per mile of a single-phase circuit using Drake conductors spaced 20 feet apart.

**Solution:**

From Table 9.1 we find that Drake has $Ds = 0.0373$ ft and $X_a = 0.399 \ \Omega/\text{mi}$. From Equation 9.30 we find

$$X_d = (2.022 \times 10^{-3})(60 \text{ Hz})\ln(20 \text{ ft}) = 0.363 \ \Omega/\text{mi}$$

so that

$$X_L = X_a + X_d = 0.399\Omega/\text{mi} + 0.363\Omega/\text{mi}$$

$$= 0.762\Omega/\text{mi}$$

Alternately, using Equation 9.28a

$$X_L = (2.022 \times 10^{-3})(60 \text{ Hz}) \ln \frac{20 \text{ ft}}{0.0373 \text{ ft}} = 0.7625\Omega/\text{mi}$$

From Table 9.2 we see that $X_d$ for 20 feet is 0.3635. Normally we would just use the table instead of calculating $X_d$.

## 9.2.7 Balanced Three-Phase Line Inductance

The flux coupling between the lines of three-phase lines must be considered in calculating the inductive reactance. For balanced loads, and most loads are reasonably well balanced, the net current of a three-phase line is zero. So for a symmetrically spaced, balanced three-phase line such as the one shown in Figure 9.8 the equation for $L$ per phase remains in the same form as for a single-phase line.

$$L_a = (2 \times 10^{-7}) \ln \frac{D}{D_s} \tag{9.31}$$

Since $r_a' = r_b' = r_c'$, $L_T = 3L_a$.

■■■ **TABLE 9.2** **Inductive reactance spacing factor $X_d$ at 60 Hz[†] (ohms per mile per conductor)**

| | Separation | | | | | | | | | | | |
|---|---|---|---|---|---|---|---|---|---|---|---|---|
| | **Inches** | | | | | | | | | | | |
| **Feet** | **0** | **1** | **2** | **3** | **4** | **5** | **6** | **7** | **8** | **9** | **10** | **11** |
| 0 | ...... | −0.3015 | −0.2174 | −0.1682 | −0.1333 | −0.1062 | −0.0841 | −0.0654 | −0.0492 | −0.0349 | −0.0221 | −0.0106 |
| 1 | 0 | 0.0097 | 0.0187 | 0.0271 | 0.0349 | 0.0423 | 0.0492 | 0.0558 | 0.0620 | 0.0679 | 0.0735 | 0.0789 |
| 2 | 0.0841 | 0.0891 | 0.0938 | 0.0984 | 0.1028 | 0.1071 | 0.1112 | 0.1152 | 0.1190 | 0.1227 | 0.1264 | 0.1299 |
| 3 | 0.1333 | 0.1366 | 0.1399 | 0.1430 | 0.1461 | 0.1491 | 0.1520 | 0.1549 | 0.1577 | 0.1604 | 0.1631 | 0.1657 |
| 4 | 0.1682 | 0.1707 | 0.1732 | 0.1756 | 0.1779 | 0.1802 | 0.1825 | 0.1847 | 0.1869 | 0.1891 | 0.1912 | 0.1933 |
| 5 | 0.1953 | 0.1973 | 0.1993 | 0.2012 | 0.2031 | 0.2050 | 0.2069 | 0.2087 | 0.2105 | 0.2123 | 0.2140 | 0.2157 |
| 6 | 0.2174 | 0.2191 | 0.2207 | 0.2224 | 0.2240 | 0.2256 | 0.2271 | 0.2287 | 0.2302 | 0.2317 | 0.2332 | 0.2347 |
| 7 | 0.2361 | 0.2376 | 0.2390 | 0.2404 | 0.2418 | 0.2431 | 0.2445 | 0.2458 | 0.2472 | 0.2485 | 0.2498 | 0.2511 |
| 8 | 0.2523 | | | | | | | | | | | |
| 9 | 0.2666 | | | | | | | | | | | |
| 10 | 0.2794 | | | | | | | | | | | |
| 11 | 0.2910 | | | | | | | | | | | |
| 12 | 0.3015 | | | | | | | | | | | |
| 13 | 0.3112 | | | | | | | | | | | |
| 14 | 0.3202 | | | | | | | | | | | |
| 15 | 0.3286 | | | | | | | | | | | |
| 16 | 0.3364 | | | | | | | | | | | |
| 17 | 0.3438 | | | | | | | | | | | |
| 18 | 0.3507 | | | | | | | | | | | |
| 19 | 0.3573 | | | | | | | | | | | |
| 20 | 0.3635 | | At 60 Hz, in Ω/mi per conductor | | | | | | | | | |
| 21 | 0.3694 | | $X_d = 0.2794 \log d$ | | | | | | | | | |
| 22 | 0.3751 | | $d$ = separation, ft | | | | | | | | | |
| 23 | 0.3805 | | For three-phase lines | | | | | | | | | |
| 24 | 0.3856 | | $d = D_{eq}$ | | | | | | | | | |
| 25 | 0.3906 | | | | | | | | | | | |
| 26 | 0.3953 | | | | | | | | | | | |
| 27 | 0.3999 | | | | | | | | | | | |
| 28 | 0.4043 | | | | | | | | | | | |
| 29 | 0.4086 | | | | | | | | | | | |
| 30 | 0.4127 | | | | | | | | | | | |
| 31 | 0.4167 | | | | | | | | | | | |
| 32 | 0.4205 | | | | | | | | | | | |
| 33 | 0.4243 | | | | | | | | | | | |
| 34 | 0.4279 | | | | | | | | | | | |
| 35 | 0.4314 | | | | | | | | | | | |
| 36 | 0.4348 | | | | | | | | | | | |
| 37 | 0.4382 | | | | | | | | | | | |
| 38 | 0.4414 | | | | | | | | | | | |
| 39 | 0.4445 | | | | | | | | | | | |
| 40 | 0.4476 | | | | | | | | | | | |
| 41 | 0.4506 | | | | | | | | | | | |
| 42 | 0.4535 | | | | | | | | | | | |
| 43 | 0.4564 | | | | | | | | | | | |
| 44 | 0.4592 | | | | | | | | | | | |
| 45 | 0.4619 | | | | | | | | | | | |
| 46 | 0.4646 | | | | | | | | | | | |
| 47 | 0.4672 | | | | | | | | | | | |
| 48 | 0.4697 | | | | | | | | | | | |
| 49 | 0.4722 | | | | | | | | | | | |

[†]From "Electrical Transmission and Distribution Reference Book," by permission of the Aluminum Association, Inc.

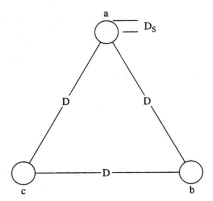

**FIGURE 9.8**   Equilateral spacing of three-phase line

Bundling of transmission lines, recall, reduces the line series impedance, which reduces line loss and improves line capacity and system stability over the use of a single conductor of the size used in the bundle. Recall also that the bundling of conductors at voltages of 230 kV and above is to reduce the loss from corona. The spacing between bundle conductors is usually 18 inches (45.72 cm) in the United States, and 45 centimeters in Canada and Europe.

To adjust for conductor bundling the Ds of equation 9.31 is replaced with the GMR of the bundle (Dsb). For a bundle of 2, used for voltages between 230 kV and 345 kV, separated by a distance d

$$D_{sb2} = \sqrt[4]{(D_s d)^2} = \sqrt{(D_s d)}$$                                          (9.32)

for a bundle of three, used at 500 kV

$$D_{sb3} = \sqrt[9]{(D_s dd)^3} = \sqrt[3]{(D_s d^2)}$$                                     (9.33)

and for a bundle of four, used from 700 kV up

$$D_{sb4} = \sqrt[16]{(D_s dd1.414d)^4} = 1.09 \sqrt[4]{(D_s d^3)}$$                          (9.34)

Figure 9.9 shows the common bundling configurations.

If the line does not have equilateral spacing, such as the one shown in Figure 9.10, the coupling between the lines is not balanced because of the asymmetrical spacing, thus the inductance of the lines is not equal. The balance of the inductive reactance of the line can be restored by transposing the lines as shown in Figure 9.11 so that each line occupies each position for one third of the line length. Very long transmission (and distribution) lines are often transposed more than three times to assure balance along the length of the line, but many are not transposed at all. The unbalance in inductive reactance caused by asymmetrical spacing is small at the common transmission line spacings. It is common practice to calculate the inductive reactance of an asymmetrically spaced line using the equivalent symmetrical spacing of the line, which is calculated using the GMD equation (9.23).

$$D_{eq} = \sqrt[6]{D_{12}D_{21}D_{23}D_{32}D_{13}D_{31}}$$

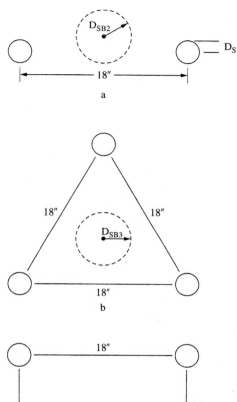

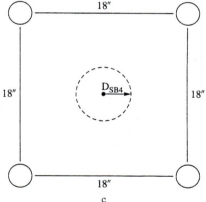

**FIGURE 9.9** Common bundles (a) Bundle of two (230 kV to 345 kV) (b) Bundle of three (500 kV) and (c) Bundle of four (700 kV and up)

However $D_{12} = D_{21}$, $D_{23} = D_{32}$, and $D_{31} = D_{13}$ so

$$D_{eq} = \sqrt[6]{D^2_{12}D^2_{23}D^2_{31}}$$

$$D_{eq} = \sqrt[6]{(D_{12}D_{23}D_{31})^2}$$

$$D_{eq} = \sqrt[3]{D_{12}D_{23}D_{31}} \tag{9.35}$$

$D_{eq}$ is used as $D$ in Equation 9.31.

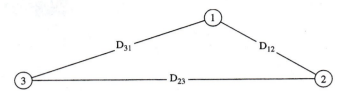

**FIGURE 9.10** Asymmetrical three-phase spacing

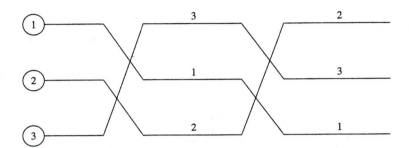

**FIGURE 9.11** Transposing phases for balance

Parallel lines are handled by calculating the combined GMD from the individual circuit GMDs, and the combined GMR of each parallel line set from the GMR of each line. The result is a single equivalent three-phase line. We will do an example in the transmission line capacitance section.

## 9.2.8 Three-Phase Inductance Examples

**Example 9.3:**

Figure 9.12 shows a very common 138 kV transmission line configuration and spacing. The H fixtures hold 336 MCM Linnet conductor at a 15 foot spacing. Using the tables calculate the series impedance per mile per phase of this line.

**Solution:**

From the tables we see that Linnet has: Al area = 336,400 CM, 26/7 stranding, $R_{ac}$ at 50°C = 0.3006 $\Omega$/mi, $D_s$ = 0.0243 ft, and $X_a$ = 0.451 $\Omega$/mi. First we solve for $D_m$

$$D_m = \sqrt[3]{(15 \text{ ft})(15 \text{ ft})(30 \text{ ft})} = 18.9 \text{ ft}$$

Interpolating from Table 9.2

$$X_d \text{ at } 18.9 \text{ ft} = [0.3507 + 0.9(0.3573 - 0.3507)] \ \Omega/\text{mi}$$
$$= (0.3507 + 0.00594) \ \Omega/\text{mi} = 0.3566 \ \Omega/\text{mi}$$

and from Table 9.1

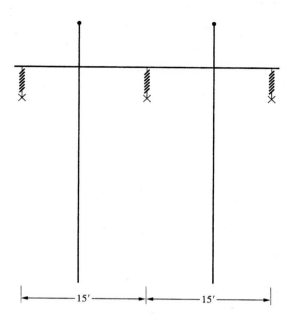

**FIGURE 9.12**   Spacing for Example 9.3 (138 kV)

$X_L = X_a + X_d = 0.451 + 0.3566 = 0.8076 \ \Omega/\text{mi}$

Finally

$Z = 0.3006 \ \Omega/\text{mi} + j0.8076 \ \Omega/\text{mi} = (0.8617 \ \Omega/\text{mi})\angle 69.58°$

**Example 9.4:**

Figure 9.13 shows a common spacing for a 345 kV line using Drake conductor in a bundle of two (18 inch bundle spacing). Calculate the series impedance per mile per phase of the line. The conductor tables may be used.

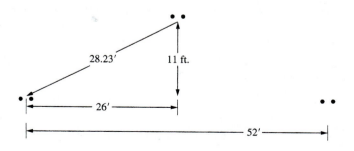

**FIGURE 9.13**   Spacing for Example 9.4 (345 kV)

**Solution:**

From Table 9.1 we find for Drake conductor: Al area = 795 MCM; 26/7 stranding; GMR = 0.0373 ft; $R_{ac}$ (50°C) = 0.1284 Ω/mi; and $X_a$ = 0.399 Ω/mi. From Equation 9.32

$$D_{sb2} = \sqrt{D_s d} = \sqrt{(0.0373 \text{ ft})(1.5 \text{ ft})} = 0.23 \text{ ft}$$

and from Equation 9.23

$$D_m = \sqrt[6]{(28.23 \text{ ft})^4 (52 \text{ ft})^2} = 34.6 \text{ ft}$$

Because the tables give only single conductor values we must calculate $X_L$ using Equation 9.28a with a 1609 m/mi factor to obtain Ω/mi.

$$X_L = 2\pi(60 \text{ Hz})(1609 \text{ m/mi})(2 \times 10^{-7}) \ln \frac{D_m}{D_{sb2}} = 0.604 \Omega/\text{mi}$$

Note from the tables that $X_L$ of a single Drake conductor is 0.829 Ω/mi, 27% more than the bundled value.

The conductors of a bundle are in parallel so

$$R_{ac} = 0.5(R_{ac(500C)}) = (0.1284 \ \Omega/\text{mi})/2 = 0.064 \ \Omega/\text{mi}$$

Thus the impedance per mile per phase is

$$Z = (0.0642 + j0.60486) \ \Omega/\text{mi} = 0.60825 \ \Omega/\text{mi} \ \angle 83.94°$$

## 9.3

## TRANSMISSION LINE CAPACITANCE

Overhead transmission line conductors are separated from each other and from ground by air, which is a dielectric. Thus transmission lines have capacitance. The effective plate area is small and the plate separation large so the leakage from line to line and line to ground because of the shunt capacitive susceptance is negligible in comparison to total power flow in lines shorter than 50 miles. However, current does flow through the shunt capacitive susceptance ($jB_C$), and this current flows even when there is no load on the line. The real part of the line shunt admittance, G, is negligible in lines with no corona. When the circuit breaker is closed to energize a line the shunt capacitance must be charged to the line voltage. The initial charging current can be quite high, especially on dc lines.

### 9.3.1 Voltage Between Two Points in an Electric Field

The electric field flux density ($\mathfrak{D}$) is defined as the electric flux pu area, normally in Coulombs per square meter. A long uniformly charged conductor has lines of electric flux extending radially out from the conductor surface. For a cylindrical conductor the flux density is given as

$$\mathfrak{D} = q/(2\pi x) \qquad C/m^2$$

where

$q$ = electric charge

$x$ = distance from the conductor

If the medium around the conductor is a dielectric other than air the electric field lines are concentrated by the dielectric constant, $\epsilon$, also called the relative permitivity. Recall that relative permitivity is a measure of the ease with which a dielectric can support electric field lines in comparison to a vacuum. The relative permitivity ($\epsilon_r$) of air is 1.00054 and is considered to be one in most calculations. The permitivity of a vacuum is $\epsilon_0 = 8.85 \times 10^{-12}$ F/m. The dielectric surrounding most overhead lines, the only type we calculate for in this chapter, is air so we use the $\epsilon_r = 1$.

The electric field intensity ($E$) is a vector quantity that is equal to the flux density divided by the permitivity with units of volts/meter.

$$E = \frac{q}{2\pi x \epsilon}$$

$$= \mathfrak{D}/\epsilon \qquad V/m \tag{9.36}$$

The voltage between two points in an electric field is equal to the work in joules per coulomb required to move charge from one point to the other. No work is required to move a charge perpendicular to an electric field because the potential is the same (equipotential) along a path perpendicular to electric field lines.

Referring to Figure 9.14, the voltage between points 1 and 2 can be found by calculating the work needed to move a charge from point 1 to point 2. The path chosen is from point 1 to point A, along which work is performed moving the charge, and from point A to point 2, along which no work is performed because this path is along an equipotential line. The polarity of the calculation depends on the polarity of the charge moved and the direction in which it is moved. The work can be calculated by summing the electric field along path x from $P_1$ to $P_2$ by integrating.

$$V_{12} = \int_{D_1}^{D_2} E \, dx = \int_{D_1}^{D_2} \frac{q}{2\pi\epsilon x} \, dx$$

$$= \frac{q}{2\pi\epsilon} \ln \frac{D_2}{D_1} \tag{9.37}$$

Note the form of this equation is reminiscent of Equation 9.16.

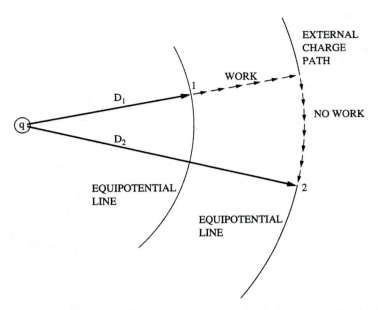

**FIGURE 9.14**  Potential between two points a distance away from a charge

## 9.3.2 Capacitance of a Two Wire Line

Capacitance is defined as charge accumulated per volt ($C = Q/V$). Along a conductor we will calculate charge/volt per meter. To calculate the capacitance of the line in Figure 9.15 we first apply Equation 9.37 to calculate the voltage between the lines.

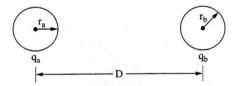

**FIGURE 9.15**  Voltage between two charged conductors

$V_a$ = V from $q_a$ + V from $q_b$.

$$V_{ab} = \frac{q_a}{2\pi\epsilon} \ln \frac{D}{r_a} + \frac{q_b}{2\pi\epsilon} \ln \frac{r_b}{D}$$

Note that with respect to conductor a there is a reversal in $D_1$ and $D_2$ in the second term. In a two wire line $q_a = -q_b$ so

$$V_{ab} = \frac{q_a}{2\pi\epsilon} \left( \ln \frac{D}{r_a} - \ln \frac{r_b}{D} \right)$$

$$= \frac{q_a}{2\pi\epsilon} \ln \frac{D^2}{r_a r_b} \tag{9.38}$$

but

$$C = \frac{q_a}{V_{ab}} = \frac{2\pi\epsilon}{\ln \dfrac{D^2}{r_a r_b}} \quad \text{F/m} \tag{9.39a}$$

and if $r_a = r_b$

$$C = \frac{\pi\epsilon}{\ln \dfrac{D}{r}} \quad \text{F/m} \tag{9.39b}$$

The actual conductor radius is used in capacitance calculations.

### 9.3.3 Line to Neutral Capacitance

The line to neutral capacitance per phase is the value usually calculated. Notice that the line to neutral point capacitance is not the line to ground capacitance. Instead it is to the neutral or zero point between conductors with equal and opposite voltages from equal and opposite charges. Basically it is the capacitance to the electric field neutral point between the conductors. The line to line capacitance is illustrated in Figure 9.16a, and the line to neutral capacitance in Figure 9.16b. Intuitively we see that the line to neutral capacitance must have half the effective plate separation with no change in plate area thus it is twice the line to line capacitance. Another way to look at it is that the two series line to neutral capacitors must have the same capacitance as the line to line capacitance, therefore the line to neutral capacitance of each line must be twice as large as the line to line capacitance. Using that approach

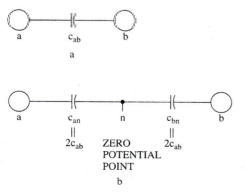

**FIGURE 9.16** Line capacitance (a) Line to line and (b) Line to neutral

$$C_{ab} = \cfrac{1}{\cfrac{1}{C_{an}} + \cfrac{1}{C_{bn}}} = \frac{C_{an}C_{bn}}{C_{an} + C_{bn}}$$

but $C_{an} = C_{bn}$ so $C_{ab} = C_{an}/2$ and $C_{an} = 2\,C_{ab}$. Therefore

$$C_{an} = \frac{2\pi\epsilon}{\ln\dfrac{D}{r}} \qquad F/m \tag{9.40}$$

recall that $X_C = 1/(2\pi f C)$ so for $\epsilon_r = 1$

$$X_C = \frac{1}{4\pi^2 f\epsilon^o}\ln\frac{D}{r} = \frac{2.862 \times 10^9}{f}\ln\frac{D}{r} \quad \Omega m \text{ to neutral}$$

Now we convert to $\Omega$ mi by dividing by 1609 m/mi to obtain

$$X_C = \frac{1.779 \times 10^6}{f}\ln\frac{D}{r} \qquad \Omega mi \tag{9.41}$$

To more easily use the tables in which are split into a 1 foot term ($X_a{}'$) in Table 9.1, and a distance correction term ($X_d{}'$) in Table 9.3, we split Equation 9.41 into two parts.

$$X_C = X_a{}' + X_d{}'$$
$$X_C = \frac{1.779 \times 10^6}{f}\ln\frac{1}{r} + \frac{1.779 \times 10^6}{f}\ln D \qquad \Omega mi \tag{9.42}$$

The reason that $X_C$ is set up to come out Ohm miles is that in the transmission line power flow and loss calculation $B_C = 1/X_C$. The reciprocal comes out S/mi as desired for most calculations.

**Example 9.4:**

A single-phase, 60 Hz, Linnet line has a spacing of 26 feet between conductors. Calculate the susceptance per mile using the conductor tables.

**Solution:**

First, from the tables, the line to neutral values are

$X_a{}' = 0.1040$ M $\Omega$ mi

$X_d{}' = 0.0967$ M $\Omega$ mi

$X_C = (0.1040 + 0.0967)$M $\Omega$ mi

$\quad = 0.2007$ M $\Omega$ mi

$B_C = 1/X_C = 4.98\ \mu S/mi$

■■■■ **TABLE 9.3** Shunt capacitive-reactance spacing factor $X'_d$ at 60 Hz[†] (megohm-miles per conductor)

| Feet | \multicolumn{12}{c}{Separation — Inches} |
|------|---|---|---|---|---|---|---|---|---|---|---|---|

| | 0 | 1 | 2 | 3 | 4 | 5 | 6 | 7 | 8 | 9 | 10 | 11 |
|------|---|---|---|---|---|---|---|---|---|---|---|---|
| 0 | ...... | −0.0737 | −0.0532 | −0.0411 | −0.0326 | −0.0260 | −0.0206 | −0.0160 | −0.0120 | −0.0085 | −0.0054 | −0.0026 |
| 1 | 0 | 0.0024 | 0.0046 | 0.0066 | 0.0085 | 0.0103 | 0.0120 | 0.0136 | 0.0152 | 0.0166 | 0.0180 | 0.0193 |
| 2 | 0.0206 | 0.0218 | 0.0229 | 0.0241 | 0.0251 | 0.0262 | 0.0272 | 0.0282 | 0.0291 | 0.0300 | 0.0309 | 0.0318 |
| 3 | 0.0326 | 0.0334 | 0.0342 | 0.0350 | 0.0357 | 0.0365 | 0.0372 | 0.0379 | 0.0385 | 0.0392 | 0.0399 | 0.0405 |
| 4 | 0.0411 | 0.0417 | 0.0423 | 0.0429 | 0.0435 | 0.0441 | 0.0446 | 0.0452 | 0.0457 | 0.0462 | 0.0467 | 0.0473 |
| 5 | 0.0478 | 0.0482 | 0.0487 | 0.0492 | 0.0497 | 0.0501 | 0.0506 | 0.0510 | 0.0515 | 0.0519 | 0.0523 | 0.0527 |
| 6 | 0.0532 | 0.0536 | 0.0540 | 0.0544 | 0.0548 | 0.0552 | 0.0555 | 0.0559 | 0.0563 | 0.0567 | 0.0570 | 0.0574 |
| 7 | 0.0577 | 0.0581 | 0.0584 | 0.0588 | 0.0591 | 0.0594 | 0.0598 | 0.0601 | 0.0604 | 0.0608 | 0.0611 | 0.0614 |
| 8 | 0.0617 | | | | | | | | | | | |
| 9 | 0.0652 | | | | | | | | | | | |
| 10 | 0.0683 | | | | | | | | | | | |
| 11 | 0.0711 | | | | | | | | | | | |
| 12 | 0.0737 | | | | | | | | | | | |
| 13 | 0.0761 | | | | | | | | | | | |
| 14 | 0.0783 | | | | | | | | | | | |
| 15 | 0.0803 | | | | | | | | | | | |
| 16 | 0.0823 | | | | | | | | | | | |
| 17 | 0.0841 | | | | | | | | | | | |
| 18 | 0.0858 | | | | | | | | | | | |
| 19 | 0.0874 | | | | | | | | | | | |
| 20 | 0.0889 | | | | | | | | | | | |
| 21 | 0.0903 | | | | | | | | | | | |
| 22 | 0.0917 | | | | | | | | | | | |
| 23 | 0.0930 | | | | | | | | | | | |
| 24 | 0.0943 | | | | | | | | | | | |
| 25 | 0.0955 | | | | | | | | | | | |
| 26 | 0.0967 | | | | | | | | | | | |
| 27 | 0.0978 | | | | | | | | | | | |
| 28 | 0.0989 | | | | | | | | | | | |
| 29 | 0.0999 | | | | | | | | | | | |
| 30 | 0.1009 | | | | | | | | | | | |
| 31 | 0.1019 | | | | | | | | | | | |
| 32 | 0.1028 | | | | | | | | | | | |
| 33 | 0.1037 | | | | | | | | | | | |
| 34 | 0.1046 | | | | | | | | | | | |
| 35 | 0.1055 | | | | | | | | | | | |
| 36 | 0.1063 | | | | | | | | | | | |
| 37 | 0.1071 | | | | | | | | | | | |
| 38 | 0.1079 | | | | | | | | | | | |
| 39 | 0.1087 | | | | | | | | | | | |
| 40 | 0.1094 | | | | | | | | | | | |
| 41 | 0.1102 | | | | | | | | | | | |
| 42 | 0.1109 | | | | | | | | | | | |
| 43 | 0.1116 | | | | | | | | | | | |
| 44 | 0.1123 | | | | | | | | | | | |
| 45 | 0.1129 | | | | | | | | | | | |
| 46 | 0.1136 | | | | | | | | | | | |
| 47 | 0.1142 | | | | | | | | | | | |
| 48 | 0.1149 | | | | | | | | | | | |
| 49 | 0.1155 | | | | | | | | | | | |

At 60 Hz, in $\Omega$/mi per conductor
$$X'_d = 0.06831 \log d$$
$d$ = separation, ft
For three-phase lines
$$d = D_{eq}$$

[†]From "Electrical Transmission and Distribution Reference Book," by permission of the Aluminum Association, Inc.

The line to line value of $B_C$ is one half this value because the capacitance is one half the line to neutral value. From Equation 9.41 we find, using the physical conductor radius

$$X_C = \frac{1.779 \times 10^6}{60\ Hz} \ln \frac{26\ ft}{0.0243\ ft} = 0.2068$$

### 9.3.4 Three-Phase Line Capacitance

Refer to Figure 9.17a. Applying Equation 9.37 to all three lines we obtain

$$V_{ab} = \frac{1}{2\pi\epsilon} \left( q_a \ln \frac{D}{r} + q_b \ln \frac{r}{D} + q_c \ln \frac{D}{D} \right)$$

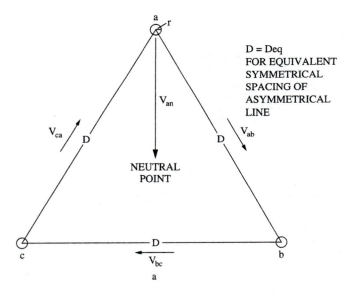

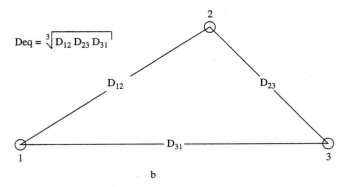

**FIGURE 9.17**   Three-phase line spacing (a) Equilateral and (b) Asymmetrical spacing for developing Equation 9.45

Similarly

$$V_{ac} = \frac{1}{2\pi\epsilon}\left(q_a\ln\frac{D}{r} + q_b\ln\frac{D}{D} + q_c\ln\frac{r}{D}\right)$$

Adding these two equations yields, with $\ln(D/D) = 0$,

$$V_{ab} + V_{ac} = \frac{1}{2\pi\epsilon}\left[2q_a\ln\frac{D}{r} + (q_b + q_c)\ln\frac{r}{D}\right]$$

In a balanced three phase circuit $q_a + q_b + q_c = 0$ so $q_b + q_c = -q_a$ thus

$$V_{ab} + V_{ac} = \frac{1}{2\pi\epsilon}\left(2q_a\ln\frac{D}{r} - q_a\ln\frac{r}{D}\right)$$

$$= \frac{1}{2\pi\epsilon}\left(2q_a\ln\frac{D}{r} + q_a\ln\frac{D}{r}\right)$$

$$= \frac{3q_a}{2\pi\epsilon}\ln\frac{D}{r}$$

But for a balanced three-phase line, $V_{an}$ is offset 30° from $V_{ab}$

$$V_{ab} = \sqrt{3}\,V_{an}(0.866 + j0.5)$$
$$V_{ac} = -V_{ca} = \sqrt{3}\,V_{an}(0.866 - j0.5)$$
$$V_{ab} + V_{ac} = \sqrt{3}\,V_{an}(0.866 + j0.5 + 0.866 - j0.5) = \sqrt{3}\,V_{an}(1.73) = 3\,V_{an}$$

thus

$$3\,V_{an} = \frac{3q_a}{2\pi\epsilon}\ln\frac{D}{r}$$

and

$$V_{an} = \frac{q_a}{2\pi\epsilon}\ln\frac{D}{r} \tag{9.42}$$

Now

$$C_{an} = \frac{q_a}{V_{an}} = \frac{2\pi\epsilon}{\ln\dfrac{D}{r}} \quad \text{F/m} \tag{9.43}$$

Note that $C_{an} = C_{bn} = C_{cn}$ for a balanced symmetrical line.

The charging current $(I_{CH})$ per phase of the line is $V/X_C$ and can be written

$$I_{CH} = V_{an}j\omega C_{an} = V_{an}B_C \quad \text{A/mi} \tag{9.44}$$

Usually the nominal phase voltage is used to calculate nominal charging current.

If the spacing of a three-phase line is asymmetrical, as illustrated in Figure 9.17b, the GMD is used as $D_{eq}$ in the calculation of line capacitance.

$$D_{eq} = \sqrt[3]{D_{12}D_{23}D_{31}}$$

resulting in

$$C_{an} = \frac{q_a}{V_{an}} = \frac{2\pi\epsilon}{\ln\dfrac{D_{eq}}{r}} \quad \text{F/m} \tag{9.45}$$

As always $X_C = 1/(j\,\omega\,C)$ thus

$$X_{cn} = \frac{1}{1609f(4\pi^2\epsilon)}\ln\frac{D_{eq}}{r}$$

$$= \frac{1.779 \times 10^6}{f}\ln\frac{D_{eq}}{r} \quad \Omega\text{mi} \tag{9.46}$$

Equation 9.46 can be broken into

$$X_{cn} = X_a' + X_d'$$

for convenient use of the conductor tables. $D_{eq}$ is used in Table 9.3.

Bundled conductors are handled identically except that $D_{sB}$ values are used in the equations for the conductor radius. Equations 9.32, 9.33, and 9.34 are used to calculate $D_{sB2}$, $D_{sB3}$, and $D_{sB4}$ respectively. The actual radius of the conductor is used instead of $0.7788r$. The distance between phase conductor bundles is measured from the geometric center of the conductor bundles. For asymmetrical bundled conductors Equation 9.45 becomes,

$$C_n = \frac{2\pi\epsilon}{\ln\dfrac{D_{eq}}{D_{sB}}} \quad \text{F/m} \tag{9.47}$$

and Equation 9.46 becomes

$$X_{cn} = \frac{1.779 \times 10^6}{f}\ln\frac{D_{eq}}{D_{sB}} \quad \Omega\text{mi} \tag{9.48}$$

Parallel three-phase circuit geometry is handled the same way as in inductance calculations, except the actual conductor radius is used for $D_s$. An example follows shortly.

The presence of the Earth causes the capacitance of the lines to be larger than it would be from the conductors alone. At normal line spacings the distance to the Earth is larger than the distance between conductors so the additional capacitance to the Earth is very small and is normally ignored in calculations of transmission line performance for balanced loads. To calculate the line to Earth capacitance the Earth is considered to be a perfect neutral between the transmission line conductors and image conductors identically spaced below the Earth as shown in Figure 9.18. This assumption allows the previous equations to be used with a distance correction term to adjust for the presence of the Earth. The distance correction term is

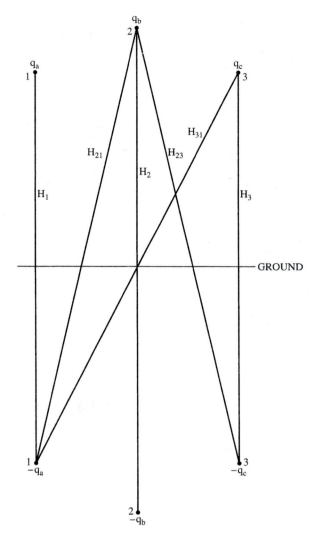

**FIGURE 9.18** Capacitance from line to ground

$$\ln \frac{\sqrt[3]{H_{12}H_{23}H_{31}}}{\sqrt[3]{H_1 H_2 H_3}}$$

and causes Equation 9.48 to become

$$C_n = \frac{2\pi\epsilon}{\ln \dfrac{D_{eq}}{D_{sB}} - \ln \dfrac{\sqrt[3]{H_{12}H_{23}H_{31}}}{\sqrt[3]{H_1 H_2 H_3}}} \qquad \text{F/m} \tag{9.49}$$

## 9.3.5 Three-Phase Capacitance Examples

**Example 9.5:**

Calculate the capacitive reactance per mile, the capacitive susceptance per mile, the charging current, and the charging MVA per mile of the line in Figure 9.19.

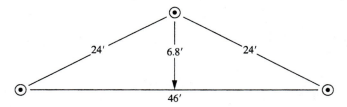

**FIGURE 9.19**　Spacing for Example 9.5. $V_l$ = 220 kV, Drake conductors.

**Solution:**

First calculate $D_{eq}$

$$D_{eq} = \sqrt[3]{D_{12}D_{23}D_{31}} = \sqrt[3]{24 \text{ ft } 24 \text{ ft } 46 \text{ ft}} = 29.812 \text{ ft}$$

From Tables 9.1 and 9.3

$$X_C = X_a' + X_d'$$

$$= 0.0912 \text{ M}\Omega\text{mi} + [0.0999 + 0.812(0.1009 - 0.0999)]\text{M}\Omega\text{mi}$$

$$= 0.0912 \text{ M}\Omega\text{mi} + 0.1007 \text{ M}\Omega\text{mi}$$

$$= 0.1919 \text{ M}\Omega\text{mi per phase to neutral}$$

Checking with Equation 9.46, where r is the actual conductor radius

$$X_C = \frac{1.779 \times 10^6}{60 \text{ Hz}} \ln \frac{29.812 \text{ ft}}{0.045 \text{ ft}} = 0.193 \quad \text{M}\Omega\text{mi}$$

Which is quite close.

Just for fun we will use Equation 9.45 to find $C$, converting to miles as we do.

$$C = \frac{2\pi\epsilon(1609 \text{ m/mi})}{\ln\dfrac{D_{eq}}{D_s}} \quad \text{F/m}$$

$$= \frac{2\pi(8.85 \times 10^{-12} \text{ F/m})(1609 \text{ m/mi})}{\ln\dfrac{29.812 \text{ ft}}{0.045 \text{ ft}}} = 13.773 \text{ n F/mi phase to neutral}$$

and the capacitive reactance is

$$X_c = \frac{1}{\omega C} = \frac{1}{\omega(13.773 \text{nF/mi})} = 0.193 \text{ M}\Omega\text{mi}$$

and the susceptance

$$B_c = \frac{1}{X_c} = \frac{1}{13.773 \text{F/mi}} = 5.189 \quad \mu\text{S/mi}$$

The charging current is

$$I_{ch} = V_{an}B_c = (220 \text{kV}/\sqrt{3})(5.189 \mu\text{S/mi}) = 0.659 \quad \text{A/mi per phase}$$

and the reactive power from charging the capacitance is

$$Q = \sqrt{3}(220 \text{ kV})(0.659 \text{ A/mi}) = 251 \text{ kVAR/mi}$$

**Example 9.6:**

Calculate the per mile $X_C$, $B_C$, $I_{CH}$ for the line shown in Figure 9.20. The line voltage is 750 kV, the conductors are Pheasant in a bundle of four per phase.

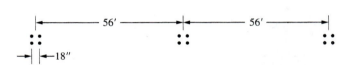

**FIGURE 9.20**   750 kV bundle of four line spacing for Example 9.6

**Solution:**

From the tables we find for Pheasant: Al area = 1272 MCM, 54/19 stranding, 1.382 in diameter, and $X_a' = 0.0897 \text{ M}\Omega\text{mi}$ per conductor.

Bundled conductors necessitate calculating C and $X_C$ because the tables are for a single conductor per phase.
First we calculate $D_{eq}$

$$D_{eq} = \sqrt[3]{56 \text{ ft } 56 \text{ ft } 112 \text{ ft}} = 70.556 \text{ ft}$$

and from Equation 9.34, where $D_s$ is the actual conductor radius

$$D_{sB4} = \sqrt[4]{0.0576 \text{ ft } 1.5 \text{ ft } 1.5 \text{ ft } (1.414 \times 1.5 \text{ ft})} = 0.724 \text{ ft}$$

Now from Equation 9.47

$$C = \frac{2\pi(8.85 \times 10^{-12} \text{ F/m})(1609 \text{ m/mi})}{\ln\dfrac{70.556 \text{ ft}}{0.724 \text{ ft}}} = 19.53 \text{ nF/mi}$$

and from Equation 9.48

$$X_{cn} = \frac{1.779 \times 10^6}{f} \ln \frac{70.556 \text{ ft}}{0.724 \text{ ft}} = 0.135 \text{ M}\Omega\text{mi}$$

and the susceptance

$$B_c = \frac{1}{0.135 \times 10^6 \ \Omega\text{mi}} = 7.365 \quad \mu\text{S/mi}$$

so that the charging current is

$$I_{ch} = \frac{750\text{kV}}{\sqrt{3}} (7.365\mu\text{S/mi}) = 3.189 \text{ A/mi}$$

At this voltage the corona loss is between 5 and 30 kW per mile for conductor diameters between 3 and 1.45 inches. The equivalent diameter of our bundle of four is 17.376 inches, which reduces corona loss to a negligible amount except in very unusual weather conditions. The insulator leakage is typically about 3 kW per mile at this voltage level. Insulator losses are lower at lower voltages.

**Example 9.7:**

Figure 9.21a shows two parallel 138 kV lines using Grosbeak conductors. Calculate $C$, $X_C$, and $I_{CH}$ per mile.

**Solution:**

From the tables for Grosbeak: Al area = 636 MCM, diameter = 0.990 in = 0.0825 ft, stranding is 26/7.

First we calculate the distance between conductors.

$$D_{ab} = D_{bc} = \sqrt{(10 \text{ ft})^2 + (2 \text{ ft})^2} = 10.198 \text{ ft}$$

$$D_{ac} = 20 \text{ ft}$$

$$D_{ab'} = D_{b'c} = D_{a'b} = D_{c'b} = \sqrt{(10 \text{ ft})^2 + (20 \text{ ft})^2} = 22.36 \text{ ft}$$

$$D_{aa'} = D_{cc'} = \sqrt{(20 \text{ ft})^2 + (18 \text{ ft})^2} = 26.91 \text{ ft}$$

Now we calculate the GMD between phases on each side, where subscript p means parallel.

$$D_{pab} = D_{pac} = \sqrt[4]{D^2_{ab}D^2_{bc}} = \sqrt{10.198 \text{ ft } 22.36 \text{ ft}} = 15.10 \text{ ft}$$

$$D_{pca} = \sqrt[4]{(20 \text{ ft})^2(18 \text{ ft})^3} = 18.97 \text{ ft}$$

Thus, the equivalent parallel spacing is that shown in Figure 9.21b. Now we need the effective $D_{eq}$.

$$D_{eq} = \sqrt[3]{(15.1 \text{ ft})^2(18.97 \text{ ft})} = 16.29 \text{ ft}$$

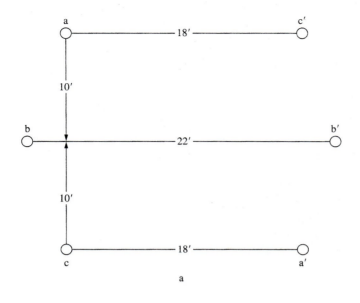

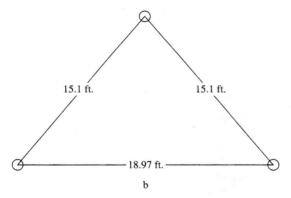

**FIGURE 9.21** Spacing of parallel circuit for Example 9.7. Grosbeak conductor. (a) Actual spacing and (b) Equivalent spacing for parallel circuit.

We must now calculate the GMR of each equivalent conductor in Figure 9.21b. $D_s$ is the actual conductor radius.

Equivalent conductor a − a′:

$$GMR = \sqrt{(26.91 \text{ ft})(0.0412 \text{ ft})} = 1.053 \text{ ft}$$

Equivalent conductor b − b′:

$$GMR = \sqrt{(22 \text{ ft})(0.0412 \text{ ft})} = 0.952 \text{ ft}$$

Equivalent conductor c − c′, the same as a − a′:

$$GMR = \sqrt{(26.91 \text{ ft})(0.0412 \text{ ft})} = 1.053 \text{ ft}$$

Now the equivalent parallel GMR is

$$D_{ps} = \sqrt[3]{(0.819\ \text{ft})^2(0.952\ \text{ft})} = 0.861\ \text{ft}$$

We now have sufficient information to calculate capacitance.

$$C_n = \frac{2\pi(8.85 \times 10^{-12}\ \text{F/m})(1609\ \text{m/mi})}{\ln\dfrac{16.29\ ft}{0.861\ ft}} = 30.43\ \text{nF/mi}$$

From which the reactance is

$$X_c = \frac{1}{2\pi(60\ \text{Hz})(30.43\text{nF/mi})} = 87.17 \quad \text{k}\Omega/\text{mi}$$

and the susceptance is

$$B = \frac{1}{87.17\text{k}\Omega/\text{mi}} = 11.47 \quad \mu\text{S/mi}$$

Finally the charging current is

$$I_{ch} = (138\text{kV}/\sqrt{3})(11.47\mu\text{S/mi}) = 0.914\ \text{A/mi per phase}$$

## 9.4

## TRANSMISSION LINE EQUIVALENT CIRCUITS

As with other circuits and networks, an equivalent circuit simplifies the calculations that are made to obtain transmission line performance estimations. The equivalent circuit that is used for transmission line calculations depends on the transmission line length; the longer the transmission line is the more complex the equivalent circuit must be for accurate results. Recall that in Chapter 2 we used the series reactance of the line for most calculation. For many short distribution lines this is entirely adequate. Generally for transmission lines less than 50 miles long the equivalent circuit of Figure 9.22a is accurate enough. Distribution lines are less than 50 miles long so this is a good equivalent circuit for overhead distribution lines. Transmission lines that are between 50 and 100 miles long can be modeled with the $\pi$ equivalent circuit of Figure 9.22b with sufficient accuracy for most purposes. The equivalent circuits of Figure 9.22a and b are called lumped parameter because the parameter represented in the equivalent circuit is lumped into one equivalent device. For example, the resistance is distributed along the line, but the equivalent circuit considers it to be a resistor in one place. Long transmission lines must take into account the distributed nature of the transmission line parameters, as illustrated in Figure 9.22c. The lumped parameter $\pi$ circuit can be used with errors less than 5% for most lines up to about 200 miles when this accuracy is sufficient.

Throughout this section the bold italic underlined letter parameters will be the pu length value. For example $\underline{\boldsymbol{Z}}$ = impedance per mile. The normaly signified values will be the lumped parameter value. For example $Z = \underline{\boldsymbol{Z}} \times 1$ = equivalent impedance.

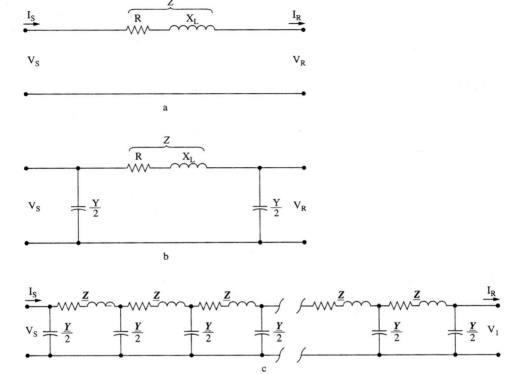

**FIGURE 9.22** Transmission line equivalent circuits including: (a) Short line (less than 50 miles) (b) Medium length line (50 miles to 120 miles) and (c) Long lines (over 120 miles)

Subscript R will refer to receiving end quantities, such as $V$ and $I$, and subscript S will refer to sending end quantities. We will assume balanced three-phase loads.

## 9.4.1 The Short Line Equivalent Circuit

The short line equivalent circuit consists of lumped series impedance parameters. This equivalent circuit, shown in Figure 9.22a, is accurate to about 50 miles. There is no shunt path so

$$V_S = V_R + I_R Z \tag{9.50}$$

The voltage regulation (VR) is calculated in the same manner as for transformers (shown in Chapter 4.7). Thus, from Equation 4.22 we obtain

$$VR = \frac{|V_{RNL}| - |V_{RFL}|}{|V_{RFL}|} \tag{9.51}$$

where the subscripts $FL$ and $NL$ are full load and no load respectively. The percent regulation is found by multiplying Equation 9.51 by 100.

### 9.4.2 Medium Length Line Equivalent Circuit

The charging current ($I_{CH}$) must be taken into account for lines over 50 miles in length. The $\pi$ equivalent circuit shown in Figure 9.22b is the preferred one for medium length lines. The capacitive susceptance is split in half with one half placed at the sending end, and the other half placed at the receiving end. The conductance is very low in most lines without corona so

$$Y = G + jB_c \approx jB_c$$

Half of $I_{CH}$ is removed at the receiving end and half at the sending end, so referring to Figure 9.22b we write

$$I_S = I_R + I_{CH} \tag{9.52}$$

$$V_S = V_R + (I_R + I_{CH}/2)Z \tag{9.53}$$

Voltage regulation is defined in Equation 9.51, but is harder to calculate with this circuit.

**Example 9.8:**

Calculate the equivalent circuit components for a medium length 138 kV, 100 mile transmission line. Use Hawk conductors with an 18 foot equilateral spacing. Also calculate $V_S$ and $VR$ for a 138 kV 60 MVA, three-phase load with a 0.9 power factor. $V_R = 138$ kV.

**Solution:**

Using Table 9.1 we find for Hawk: Al area = 477 MCM, OD = 0.857 in, $D_s = 0.0289$ ft. $X_a = 0.43$ $\Omega$/mi, $X_a' = 0.0988$ M$\Omega$mi, and $R_{ac}$ at 50° = 0.2120 $\Omega$/mi. From Tables 9.2 and 9.3 we find that for our spacing $X_d = 0.3507$ $\Omega$/mi, and $X_d' = 0.0858$ M$\Omega$mi. Now we can solve for the equivalent circuit components.

$$\underline{X}_L = X_a + X_d = (0.430 + 0.3507)\text{M}\Omega\text{mi} = 0.780 \ \Omega/\text{mi}$$

$$X_L = 100\underline{X}_L = 78 \ \Omega$$

$$Z = R + j\underline{X}_L = 21.2\Omega + J78\Omega = 80.83\Omega\angle74.8°$$

$$\underline{X}_c = X_a' + X_d' = (0.0988 + 0.0858)\text{M}\Omega\text{mi} = 0.1846 \ \text{M}\Omega/\text{mi}$$

$$\underline{Y} \approx \underline{B}_c = \frac{1}{\underline{X}_c} = 5.417 \ \mu\text{S}/\text{mi}$$

$$B = 100(5.417\mu\text{S}/\text{mi}) = 541 \ \mu\text{S}$$

$$B/2 = 270.85 \ \mu\text{S}$$

Calculating $I_{CH}$ and the line current we obtain

$$I_{CH} = (138\text{kV}/\sqrt{3})(541\mu\text{S}) = 43.1 \ \text{A}$$

$$I_R = \frac{60\text{MVA}}{\sqrt{3}\,(138\text{kV})} = 251 \text{ A}$$

The load impedance is equal to

$$Z_L = \frac{138\text{kV}/\sqrt{3}}{251\text{A}} \angle\arccos(0.9) = 317\Omega\angle25.8° \text{ per phase}$$

$V_R$ is 138 kV, 79.67 kV phase voltage, at the reference angle, so from Equation 9.53 we see

$$V_S = 79.67\text{kV} + (251\text{A}\angle{-25.8°} + 21.6\text{A}\angle90°)(80.83\Omega\angle74.8°)$$

$$= 79.67\text{kV} + (226 - j109.24 + j21.6)\text{A}(80.83\Omega\angle74.8°)$$

$$= 79.67\text{kV} + 19.592\text{kV}\angle53.6° = 92.65\text{kV}\angle9.8°$$

The source voltage is not equal to the no load voltage because one half the charging current flows through the series impedance. The no load voltage is

$$V_{RNL} = V_S - (I_{CH}/2)Z = 92.65\text{kV}\angle9.8° - (21.6\text{A}\angle90°)(80.84\Omega\angle74.80°$$

$$= 92.65\text{kV}\angle9.8° - 1.76\text{kV}\angle164.8° = 94.23\text{kV}\angle9.35°$$

Finally, from Equation 9.51

$$\text{VR} = \frac{94.23\text{kV} - 79.67\text{kV}}{79.67\text{kV}} = 0.183 \text{ or } 18.3\%$$

The reader will please note that this problem could have easily been solved with pu values if an appropriate base is chosen. If the calculations are for a larger system of which the line is only a part, especially with one or more transformers, pu expressions would be almost a necessity for reasonably easy calculations.

### 9.4.3 Generalized Circuit Constants

Referring again to Figure 9.22b we can write

$$V_S = (I_R + I_{CH}/2)Z + V_R = \left(I_R + \frac{V_R Y}{2}\right)Z + V_R$$

Rearranging terms we obtain

$$V_S = \left(\frac{YZ}{2} + 1\right)V_R + ZI_R \tag{9.54a}$$

and we write the sending current as

$$I_S = V_S\left(\frac{Y}{2}\right) + V_R\left(\frac{Y}{2}\right) + I_R \tag{9.55}$$

Substituting Equation 9.54a into Equation 9.55

$$I_S = V_R\left(\frac{YZ}{2} + 1\right)\left(\frac{Y}{2}\right) + V_R\left(\frac{Y}{2}\right) + I_R + ZI_R\left(\frac{Y}{2}\right)$$

$$= V_R\left(\frac{Y}{2}\right)\left(\frac{YZ}{2} + 2\right) + I_R\left(1 + \frac{ZY}{2}\right)$$

Thus

$$I_S = V_R Y\left(1 + \frac{YZ}{4}\right) + I_R\left(1 + \frac{ZY}{2}\right) \tag{9.54b}$$

Equations 9.54a and b are in the form of generalized four terminal network equations. The general form is

$$V_S = AV_R + BI_R \tag{9.56}$$

$$I_S = CV_R + DI_R \tag{9.57}$$

where

$$A = D = YZ/2 + 1 \tag{9.58a}$$

$$B = Z \tag{9.58b}$$

$$C = Y(YZ/4 + 1) \tag{9.58c}$$

The **ABCD** constants are called generalized circuit constants. They are usually complex numbers. They can be evaluated as follows.

Let $I_R = 0$, in other words no load, then

$$A = V_S/V_R \tag{9.59a}$$

$$C = I_S/V_R \tag{9.59b}$$

Now let $V_R = 0$, the output terminal short circuited, and

$$B = V_S/I_R \tag{9.59c}$$

$$D = I_S/I_R \tag{9.59d}$$

A bit of algebra allows us to calculate the receiving current and voltage in terms of the sending current and voltage

$$V_R = DV_S - BI_S \tag{9.60}$$

$$I_R = {-}CV_S + AI_S \tag{9.61}$$

where $(AD - BC) = 1$

The general four terminal network is shown in Figure 9.23a. The **ABCD** constants vary with the contents of the four terminal network.

The use of the **ABCD** constants has subsided with the advent of computers because a computer program can use the most precise equivalent circuit available for long, medium, and short lines with equal ease. The **ABCD** constants are very useful for calculator and general purpose mathematics solver programs computations, though.

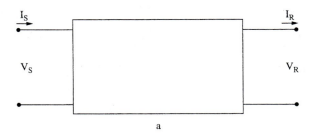

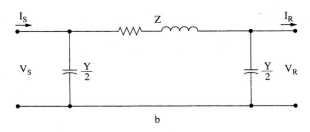

**FIGURE 9.23** Generalized circuit constants (a) Four terminal network and (b) $\pi$ network

### Example 9.9:

Derive the **ABCD** constants of Figure 9.23b, and calculate their value for the line in Example 9.8.

**Solution:**

Using Equations 9.59a–d and letting

$$Z' = 1/(Y/2) = 2/Y$$

we solve for each constant. Let $I_R = 0$.

First **A**

$$V_R = V_S\left(\frac{Z'}{Z + Z'}\right)$$

$$A = \frac{V_S}{V_R} = \frac{Z + Z'}{Z'} = \frac{Z + 2/Y}{2/Y} = \frac{YZ}{2} + 1$$

Now for **C**

$$I_S = I_R + \frac{V_R}{Z'} + \frac{V_S}{Z'} = 0 + \frac{V_R}{Z'} + V_R\left(\frac{Z + Z'}{Z'}\right)\left(\frac{1}{Z'}\right)$$

$$= V_R\left(\frac{1}{Z'}\right)\left(1 + \frac{Z + Z'}{Z'}\right) = V_R\left(\frac{Y}{2}\right)\left(\frac{YZ}{2} + 2\right)$$

$$C = \frac{I_S}{V_R} = Y\left(\frac{YZ}{4} + 1\right)$$

Now let $V_R = 0$. Solving for **B**

$$B = \frac{V_S}{I_R} = \frac{V_S}{\left(\frac{V_S}{Z}\right)} = Z$$

Now for **D**. The receiving voltage is zero so

$$I_R = \frac{V_S}{Z}$$

$$I_S = \frac{V_S}{Z} + \frac{V_S}{Z'}$$

$$D = \frac{I_S}{I_R} = \frac{\dfrac{V_S}{Z} + \dfrac{V_S}{Z'}}{\dfrac{V_S}{Z}} = 1 + \frac{Z}{Z'}$$

$$D = \frac{YZ}{2} + 1 = A$$

The values of the **ABCD** constants are evaluated by multiplying the value pu distance by the distance and using Equations 9.58a, b, and c. In Example 9.8 we found the lumped parameter values. Evaluating for **ABCD** we obtain

**B** $= Z = 80.83 \; \Omega \; \angle 74.8°$

$Y/2 = 271 \; \mu S \; \angle 90°$

**A** $=$ **D** $= (YZ)/2 + 1 = 21.905 \times 10^{-3} \; \angle 164.8° + 1 = 0.978 \; \angle 0.336°$

**C** $= Y[(YZ)/4 + 1]$

$\quad = 542 \; \mu S \; \angle 90°[(135.5 \; \mu S \; \angle 90°)(80.83 \; \Omega \; \angle 74.8°) + 1]$

$\quad = {}^{-}1.556 \times 10^{-6} + j(536.27 \times 10^{-6}) = 536.28 \times 10^{-6} \; \angle 90.16°$

Appendix C gives some generalized constants for some useful four terminal networks.

## 9.4.4 Long Transmission Line Equivalent Circuit, Traveling Waves

To accurately calculate the current and voltage along a long line we must take into account that the series impedance $(R + jX_L)$ and the shunt admittance $(G + jB_C)$ are distributed along the line. In Figure 9.24 the long line distributed parameters are illustrated as lumped parameters per unit distance, as is commonly done. Each section, $\Delta d$, represents a section of line, usually a mile or a kilometer.

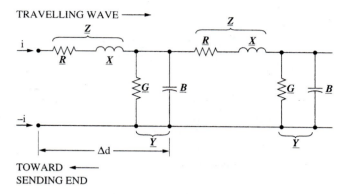

TOWARD ◄───
SENDING END

**FIGURE 9.24** Generalized long line section

If a switch is closed at the sending end of the line in Figure 9.22c the line must energize by the line capacitance and inductance charging one section at a time, from left to right. The voltage and current then propagate by one small element of the line at a time. The charging current of the segment of the line the wave front is at is the voltage at that section divided by the impedance of that section of line. The impedance of a section of the line is called the line characteristic impedance ($Z_C$). Referring to Figure 9.24 we see

$$\frac{\Delta v}{\Delta d} = -L\frac{\Delta i}{\Delta t} - Ri \tag{9.62}$$

$$\frac{\Delta i}{\Delta d} = -C\frac{\Delta v}{\Delta t} - Gv \tag{9.63}$$

Letting R and G equal zero, a lossless line, and dividing Equation 9.62 by Equation 9.63 we find

$$\frac{-\dfrac{\Delta v}{\Delta d}}{-\dfrac{\Delta i}{\Delta d}} = \frac{L\dfrac{\Delta i}{\Delta t}}{C\dfrac{\Delta v}{\Delta t}}$$

$$\left(\frac{\Delta v}{\Delta i}\right)^2 = Z_C^2 = \frac{L}{C}$$

$$Z_C = \sqrt{\frac{L}{C}} \tag{9.64}$$

Typical transmission line characteristic impedance values are from around 400Ω to around 800Ω. If the loss components are not zero a somewhat more complicated process of mathematics gives the following equation for $Z_C$.

$$Z_c = \sqrt{\frac{\underline{Z}}{\underline{Y}}} \qquad (9.65)$$

Where $\underline{Z}$ is the complex line impedance per unit length, and $\underline{Y}$ is the complex line admittance pu length.

Equation 9.65 yields the complex characteristic impedance, $Z_C = R + jX$, which reduces to Equation 9.64 if $R$ and $G$ are zero. Try it. $G$ is normally nearly zero and can be neglected, but $R$ cannot. Because of the power loss in R the voltage wave front decreases in amplitude as it travels from the sending to the receiving end of a line.

As $\Delta d \rightarrow 0$ Equations 9.62 and 9.63 become differential equations. The solutions to them are in section 9.4.5.

If we multiply Equations 9.62 and 9.63 together, once again with R and G equal to zero, and solve for t we have the time it takes the voltage (or current) waveform to travel $\Delta d$ in Figure 9.24.

$$\frac{\Delta v}{\Delta d} \frac{\Delta i}{\Delta d} = LC \frac{\Delta i \Delta v}{(\Delta t)^2}$$

$$\frac{(\Delta t)^2}{(\Delta d)^2} = \frac{1}{v^2} = LC$$

If the $\Delta d = 1$, a unit distance then

$$\Delta t = \sqrt{LC} \text{ per unit length} \qquad (9.66)$$

$\Delta t \times l =$ time to propagate the length of the line. $\qquad (9.67)$
and the velocity of propagation, $v$, is,

$$v = \frac{1}{\sqrt{LC}} \qquad \text{unit distance per second} \qquad (9.68)$$

The square root of the complex impedance times the complex admittance is not the complex time, instead it is the propagation constant, $\gamma$.

$$\gamma = \sqrt{\underline{ZY}} = \sqrt{(\underline{R} + j\underline{X}_L)(\underline{G} + j\underline{B}_C)} \qquad (9.69)$$

Which, after some complex math can be expressed as

$$\gamma = \alpha + j\beta \qquad (9.70)$$

where

$\alpha =$ attenuation constant per unit length (1 neper = 8.68 dB)

$\qquad \approx \underline{R}/(2\underline{Z})$

and

$\beta =$ phase constant in radians pu length

$\qquad = (2\pi)/\lambda \qquad (9.71)$

where $\lambda =$ wavelength in the same units as $\beta$

The velocity of the wave on the line is

$$v = f\lambda \tag{9.72}$$

The total attenuation (reduction in amplitude) along the line is $\alpha l$, and the total phase shift is $\beta l$.

Summing up, after the sending end switch is closed energizing the line the voltage and current move down the line as a wave front. The impedance of the line to the travelling wave is $\sqrt{L/C}$, and the velocity of the wave front is $1/\sqrt{LC}$. The voltage amplitude is reduced along the line by $\alpha l$, and the total phase shift of the wave front along the line is $\beta l$. At the end of the line all of the power that reaches the receiving end is absorbed if $Z_R = Z_C$. In transmission lines the object is to transport power with a minimal voltage loss so normally $Z_R > Z_C$.

### 9.4.5 Long Transmission Line: Reflections

The portion of the incident traveling wave front that is not absorbed by the receiving impedance is reflected and travels back toward the sending end where, if it is not all absorbed, a portion is re-reflected toward the receiving end. This process continues until all an equilibrium condition is established. If $Z_R > Z_C$ the reflected voltage ($V_{rfl}$) has the same instantaneous polarity as the incident voltage ($V_{inc}$). If $Z_R < Z_C$ the polarity of the incident voltage is reversed at reflection.

Let us first consider a line energized with a dc voltage, as shown in Figure 9.25a. Let the line resistance and conductance be negligible and the source and load impedances be pure resistance. The voltage wave begins to travel at $t = 0$, when the switch is closed. We will monitor a point just on the source side of the switch. Normally we would monitor just on the line side of the switch. As the wave front moves across the switch it encounters $Z_c = \sqrt{L/C}$. The incident voltage divides between the source impedance and the characteristic impedance at this point, with the amplitude of the wave traveling on down the line being the voltage dropped across $Z_C$. This is shown in Figure 9.25b. The voltage at the observation point does not change until the reflected wave returns from the load. This is twice the time it takes the wave to travel from the source to the load, $t = 2l\sqrt{LC}$.

At the end of the line the incident wave encounters the load impedance. If $Z_C \neq Z_R$ then a portion is reflected. To calculate the reflected voltage refer to Figure 9.26a. The maximum available voltage at the end of the line occurs if all of the incident wave is reflected. This can occur if none is absorbed, or $Z_R = \infty$. In this case all of the incident voltage is reflected in the same polarity as the incident voltage and the voltage on the line is twice the incident voltage, $V_{inc} + V_{rfl} = 2V_{inc}$. If we close the switch just at the point the voltage at the end of the line reaches maximum, this voltage divides between $Z_C$ and $Z_R$. Now, we close the switch in Figure 9.26a at $t = l\sqrt{LC}$. After closure

$$V_{rfl} = 2V_{inc}\frac{Z_R}{Z_C + Z_R} - V_{inc} = \frac{2V_{inc}Z_R}{Z_R + Z_C} - \frac{V_{inc}(Z_R + Z_C)}{Z_R + Z_C}$$

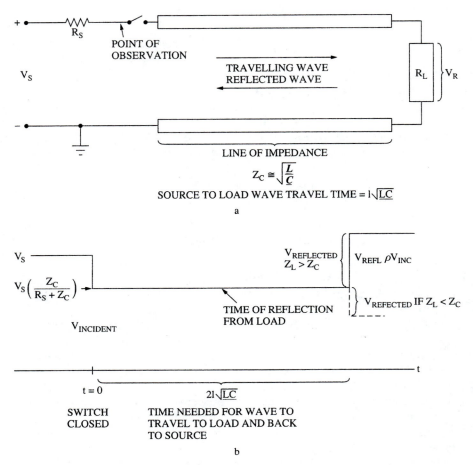

**FIGURE 9.25** Reflections on a dc line (a) Line and source and (b) Voltage at observation point

$$= V_{inc} \frac{2Z_R - Z_R - Z_c}{Z_R + Z_c} = V_{inc} \frac{Z_R - Z_c}{Z_R + Z_c}$$

The ratio of the reflected voltage to the incident voltage is the reflection coefficient ($\Gamma$). From the equation above we see

$$\Gamma = \frac{V_{rfl}}{V_{inc}} = \frac{Z_R - Z_c}{Z_R + Z_c} \tag{9.73}$$

If $Z_R > Z_C$ then $\Gamma$ is $+$ and $V_{rfl}$ adds to $V_{inc}$

If $Z_R < Z_C$ then $\Gamma$ is $-$ and $V_{rfl}$ subtracts from $V_{inc}$

To keep the volt amperes constant at a reflection the reflected current polarity is opposite to the voltage so

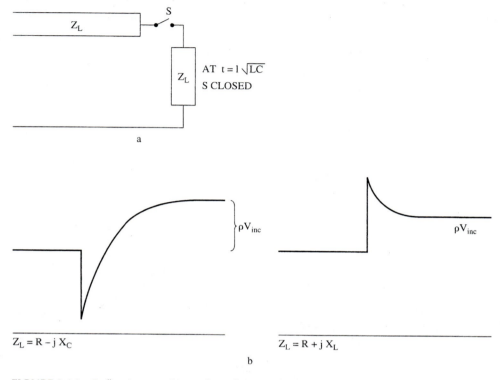

**FIGURE 9.26** Reflections (a) Circuit for reflection coefficient equation development and (b) DC reflections from complex loads

$$\Gamma_I = \frac{I_{inc}}{I_{rfl}} = \frac{Z_R - Z_c}{Z_R + Z_c} \qquad (9.74)$$

Notice that the reflection coefficient magnitude is the same.

The reflections of dc voltage from a complex impedance are shown in Figure 9.26b. The capacitive load pulls the voltage down while it charges and the inductive load causes voltage to peak until the incident current begins to rise. The final value of the reflected wave is the reflection coefficient from the resistances and characteristic impedance, $\sqrt{L/C}$, of the line times the incident voltage for a dc wavefront. AC waves reflect according to the complex receiving and line impedances.

### Example 9.10:

Draw the voltage reflection waveform for the dc line shown in Figure 9.27a for the first three reflections. Also calculate the time for the voltage wave to travel from one end of the line to the other. The dc line uses Hawk conductors spaced 49 feet apart. Ignore $R_{line}$ in $Z_C$ calculation.

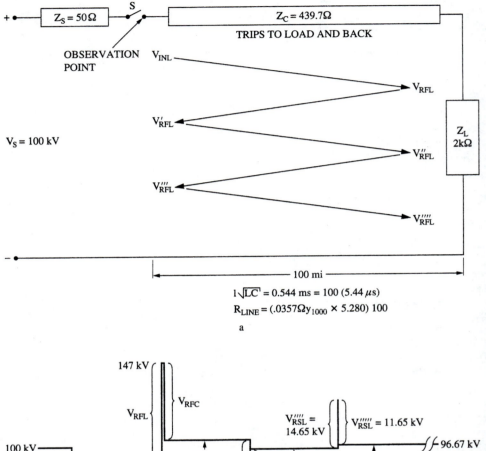

$$l\sqrt{LC'} = 0.544 \text{ ms} = 100 \ (5.44 \ \mu s)$$
$$R_{LINE} = (.0357\Omega y_{1000} \times 5.280) \ 100$$

a

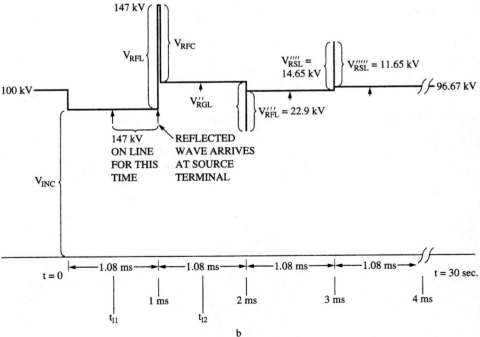

b

**FIGURE 9.27** Example 9.10 circuit and waveform (a) Circuit and (b) Voltage waveform at observation point. $t_l$ = time of reflection from load

**Solution:**

We find in the conductor tables that for Hawk: $X_a = 0.430\ \Omega/\text{mi}$, $X_d = 0.4722\ \Omega/\text{mi}$, $X_a' = 0.0988\ \text{M}\Omega\text{mi}$, and $X_d' = 0.1155\ \text{M}\Omega\text{mi}$. We now calculate L and C per mile.

$$X_L = X_a + X_d = 0.430\ \Omega/\text{mi} + 0.4722\ \Omega/\text{mi} = 0.902\ \Omega/\text{mi}$$

$$L = \frac{X_L}{\omega} = 2.39\ \text{mH/mi}$$

$$X_C = X'_a + X'_d = 0.0988\ \text{M}\Omega/\text{mi} + 0.1155\ \text{M}\Omega/\text{mi} = 0.2143\ \text{M}\Omega/\text{mi}$$

$$C = \frac{1}{X_C} = 12.378\ \text{nF/mi}$$

so the characteristic impedance is

$$Z_c \approx \sqrt{\frac{L}{C}} = \sqrt{\frac{2.39\ \text{mH/mi}}{12.378\ \text{nF/mi}}} = 439.7\Omega$$

The time for the wave to travel a mile is

$$t = \sqrt{LC} = 5.4425\mu\text{S/mi}$$

so the time needed to travel 100 miles is

$$t_{100\ mi} = 0.544\ \text{mS}$$

The incident voltage is

$$V_{inc} = V_S \frac{Z_c}{Z_S + Z_c} = 100\text{kV}\frac{439.7\Omega}{439.7\Omega + 50\Omega} = 89.7\ \text{kV}$$

After travelling 100 miles the incident wave reaches the load and reflects.

$$V_{rfl} = V_{inc}\Gamma = V_{inc}\frac{Z_R - Z_c}{Z_R - Z_c}$$

$$= 89.7\text{kV}\frac{2\text{k}\Omega - 439.7\Omega}{2.4397\text{k}\Omega} = 89.7\text{kV}(0.6397) = 57.36\text{kV}$$

The reflected voltage adds to the incident voltage because it is the same polarity, since $Z_R > Z_C$. Notice the total voltage on the line behind the reflected wave is 89.7 kV + 57.36 kV = 147 kV. This voltage must be anticipated when selecting the equipment BIL levels.

The reflected voltage now travels back to the source, and is reflected. $Z_S < Z_C$ so the voltage reflected back from the source is opposite in polarity to the voltage incident on the source. The source side reflection coefficient is

$$\Gamma_S = \frac{Z_S - Z_c}{Z_S + Z_c} = \frac{50\Omega - 439.7\Omega}{50\Omega + 439.7\Omega} = -0.7958$$

and the reflected voltage is

$$V'_{rfl} = (-0.7958)(57.36 \text{ kV}) = -45.6 \text{ kV}$$

The reflected voltage from the source causes an immediate change in our observed total line voltage because our observation point is at the source, as shown in Figure 9.27b. The total line voltage is now $147 \text{ kV} - 45.6 \text{ kV} = 101 \text{ kV}$. The $-45.6 \text{ kV}$ reflected voltage wave travels back toward the load subtracting from the 147 kV as it goes.

When it reaches the load an amount governed by the reflection coefficient reflects again.

$$V''_{rfl} = (-45\text{kV})(0.6397) = -28.78 \text{ kV}$$

This reflected voltage is of the same polarity as the incident voltage that just arrived, not the total line voltage. Now the total line voltage at the load is $102 \text{ kV} - 28.78 \text{ kV} = 73.22 \text{ kV}$. As the reflected wave $V_{rfl}''$ travels back toward the source the total line voltage adjusts to 73.22 kV as it travels. After 0.544 ms the voltage reflected from the load arrives back at the source and is reflected again.

$$V'''_{rfl} = (-0.7958)(-28.78\text{kV}) = 22.9 \text{ kV}$$

This reflection brings the line voltage to $73.22 \text{ kV} + 22.9 \text{ kV} = 96.12 \text{ kV}$ as $V'''_{rfl}$ travels back toward the load. The reflections continue until the dc steady state conditions

$$V_R = V_S \frac{Z_L}{Z_S + Z_L + R_{DC}} = \frac{100\text{kV} \times 2\text{k}\Omega}{50\Omega + 2\text{k}\Omega + 18.85\Omega} = 96.67\text{kV}$$

are met.

$Z_C$, $Z_S$, and $Z_L$ are usually complex in transmission line circuits, but other than the waveform adjustments at the line ends, as shown in Figure 9.26b, and some loss along the line causing $V_{inc}$ and $V_{rfl}$ drop along the line, the reflection process is the same for any dc line.

The same reflection process occurs at the line ends with an ac voltage applied but the reactance causes the phase of the continuously changing ac wave to either retard, inductive $Z_R$, or advance, capacitive $Z_R$, as it reflects. Because an ac $V_{inc}$ is continuously changing, the voltage reflected from it is continuously varying. The incident and reflected ac voltages add and subtract from each other along the line, just as in the dc case, thus the total line voltage at any point is an ac voltage whose instantaneous voltage is the algebraic sum of the waves travelling along the line in opposite directions. This process, illustrated in Figure 9.28, causes the ac voltage to be higher at some locations on a line than at others. The ratio of the maximum to the minimum voltage on the line is the voltage standing wave ratio (VSWR).

$$\text{VSWR} = \frac{V_{max}}{V_{min}} \qquad (9.75)$$

It is useful to express the VSWR in terms of the reflection coefficient. Note that

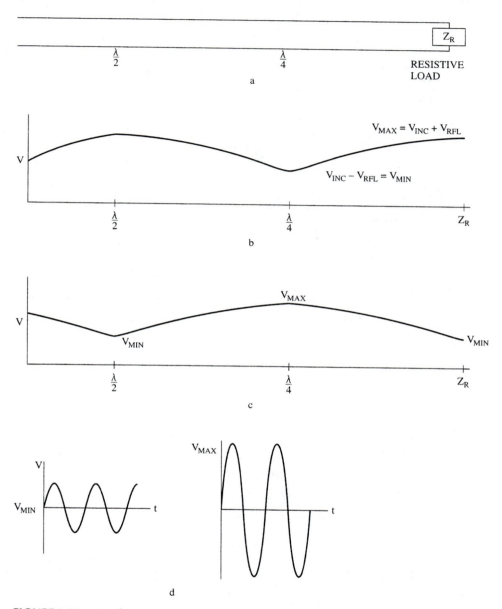

**FIGURE 9.28** Standing waves (a) Line (b) rms voltage along line if $Z_R$ is resistive and larger than $Z_C$ (c) rms voltage along line if $Z_R$ is resistive and smaller than $Z_C$ and (d) Waveform of voltage at $V_{min}$ and $V_{max}$

$$V_{max} = V_{inc} + V_{rfl} = V_{inc} + V_{inc}\Gamma = V_{inc}(1 + \Gamma)$$

and

$$V_{min} = V_{inc} - V_{rfl} = V_{inc} - V_{inc}\Gamma = V_{inc}(1 - \Gamma)$$

Using the definition of VSWR

$$\text{VSWR} = \frac{V_{max}}{V_{min}} = \frac{1 + \Gamma}{1 - \Gamma}$$

The VSWR of the circuit of Example 9.10 would be 4.549 if the voltage source was ac. The waves are called standing waves because the maximum and minimum voltages occur at the stationary locations on the line, as illustrated in Figure 9.28. If the load is complex the maximum and minimum voltage points are not at integer multiples of λ/4 from the end of the line, but are still spaced λ/4 apart. The current wave reflects, as does the voltage wave, but in the opposite polarity.

The characteristic impedance of a line with the series R neglected ($\sqrt{L/C}$) is called the surge impedance. This impedance is sometimes used to estimate the maximum power that can be transferred by a line into a matched load ($Z_R = Z_C$).

A transient voltage (such as that often induced by a lightning strike) propagates down a line at a velocity of $1/\sqrt{LC}$, as shown in Figure 9.29. The transient voltage adds to, or subtracts from depending on polarity, the voltage already on the line. The line losses attenuate the transient wave as it travels along the line, and at the end of the line it reflects from an unmatched load like any other wave.

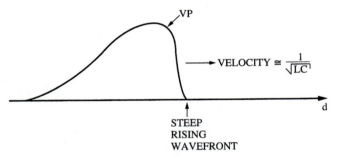

**FIGURE 9.29**   Transient from lightning on long line

## 9.4.6 Long Transmission Line Equivalent Circuit: Equation Solution

The solution of Equations 9.62 and 9.63 involve replacing the deltas with derivatives (for example ΔV/Δt with dV/dt) and solving the differential equations. The solution steps are somewhat lengthy and are presented in several of the references in Appendix E. We will present the solutions here. The current and voltage at any point along the line, referenced to the receiving current and voltage, are given in the following two equations.

$$V = \frac{V_R + I_R Z_c}{2} e^{\gamma l_R} + \frac{V_R - I_R Z_c}{2} e^{-\gamma l_R} \qquad (9.76)$$

$$I = \frac{(V_R/Z_c) + I_R}{2} e^{\gamma l_R} - \frac{(V_R/Z_c) - I_R}{2} e^{-\gamma l_R} \qquad (9.77)$$

where

$\gamma = \sqrt{ZY}$     the propagation constant

$\gamma = \alpha + j\beta$

$l_R$ = distance from the receiving end of the line

The first term of each equation is the incident voltage (9.76) or current (9.77), and the second term is the reflected voltage or current, Equations 9.76 and 9.77 respectively. The first terms increase in magnitude toward the source and the second terms decrease, generally. The voltage, current, impedance, and propagation constant are all complex quantities.

Equations 9.76 and 9.77 are sometimes easier to solve if $\gamma$ is broken into its real and imaginary parts as follows.

$$\sinh \theta = \frac{e^{\theta} - e^{-\theta}}{2} \qquad (9.80)$$

$$\cosh \theta = \frac{e^{\theta} + e^{-\theta}}{2} \qquad (9.81)$$

Recall that if a quantity $\theta$ is complex that

$$e^{\theta} = e^{\theta_R} e^{j\theta_I} \qquad (9.82)$$

where

$\theta_R$ = real part

$\theta_I$ = imaginary part

Some algebraic manipulation of Equations 9.76 and 9.77 result in the following forms

$$V = V_R \frac{e^{\gamma l_R} + e^{-\gamma l_R}}{2} + I_R Z_c \frac{e^{\gamma l_R} - e^{-\gamma l_R}}{2}$$

$$I = I_R \frac{e^{\gamma l_R} + e^{-\gamma l_R}}{2} + \frac{V_R}{Z_c} \frac{e^{\gamma l_R} - e^{-\gamma l_R}}{2}$$

which can be written as

$$V = V_R \cosh \gamma l_R + I_R Z_c \sinh \gamma l_R$$

$$I = I_R \cosh \gamma l_R + \frac{V_R}{Z_c} \sinh \gamma l_R$$

If we let $l_R = l$, the entire line distance, in the preceding two equations we obtain the source voltage and current.

$$V_S = V_R \cosh\gamma l + I_R Z_c \sinh\gamma l \qquad (9.83)$$

$$I_S = I_R \cosh\gamma l + \frac{V_R}{Z_c} \sinh\gamma l \qquad (9.84)$$

Referring back to Equations 9.56 and 9.57 we notice that they are the same as Equations 9.83 and 9.84 if the general constants are as follows

$$A = D = \cosh \gamma l \qquad (9.85)$$

$$B = Z_C \sinh \gamma l \qquad (9.86)$$

$$C = (1/Z_C) \sinh \gamma l \qquad (9.87)$$

The long line equivalent circuit can be represented with lumped parameters, as shown in Figure 9.30. The long line lumped parameters are primed. To be equivalent to the medium length line the coefficients of Equations 9.83 and 9.84 and Equations 9.56 and 9.57 must be equal. Thus from B

$$Z' = Z_c \sinh\gamma l \qquad (9.88)$$

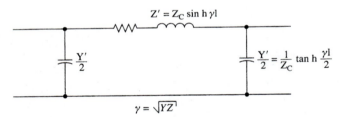

**FIGURE 9.30**   Lumped parameter long line $\pi$ equivalent circuit

For A and D

$$\frac{Z'Y'}{2} + 1 = \cosh\gamma l$$

$$\frac{Y' Z_c \sinh\gamma l}{2} + 1 = \cosh\gamma l$$

$$\frac{Y'}{2} = \frac{\cosh\gamma l - 1}{Z_c \sinh\gamma l}$$

$$\frac{Y'}{2} = \frac{1}{Z_c} \frac{\cosh\gamma l - 1}{\sinh\gamma l}$$

From the hyperbolic function identities (check these with the exponential forms if you wish)

$$\tanh \frac{\gamma l}{2} = \frac{\cosh \gamma l - 1}{\sinh \gamma l}$$

Thus

$$\frac{Y'}{2} = \frac{1}{Z_c} \tanh \frac{\gamma l}{2} \qquad (9.89)$$

Equations 9.62 and 9.63 can be solved for receiving values in terms of sending values yielding

$$V_R = V_S \cosh \gamma l - I_S Z_c \sinh \gamma l \qquad (9.90)$$

$$I_R = I_S \cosh \gamma l - \frac{V_S}{Z_c} \sinh \gamma l \qquad (9.91)$$

Many calculators and mathematics-solving computer programs have hyperbolic function tables or evaluation programs built in, but few can evaluate the hyperbolic functions of complex arguments. The following two identities make the evaluation of complex arguments in terms of real numbers possible.

$$\cosh \gamma l = \cosh(\alpha l + j\beta l)$$
$$= \cosh \alpha l \cos \beta l + j\sinh \alpha l \sin \beta l \qquad (9.92)$$
$$\sinh \gamma l = \sinh(\alpha l + j\beta l)$$
$$= \sinh \alpha l \cos \beta l + j\cosh \alpha l \sin \beta l \qquad (9.93)$$

where $\alpha$ and $\beta$ are found by computing

$$\gamma l = \sqrt{ZYl} = (\alpha + j\beta)l = \alpha l + j\beta l \qquad (9.94)$$

Examples follow to illustrate the use of the equivalent circuits.

## 9.5

## TRANSMISSION LINE LOSS EXAMPLE

The purpose of this example is to show the effect of voltage, length, and bundling on line loss for 69 kV, 138 kV, 345 kV, and 500 kV lines in lengths of 50, 100, and 200 miles. To make the example reasonable to do with a scientific calculator (programmable type preferable), or a mathematics-solving program for PCs, some simplifying assumptions were made.

1. The medium length lumped parameter equivalent circuit was used and considered sufficient in accuracy for the 200 mile line. The error is normally less than 5% for overhead lines with this equivalent circuit.
2. The source has no impedance and infinite capacity.

3. The load and load power factor for each line is fixed. The standard assumed power factor of 80% lagging is used.
4. The conductor used is Linnet, which is normally used only for 115 kV and 138 kV lines in the United States. To have a comparison basis Linnet is used for the 69 kV, 138 kV, 345 kV line (in a bundle of two), and the 500 kV line (in a bundle of three) of the example.
5. To have a basis for comparison the current per conductor of each line is 100 A. This causes the loads to be a bit light, but still reasonable. 100 A is a reasonable current for Linnet.
6. Common line spacings, shown in Figure 9.31, are used.
7. The resistance at 50°C is used.
8. The line lengths are to be the same for all the example lines for comparison. Actually a 69 kV line seldom exceeds 30 miles, a 138 kV line seldom exceeds 100 miles, and 345 kV and 500 kV lines are seldom built for distances less than 100 miles.

**Example 9.11:**

Calculate the line loss for a 50 mile, a 100 mile, and a 200 mile line for each voltage of 69 kV, 138 kV, 345 kV, and 500 kV. Use the spacings of Figure 9.31, and Linnet conductor. The current per conductor is 100 amperes. The 345 kV line is to use a bundle of two, and the 500 kV line a bundle of three conductors per phase. The medium length lumped parameter equivalent circuit is to be used throughout the example.

**Solution:**

From Table 9.1 we find for Linnet: Al area = 336,400 CM, 26/7 stranding, 0.721 in OD, GMR = 0.0243 ft, $X_a$ = 0.451 $\Omega$/mi, $X_a'$ = 0.1040 M $\Omega$mi, and $R_{ac}$ at 50°C = 0.3006 $\Omega$/ mi.

The resistances per mile are

69 kV and 138 kV—0.3006 $\Omega$/mi

345 kV bundle of two—0.3006/2 = 0.1503 $\Omega$/mi

500 kv bundle of three—0.3006/3 = 0.1002 $\Omega$/mi

The inductive reactances are dependent on the spacing as well as the diameter of the conductor. For each conductor $X_L$ is

69 kV $X_L$: 10 ft – 10 ft – 20 ft spacing

$$D_{eq} = \sqrt[3]{10 \text{ ft } 10 \text{ ft } 20 \text{ ft}} = 12.6 \text{ ft}$$

$$X_L = (2.022 \times 10^{-3}) f \ln \frac{D_{eq}}{D_s}$$

$$X_L = (2.022 \times 10^{-3})(60 \text{ Hz}) \ln \frac{12.6 \text{ ft}}{0.0243 \text{ ft}} = 0.75825 \text{ } \Omega/\text{mi}$$

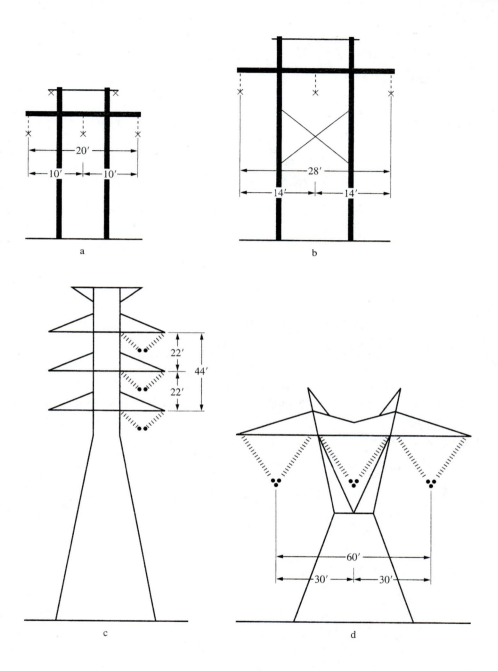

**FIGURE 9.31** Towers and spacings for Example 9.11 (approximately to scale) (a) 69 kV H fixture (b) 138 kV H fixture (c) 345 kV restricted R.O.W. type tower and (d) 500 kV cross country type tower

$138KV$ $X_L$: 14 ft $-$ 14 ft $-$ 28 ft spacing

$D_{eq} = \sqrt[3]{14 \text{ ft } 14 \text{ ft } 28 \text{ ft}} = 17.64 \text{ ft}$

$X_L = (2.022 \times 10^{-3})(60 \text{ Hz})\ln \dfrac{17.64 \text{ ft}}{0.0243 \text{ ft}} = 0.79919 \ \Omega/\text{mi}$

$D_s$ for the bundled conductors Bundle spacing—18"

$D_{sB2} = \sqrt{0.0243 \text{ ft } 1.5 \text{ ft}} = 0.1909 \text{ ft}$

$D_{sB3} = \sqrt[3]{0.0243 \text{ ft}(1.5 \text{ ft})^2} = 0.3795 \text{ ft}$

345 KV $X_L$: 22 ft – 22 ft – 44 ft spacing

$D_{eq} = \sqrt[3]{22 \text{ ft } 22 \text{ ft } 44 \text{ ft}} = 27.72 \text{ ft}$

$X_L = (2.022 \times 10^{-3})(60 \text{ Hz})\ln \dfrac{27.72 \text{ ft}}{0.1909 \text{ ft}} = 0.60385 \ \Omega/\text{mi}$

500 kV $X_L$: 30 ft – 30 ft – 60 ft spacing

$D_{eq} = \sqrt[3]{30 \text{ ft } 30 \text{ ft } 60 \text{ ft}} = 37.8 \text{ ft}$

$X_L = (2.022 \times 10^{-3})(60 \text{ Hz})\ln \dfrac{37.8 \text{ ft}}{0.3795 \text{ ft}} = 0.55813 \ \Omega/\text{mi}$

We now have enough information to calculate the series impedance pu length. Now for the capacitive reactance. $D_S$ of the bundled conductors for $X_C$ using actual radius

$D_{sB2} = \sqrt{0.0300 \text{ ft } 1.5 \text{ ft}} = 0.2121 \text{ ft}$

$D_{sB3} = \sqrt[3]{0.0300 \text{ ft}(1.5 \text{ ft})^2} = 0.4072 \text{ ft}$

69 kV $X_C$ from Equation 9.46

$$X_C = \frac{1.779 \times 10^6}{f} \ln \frac{D_{eq}}{r}$$

$X_C = (2.965 \times 10^4)\ln \dfrac{12.6 \text{ ft}}{0.030 \text{ ft}} = 0.1791 \text{ M}\Omega\text{mi}$

138 kV $X_C$

$X_C = (2.965 \times 10^4)\ln \dfrac{17.64 \text{ ft}}{0.030 \text{ ft}} = 0.1891 \text{ M}\Omega\text{mi}$

345 kV $X_C$ bundle of two

$X_C = (2.965 \times 10^4)\ln \dfrac{27.72 \text{ ft}}{0.2121 \text{ ft}} = 0.1445 \text{ M}\Omega\text{mi}$

500 kV $X_C$ bundle of three

$$X_C = (2.965 \times 10^4)\ln \frac{37.8 \text{ ft}}{0.4072 \text{ ft}} = 0.1344 \text{ M}\Omega\text{mi}$$

Table 9.4 lists the impedance per mile, the admittance per mile, and the ABCD constants needed for the calculation of $V_S$ and $I_S$.

We proceed for each line length as follows. First the equivalent circuit parameters are calculated, except in Table 9.4 $Y$ rather than $Y/2$ is listed. Second we calculate the ABCD constants from Equations 9.58a, b, and c.

$$A = D = \frac{YZ}{2} + 1$$

$$B = Z$$

$$C = Y\left(\frac{YZ}{4} + 1\right)$$

### TABLE 9.4 ABCD Values

| Z = B(Ω) | Y(S) | A = D | C(S) |
|---|---|---|---|
| **69 kv** | | | |
| $Z = 0.3006 + j0.7582\ \Omega/\text{mi}$<br>$Y = j5.583 \times 10^{-6}\ \text{s/mi}$ | | | |
| 50 mi   15.03 + j 37.91 | $297.2 \times 10^{-6}$ | $0.9947 + j2.098 \times 10^{-3}$ | $-292.9 \times 10^{-9} + j278.4 \times 10^{-6}$ |
| 100 mi   30.06 + j75.828 | $558.3 \times 10^{-6}$ | $0.9788 + j8.391 \times 10^{-3}$ | $-2.342 \times 10^{-6} + j0.5524 \times 10^{-3}$ |
| 200 mi   60.12 + j151.65 | $1.118 \times 10^{-3}$ | $0.9153 + j33.57 \times 10^{-3}$ | $-18.74 \times 10^{-6} + j1.069 \times 10^{-3}$ |
| **138 kv** | | | |
| $Z = 0.3006 + j0.7992\ \Omega/\text{mi}$<br>$Y = j5.288 \times 10^{-6}\ \text{s/mi}$ | | | |
| 50 mi   15.03 + j39.96 | $264.4 \times 10^{-6}$ | $0.9947 + j1.987 \times 10^{-3}$ | $-262.7 \times 10^{-9} + j263.7 \times 10^{-6}$ |
| 100 mi   30.06 + j79.92 | $528.8 \times 10^{-6}$ | $0.9789 + j7.948 \times 10^{-3}$ | $-2.101 \times 10^{-6} + j0.5232 \times 10^{-3}$ |
| 200 mi   60.12 + j159.8 | $1.058 \times 10^{-3}$ | $0.9155 + j31.79 \times 10^{-3}$ | $-16.81 \times 10^{-6} + j1.013 \times 10^{-3}$ |
| **345 kv** | | | |
| $Z = 0.1503 + j0.6038\ \Omega/\text{mi}$<br>$Y = j6.923 \times 10^{-6}\ \text{s/mi}$ | | | |
| 50 mi   7.515 + j30.19 | $346.2 \times 10^{-6}$ | $0.9948 + j1.301 \times 10^{-3}$ | $-225.1 \times 10^{-9} + j345.2 \times 10^{-6}$ |
| 100 mi   15.30 + j60.38 | $692.3 \times 10^{-6}$ | $0.9791 + j5.203 \times 10^{-3}$ | $-1.801 \times 10^{-6} + j0.6851 \times 10^{-3}$ |
| 200 mi   30.06 + j120.8 | $1.385 \times 10^{-3}$ | $0.9164 + j20.81 \times 10^{-3}$ | $-14.41 \times 10^{-6} + j1.327 \times 10^{-3}$ |
| **500 kv** | | | |
| $Z = 0.1002 + j0.5581\ \Omega/\text{mi}$<br>$Y = j7.445 \times 10^{-6}\ \text{s/mi}$ | | | |
| 50 mi   5.010 + j27.91 | $372.3 \times 10^{-6}$ | $0.9948 + j.9325 \times 10^{-3}$ | $-173.6 \times 10^{-9} + j371.2 \times 10^{-6}$ |
| 100 mi   10.02 + j55.81 | $744.5 \times 10^{-6}$ | $0.9792 + j3.730 \times 10^{-3}$ | $-1.388 \times 10^{-6} + j0.7368 \times 10^{-3}$ |
| 200 mi   20.04 + j111.62 | $1.489 \times 10^{-3}$ | $0.9169 + j14.92 \times 10^{-3}$ | $-11.11 \times 10^{-6} + j1.427 \times 10^{-3}$ |

Third we calculate $V_S$ and $I_S$ from Equations 9.56 and 9.57. $V_S$ is the phase voltage.

$$V_S = AV_R + BI_R$$

$$I_S = CV_R + DI_R$$

Fourth we find the power received

$$P_R = \sqrt{3}\, V_R I_R \cos(-36.8°)$$

and the power sent

$$P_S = \sqrt{3}\, V_S I_S^* \cos\angle V_{S'} I_S^*$$

where $V_S$ has been converted back to the line value. $S_S$ may at times be smaller than $S_R$ ($S_R = \sqrt{3}\, V_R I_R \angle -36.8°$) in magnitude, but the $S$ values are vector values and $P_S > P_R$. Finally the line loss, $\Delta P = P_S - P_R$. Appendix D gives the received voltage, sending end voltage, and line loss ($\Delta P$) per phase for both the medium and long line equivalent circuit at 100A and 300A.

### Example calculation of loss 69 kV 50 mile line

The values pu length are

$$Z = (0.3006 + j0.75825)\ \Omega/\text{mi}$$

$$Y = B_C = j5.583\ S/\text{mi}$$

so the lumped parameters are 50 miles times the value per mile

$$Z = (15.03 + j37.91)\ \Omega$$

$$Y = j279.15\ \mu S$$

The ABCD constants are

$$A = D = \frac{ZY}{2} + 1 = 0.9947 \times 10^{-3} + j2.098 \times 10^{-3}$$

$$B = Z$$

$$C = Y\left(\frac{YZ}{2} + 1\right) = -292.9 \times 10^{-9} + j278.4 \times 10^{-6}$$

and the sending voltage and current are, from Equations 9.56 and 9.57

$$V_S = 3.855 \times 10^4 + j4.018 \times 10^3 = 38.76\ kV\underline{/5.94°}\quad V$$

The sending line voltage is

$$\sqrt{3}\,38.76\ kV = 67.14\ kV\underline{/5.94°}\quad V$$

$$I_S = 79.44 + j70.94 = 106.51\ \underline{/41.76°}\quad A$$

The received power is

$P_R = \sqrt{3}$ 69 kV 100A cos 36.87° = 9.561MW

and the sent power is the real part of $S_S = \sqrt{3}\, V_S\, \overline{I}_S$

$S_S = \sqrt{3}$ (67.14 kV∠5.94°)(1065A∠-41.76°) = 10.043 MW − j7.247 MVAR

Thus $P_S$ = 10.043 MW

so the line loss is

$\Delta P = P_S - P_R = 482$ kW

Which is 4.41% of the received power. The results of the calculations for all of the lines are in Appendix D.

## 9.6

### SUMMARY

Transmission lines have resistance, inductance, and capacitance. These parameters, R, L, and C, are distributed along the line, a fact that makes the calculation of the parameters more complicated. Additionally, the line electric and magnetic fields couple to each other, which complicates the calculation task even more. Field interactions with the shield lines and ground are additional complicating factors. The line parameters are usually represented on a pu length basis, usually R, L, or C per mile or kilometer. Tables are available that allow easy calculation of the line parameters. The equations for the line electrical parameters are presented in sections 9.1 through 9.4.

Transmission line distributed calculations are very complicated, so equivalent circuits for transmission lines are used to simplify both discussion and calculations. As transmission line length is increased the equivalent circuit needed to accurately model the line becomes more complex. The equivalent circuits of transmission lines are covered in section 9.4.

The load impedance is always more than the line impedance on unfaulted electrical power transmission lines. If the load impedance differs from the line impedance when an electrical wave travels down a line reflections occurs. If the load impedance is larger than the line impedance the voltage is reflected in phase and the total line voltage rises until the line stabilizes. This phenomena can result in switching transients of almost twice the line voltage. These transients must be taken into account when transmission line protection systems are designed.

## 9.7

### QUESTIONS

1. Calculate the percent resistance increase from skin effect at 60 Hz and 200C for Waxwing, Ortulan, and Bluebird ACSR conductors. Use Table 9.1.

2. Calculate the total inductance of a single-phase, 60 Hz, circuit with Waxwing conductor spaced 9 feet apart.

3. Three conductor filaments are 1 cm in diameter and separated from two 2 cm diameter conductor filaments, as shown in Figure 9.32. Calculate the GMR of each conductor and the GMD between the conductors.

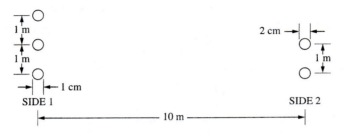

**FIGURE 9.32**   Conductor arrangement for Problem 9.3

4. Calculate the total inductance of the equivalent geometry of the conductors in problem 3. Assume single-phase line voltage.

5. Calculate, with the aid of Tables 9.1 and 9.2, the inductance per mile of Osprey conductors spaced 22 feet apart.

6. The circuit shown in Figure 9.12 uses Osprey conductors. Calculate the series impedance per mile.

7. The 345 kV line of Figure 9.13 uses Ortulan conductors. Calculate the series impedance per mile.

8. Calculate the line to neutral capacitance per meter of the conductors of Figure 9.15 if $r_a = 2$ cm, $r_b = 5$ cm, and D = 2 m.

9. Calculate the capacitance per meter of the equivalent geometry of the conductors in Figure 9.32.

10. Calculate the shunt admittance per mile of the circuit of problem 6.

11. Calculate the shunt admittance per mile of the circuit of problem 7.

12. Calculate the series impedance, shunt admittance, and charging current per mile of the line shown in Figure 9.19 if the conductor is Hawk.

13. The conductor of Figure 9.20 is changed to Drake. Calculate the series impedance, shunt admittance, and charging current per mile.

14. Calculate the $\pi$ lumped parameter medium length line equivalent circuit for the line of problem 12 if the line is 60 miles long.

15. Calculate the $\pi$ lumped parameter medium length line equivalent circuit for the line of problem 13 if the line is 120 miles long.

16. Calculate the ABCD generalized circuit parameters for a single capacitor, as shown in Figure 9.33.

17. Calculate the power loss on the line of problem 14 if the receiving voltage, current, apparent power, and power factor are 220 kV, 140 A, 53.347 MVA, and 0.9 respectively.

18. Repeat problem 17 with a 120 mile line length.

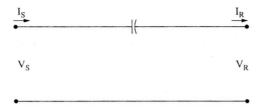

**FIGURE 9.33** Circuit for Problem 9.16

19. Calculate the power loss on the line of problem 15 if $V_R = 750$ kV, $I_R = 800$ A, SR = 1,039 MVA, and the receiving power factor is 0.8.
20. If electric power costs $0.08 per kW hr, what is the difference in the cost of line power loss between the lines of problem 18 and problem 19?

*Optional*

21. Repeat problem 19 using the long line equivalent circuit.

# CHAPTER 10

# Transmission Line Fault Current Calculation, Protection, and Bulk Power Substations

This chapter deals with three major topics. The first is a method of accurately calculating fault currents on transmission lines. The method is called *symmetrical components*. The second topic is *transmission line protection*. Protective relays are used to protect transmission lines. The relays are housed in the substations at which the transmission lines terminate. The third topic is *bulk power substations*. We have already studied the equipment used in bulk power substations since it varies only in capacity and size from that used in distribution substations. So we will contrast bulk power and distribution substations, and take a closer look at grounding.

## 10.1

---

## TRANSMISSION LINE FAULT CALCULATIONS

This section classifies faults, introduces symmetrical components, and shows an example fault current calculation.

### 10.1.1 Fault Classifications

A transmission system fault is defined as any abnormal condition. In this section we are concerned only with faults caused by a transmission line short circuit. The short circuit faults are classified as:

1. Line to ground. Line to ground faults are caused by a transmission line touching the ground. Wind, ice loading, or an accident such as a tree limb falling on a line can cause a line to ground fault. This category accounts for about 70% of all transmission line short circuit faults.
2. Line to line. These faults are normally caused by high winds blowing one line into another, or by a line breaking and falling on a line below it. These account for about 15% of transmission line faults.
3. Double line to ground. This category is caused by the same things that cause single line to ground faults, except two lines are involved instead of one. These account for about 10% of transmission line faults.
4. Three-phase faults. If a line condition occurs in which all three phases are shorted together, either by something falling on the phase conductors, an equipment failure, or all three lines falling to the ground, it is called a three-phase fault. These are relatively rare, accounting for only about 5% of all transmission line faults.

These faults are summarized in Figure 10.1. Open circuits also occur but they are normally caused by relays responding erroneously to a switching transient or load change. This accounts for about 3% of all relay operations.

The point at which a conductor touches ground or another conductor during a fault is usually accompanied by an arc. The arc is resistive, but arc resistance varies widely. The usual utility practice is to consider the fault resistance zero to calculate the maximum fault current that can occur at a point of interest on a line, and the fault resistance 20 Ω (40 Ω for REA) to calculate minimum fault current. The fault resistance consists of the arc resistance, plus the ground resistance in the event a fault is to ground.

The fault current that flows depends on the source, line, and fault impedances, as illustrated in Figure 10.2.

$$I_{FAULT} = \frac{V_{SOURCE}}{Z_S + Z_L + Z_f} \qquad (10.1)$$

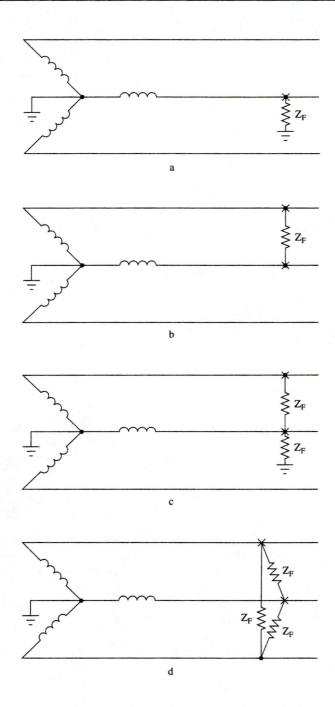

**FIGURE 10.1** Fault classifications (a) Line to ground  (b) Line to line
(c) Double line to ground and (d) Three-phase fault

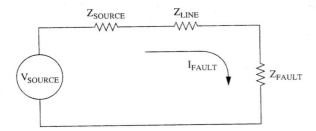

**FIGURE 10.2** Simple equivalent circuit for fault current calculation

Recall from Chapter 5.1.1 that the fault current is normally asymmetrical, as shown in Figure 5.2.

A shorthand, per phase, method of calculating line current and voltages called per unit (pu) was introduced in Chapter 2. We can solve nearly all power problems using per phase values, although when no base change is necessary, such as in the single line loss problems of Chapter 9, actual values may be as convenient as pu. Our methods of calculation to this point have depended on the load fed by the system being reasonably well balanced. The line voltages and currents during a fault are very seldom balanced.

A method of calculating voltage and current for unbalanced loads was published by C. L. Fortescue in 1918. It is called the method of symmetrical components. The method is a powerful technique for analyzing ac power systems with unbalanced loads.

## 10.1.2 Definition of Symmetrical Components

An unbalanced system of N phasors can be resolved into N systems of balanced phasors, each of which has phasors of equal amplitude with equal angles between each phasor. The systems of balanced phasors are called symmetrical components. The vector sum of the symmetrical components is equal to the original system of unbalanced phasors. Thus an unbalanced three-phase system is the sum of three sets of balanced phasors, or symmetrical components.

The symmetrical components of a three-phase system, illustrated in Figure 10.3a, are:

1. *Positive sequence components.* These are three phasors of equal magnitude, offset from each other by 120°, and rotating in the same direction as the original phasors, usually signified counterclockwise in a phasor diagram. Positive sequence quantities are signified by the subscript 1, for example $V_{A1}$ or $I_{B1}$.
2. *Negative sequence components.* Three phasors of equal magnitude, offset from each other 120° in phase, and rotating in the opposite direction to the original phasor, make up the negative sequence components. Normally the rotation of the phasors is considered to be in the same direction as the original phasors, but two of the phase vectors, b and c, are reversed, as shown in Figure 10.3a. Thus the negative sequence components phase rotation is acb instead of abc. Negative sequence quantities are given subscript 2 identification as in $Z_2$ or $V_2$.

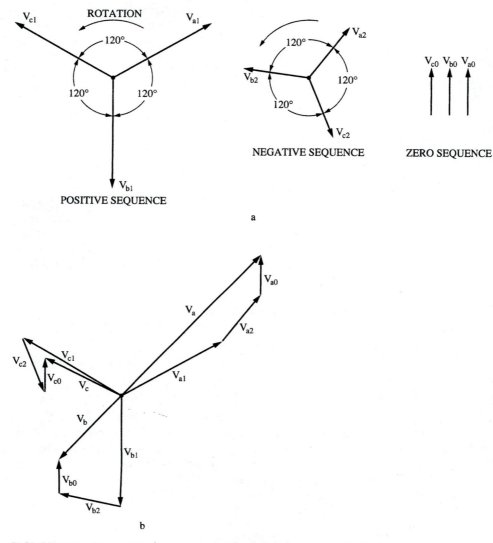

**FIGURE 10.3** Symmetrical components of unbalanced phasors (a) Symmetrical components (b) Unbalanced phasor, sum of symmetrical components

3. *Zero sequence components.* These are three phasors of equal magnitude with no phase offset between them. They are in phase. The zero sequence components are often considered to be nonrotating, like a dc voltage. However, zero sequence impedance is complex, so the zero sequence voltage is treated as a single-phase source for calculating zero sequence impedance in relaying problems. Zero sequence quantities are identified with a subscript zero as in $I_0$ or $Z_0$.

Figure 10.3b shows the asymmetrical phasors of which the phasors of Figure 10.3a are the symmetrical components.

The defining equations for the symmetrical components, where $V_a$, $V_b$, and $V_c$ are the original unbalanced phasors, are

$$V_a = V_{a1} + V_{a2} + V_{a0} \tag{10.2}$$

$$V_b = V_{b1} + V_{b2} + V_{b0} \tag{10.3}$$

$$V_c = V_{c1} + V_{c2} + V_{c0} \tag{10.4}$$

The quantities in the symmetrical components are generally complex. These equations are illustrated in Figure 10.3b. The current equations are identical in form.

$$I_a = I_{a1} + I_{a2} + I_{a0} \tag{10.5}$$

$$I_b = I_{b1} + I_{b2} + I_{b0} \tag{10.6}$$

$$I_c = I_{c1} + I_{c2} + I_{c0} \tag{10.7}$$

## 10.1.3 Operators

The j operator, which we have used since ac circuits, signifies a 90° phase shift. The $a$ operator is used to indicate a 120° phase shift.

$$a = 1\angle120° = -0.5 + j0.866$$

$$a^2 = 1\angle240° = -0.5 - j0.866$$

$$a^3 = 1\angle360° = 1\angle0°$$

The relative angular position of the three phasors for a three-phase voltage can be expressed as the product of the amplitude of the phasor and the operator, $a$. Referring to Figure 10.3a we see, since $V_{a1} = V_{b1} = V_{c1}$

$$V_{b1} = a^2 V_{a1}$$

$$V_{c1} = a V_{a1}$$

and

$$V_{b2} = a V_{a2}$$

$$V_{c2} = a^2 V_{a2}$$

and

$$V_{c0} = V_{b0} = V_{a0}$$

Now Equations 10.2 through 10.4 can be written

$$V_a = V_{a1} + V_{a2} + V_{a0} \tag{10.8}$$

$$V_b = a^2 V_{a1} + a V_{a2} + V_{a0} \tag{10.9}$$

$$V_c = a V_{a1} + a^2 V_{a2} + V_{a0} \tag{10.10}$$

### 10.1.4 Symmetrical Components of Asymmetrical Phasors

The magnitudes of the phasors of each sequence of symmetrical components are equal. Thus we need only find the magnitude of $V_{a1}$, $V_{a2}$, and $V_{a0}$ to calculate $V_a$, $V_b$, and $V_c$, as shown in Equations 10.8, 10.9, and 10.10. If we add the terms in Equations 10.8, 10.9, and 10.10 we find

$$V_{a1} + a^2 V_{a1} + a V_{a1} = |V_{a1}| + (-0.5 - j0.866) |V_{a1}| + (-0.5 + j0.866) |V_{a1}| = 0$$

and similarly

$$V_{a2} + a V_{a2} + a^2 V_{a2} = 0$$

thus

$$V_{a0} = \frac{1}{3} (V_a + V_b + V_c) \tag{10.11}$$

We can eliminate the Va1 column coefficients by multiplying Equation 10.9 by a and Equation 10.10 by $a^2$ yielding

$$V_{a1} = \frac{1}{3} (V_a + a V_b + a^2 V_c) \tag{10.12}$$

Multiplying Equation 10.10 by a and Equation 10.9 by $a^2$ and adding results in

$$V_{a2} = \frac{1}{3} (V_a + a^2 V_b + a V_c) \tag{10.13}$$

As before, the form of the current equations is the same as the voltage equations.

$$I_{a0} = \frac{1}{3} (I_a + I_b + I_c) \tag{10.14}$$

$$I_{a1} = \frac{1}{3} (I_a + a I_b + a^2 I_c) \tag{10.15}$$

$$I_{a2} = \frac{1}{3} (I_a + a^2 I_b + a I_c) \tag{10.16}$$

Recall that for an unbalanced three-phase system with a neutral

$$I_n = I_a + I_b + I_c$$

From Equation 10.14 we see that

$$I_n = 3 I_{a0}$$

### Example 10.1:

Calculate the symmetrical components of the currents shown in Figure 10.4.

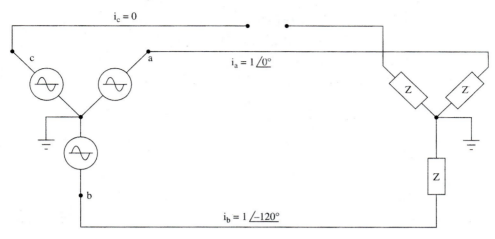

**FIGURE 10.4** Circuit for Example 10.1

**Solution:**

First we express the currents in component form for future use

$I_a = 1\angle0° = 1 + j0$

$I_b = 1\angle-120° = -0.5 - j0.866$

$I_c = 0$

From Equation 10.14 we calculate the zero sequence current

$$I_{a0} = \frac{1}{3}(I_a + I_b + I_c) = \frac{1}{3}(1 - 0.5 - j0.866) = \frac{1}{3}(0.5 - j0.866)$$

$$= 0.\frac{999}{3}A\angle-60° = 0.333A\angle-60°$$

Using Equation 10.15 we calculate the positive sequence current.

$$I_{a1} = \frac{1}{3}(I_a + aI_b + a^2I_c) = \frac{1}{3}[1 + (1\angle120°)(1\angle-120°)]$$

$$= \frac{1}{3}(1 + 1) = 0.667A\angle0°$$

Using Equation 10.16 to calculate the negative sequence current

$$I_{a2} = \frac{1}{3}(I_a + a^2I_b + aI_c) = [1 + (1\angle240°)(1\angle-120°)]$$

$$= \frac{1}{3}[1 + (-0.5 + j0.866)] = 0.333A\angle60°$$

These are the a phase symmetrical components. The b and c phase symmetrical components are the same magnitude but offset in phase.

$$I_{b1} = \alpha^2 I_{a1} = 0.667 A \angle 240°$$

$$I_{c1} = \alpha I_{a1} = 0.667 A \angle 120°$$

$$I_{b2} = \alpha I_{a2} = 0.333 A \angle 180°$$

$$I_{c2} = \alpha^2 I_{a2} = 0.333 A \angle 300°$$

Of course $I_{b0}$ and $I_{c0}$ are both equal to $I_{a0}$.

$$I_{a0} = 0.333 A \angle \text{-}59.9°.$$

The symmetrical components of Example 10.1 are shown in Figure 10.5.

Let us now check our work by calculating the original phasors from the symmetrical components. The sums of the components are:

$$I_a = I_{a0} + I_{a1} + I_{a2} = 0.333 A \angle \text{-}60° + 0.667 A \angle 0° + 0.333 A \angle 60°$$

$$= (0.1667 - j0.2887)A + 0.667A + (0.1667 + j0.2887) = 1 A \angle 0°$$

$$I_b = I_{b0} + I_{b1} + I_{b2} = 0.333 A \angle \text{-}60° + 0.667 A \angle 240° + 0.333 A \angle 180°$$

$$= (0.1667 - j0.2887)A + (\text{-}0.333 - j0.557)A - 0.333A = 1 A \angle 120°$$

$$I_c = I_{c0} + I_{c1} + I_{c2} = 0.333 A \angle \text{-}60° + 0.667 A \angle 120° + 0.333 A \angle 300°$$

$$= (0.1667 - j0.2887)A + (\text{-}0.333 + j0.557)A + (0.1667 - j0.2887)A = 0$$

Figure 10.6 shows the sums of the symmetrical components of Example 10.1.

## 10.1.5 Sequence Impedances

The sequence impedance is defined as that obtained by providing unit current of that sequence in the three phases of the apparatus in question and writing the equation for the voltage drop. The sequence impedance, which can also be found by measurement, is normally expressed in pu for calculations. The impedance of one sequence need not be that of another. For example, linear networks such as transmission lines normally have equal positive and negative sequence impedances, but the zero sequence impedance, which must take into account the neutral return line impedance and coupling, is different. Transformers also have identical positive and negative sequence impedances, but different zero sequence impedance. Rotating machines may have different positive and negative sequence impedances as well. The sequence impedances for rotating machines is the transient $(X'_d)$ or subtransient impedances $(X''_d)$ depending on the time of interest. $X''_d$ is normally used for calculating fault currents to set relays because it is the effective machine impedance during the relay sensing time. $X'_d$ may be used to calculate currents used to set time delay relays and other slower devices. Frequently only the sequence reactance is used to calculate fault currents when the resistance of a component is much less than its reactance. Computer programs normally include the resistance to obtain the added accuracy. Computers do not tire of tedious calculations as easily as do people.

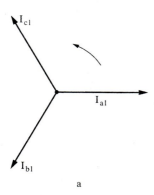

a

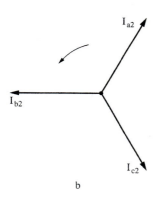

b

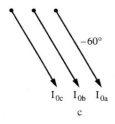

c

**FIGURE 10.5** Symmetrical components of Example 10.1 (a) Positive sequence (b) Negative sequence and (c) Zero sequence

## 10.1.6 Sequence Networks

Sequence networks are one phase to neutral diagrams showing the elements of interest for each sequence of the network under consideration. The sequence networks are called the positive sequence network, the negative sequence network,

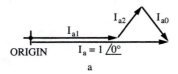

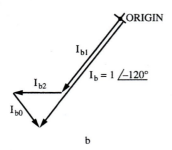

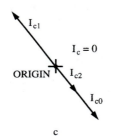

**FIGURE 10.6**   Phase sums of symmetrical components in Figure 10.5
(a) Phase a (b) Phase b and (c) Phase c

and the zero sequence network. In a balanced three-phase network the sequence components are independent of each other until a fault occurs. The sequence impedance networks can therefore be computed independently and combined appropriately for calculating a fault current.

The positive sequence network contains all of the generator voltages and the positive sequence impedance of all generators, transformers, lines, and motors in the network under consideration. The negative sequence network consists of the negative sequence impedances of all the elements of the network. The zero sequence impedance contains the zero sequence impedances of all the elements of the network. All three of the networks are not needed for every fault. For example, a system with no neutral return has an infinite zero sequence impedance, so the zero sequence network is not used in calculating the current for the faults that can occur in the system. The configuration of the sequence networks varies with the type of fault on the system. Figure 10.7 shows the sequence networks of a grounded wye connected generator with a line to ground fault at one terminal. The fault impedance may be drawn outside of the sequence networks.

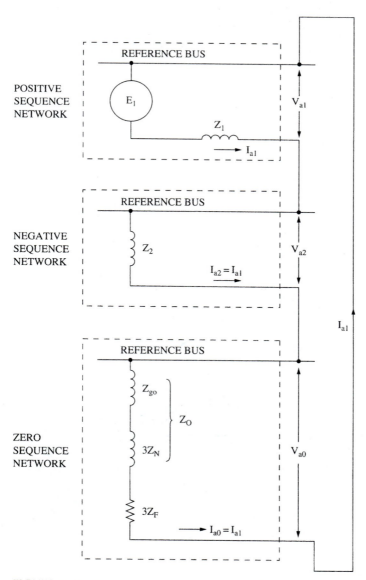

**FIGURE 10.7** Sequence networks, connected for line to ground fault

## 10.1.7 Generator Sequence Networks

Figure 10.8 shows the equivalent circuit of a generator for transient conditions. Let $E_a = E_b = E_c$ and $Z_1 = Z_2 = X''_d$. The positive sequence network will contain $E_a$ and $Z_1$ as shown in Figure 10.9a. The sequence impedances must be obtained from the machine manufacturer or from testing. The negative sequence network, shown in

Figure 10.9b, consists only of the negative sequence impedance, which is often approximately equal to the positive sequence impedance.

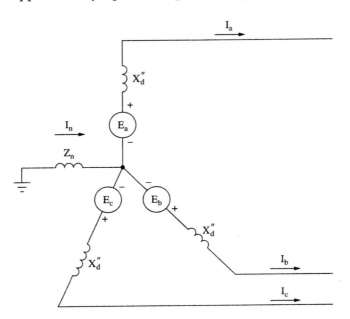

**FIGURE 10.8**   Equivalent circuit of a generator

The zero sequence network includes the neutral impedance as well as the machine zero sequence impedance. The machine zero sequence impedance varies with the type of machine between 0.1 and 0.7 $X''_d$. Often $Z_{g0} = 0.5X''_d$. Note that if zero sequence current is flowing in each phase of the generator in Figure 10.8 that the neutral current must be the sum of the phase currents. Since the zero sequence current phasors have no angular offset between them (or are effectively in phase) the neutral current is three times the current in each phase. Thus in the zero sequence network shown in Figure 10.9c,

$$V_{a0} = 3I_{a0}Z_R + I_{a0}Z_{g0}$$

$$Z_0 = \frac{V_{a0}}{I_{a0}} = 3Z_n + Z_{g0} \qquad (10.17)$$

Thus the effect is like tripling $Z_n$ at the zero sequence phase current.

The sequence networks for a large synchronous motor during a fault is the same as that for a generator. In the event of a fault on a line supplying a motor, the motor becomes a source until the protective relays open or the motor rotor stops.

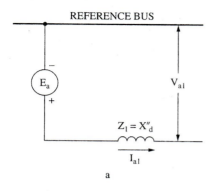

a

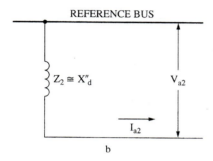

b

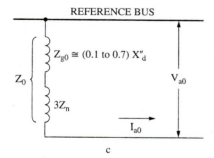

c

**FIGURE 10.9** Sequence networks of a generator (sequence impedances are obtained from machine specifications) (a) Positive sequence (b) Negative sequence and (c) Zero sequence

## 10.1.8 Transformer Sequence Impedance

Transformers are static machines so their positive and negative sequence impedances are the same. Recall that the pu impedance of a transformer is the same from the high or the low side of a transformer.

The 30° phase shift in a delta-wye transformer must be considered in sequence networks. Recall that the convention in the United States is to connect the high side, whether it is the delta or the wye connected side, to lead the low side by 30°

as shown in Figure 10.10a. The phase shift is taken care of in sequence networks by adding 30° to the transformer impedance, or subtracting 30° from it, as shown in Figures 10.10b and c.

The T equivalent circuit of three winding transformers is used in sequence networks. The equations for the equivalent T impedances of a three winding transformer are given in Chapter 4.10. Transformer impedances must, of course, be referred to the system base values.

The zero sequence impedance of a transformer differs from the positive and negative sequence impedances. No significant current can flow to ground from an ungrounded system. Thus for an ungrounded wye or a delta connected transformer $Z_0 = \infty$. For shell constructed three-phase transformers $Z_0 = 3Z_1$. For a core constructed three-phase transformer $Z_0$ ranges from about $0.8Z_1$ to $0.85Z_1$ because there is no path for zero sequence flux return in the core. Figure 10.11 shows the zero sequence impedance for some two winding three-phase transformer configurations. Some electrical power handbooks contain several pages of equivalent sequence impedances for just about any conceivable two or three winding transformer connection.

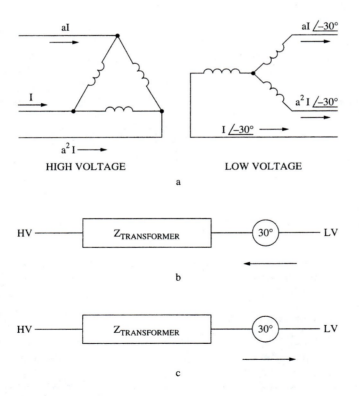

**FIGURE 10.10** Sequence phasor phase shift through delta-wye or wye-delta transformer connection (a) Transformer connection, current phase shift (b) From low voltage to high voltage add 30° to impedance angle (c) From high voltage to low voltage subtract 30° from impedance angle

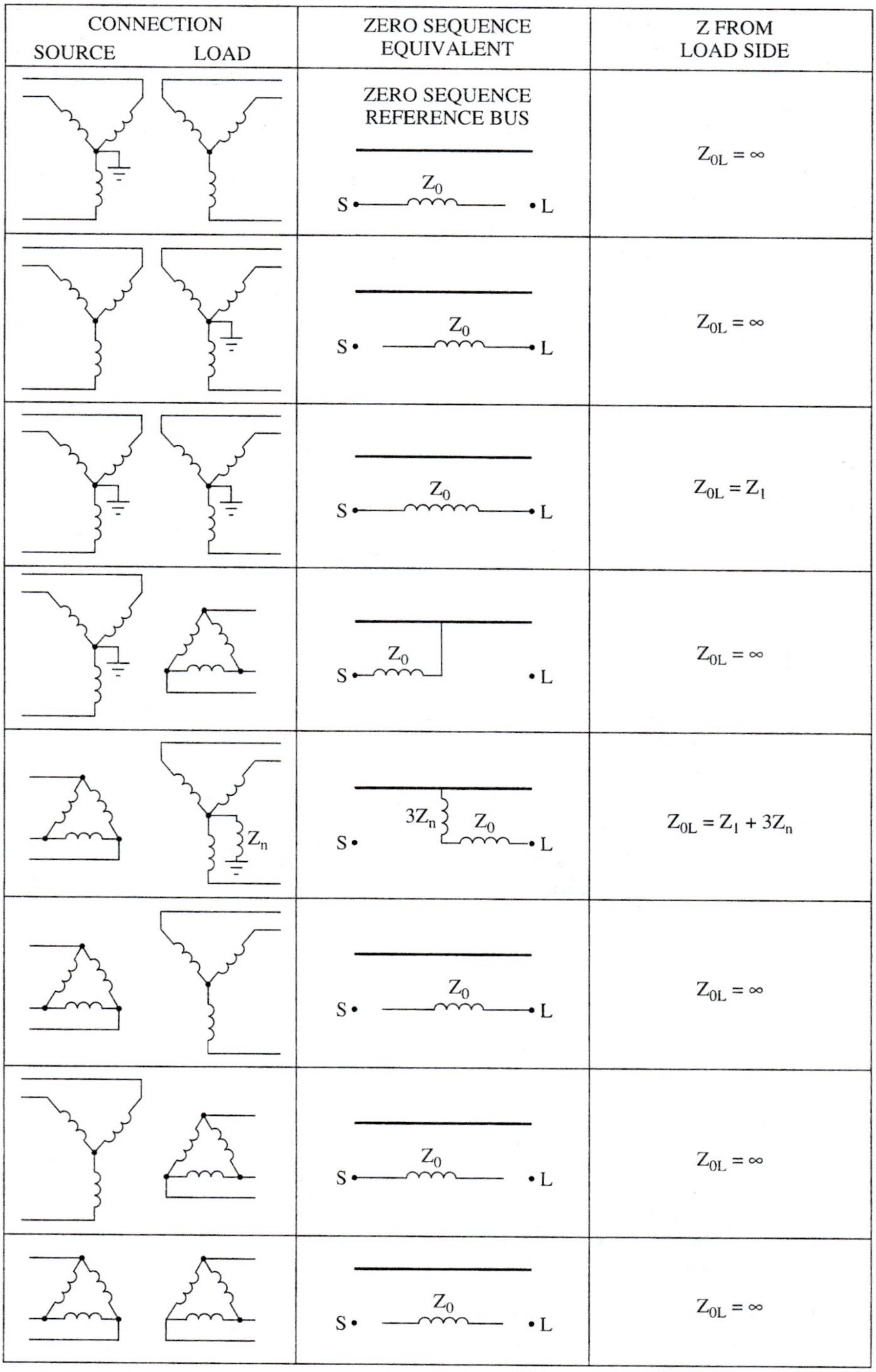

| CONNECTION | | ZERO SEQUENCE EQUIVALENT | Z FROM LOAD SIDE |
|---|---|---|---|
| SOURCE | LOAD | | |

FIGURE 10.11 Zero sequence equivalent circuits for three-phase transformer banks

## 10.1.9 Transmission Line Sequence Impedance

The positive and negative sequence impedances of a transmission line are equal, and they are the line impedances as calculated or found in the tables in Chapter 9. The zero sequence impedance of the transmission line must take into account the resistance and inductance of the neutral return conductors, including the flux coupling of each ground wire with each phase conductor. This is a complicated and tedious calculation that is carried out like those of Chapter 9. Each conductor is treated like a single phase conductor for the calculations. A useful, but often violated, rule of thumb is $Z_0 = 3Z_1$.

The sequence impedances of cables is very difficult to calculate because of the capacitive effect of the insulating dielectric, and the magnetic coupling from induced currents in near-by conduit, water pipes, and soil. Steel conduit, which is magnetic, complicates calculations even further. Sequence impedances for cables is available from cable manufacturers. Usually the positive and negative sequence impedances are equal and equal to the cable impedance. The zero sequence impedance can range from two to 50 times the positive sequence impedance.

Accurate tables of sequence impedances for commonly used overhead distribution conductors and spacings, as well as some underground cable, is available in the McGraw Edison (now Cooper Power Systems) book listed in the reference list.

## 10.1.10 Motor Sequence Impedance

Motors are a factor in fault current for systems in which there are large motors. When a fault occurs the motors connected on the system continue spinning for a short time because of the load inertia. The motor rotors cause the motors to operate as generators driven by their loads until the rotors stop or they are disconnected from the system by relays. Thus both system motors and generators deliver fault current initially, as shown in Figure 10.12. Only very large motors need to be included in fault current calculations. In many systems all motors under 50 hp are neglected.

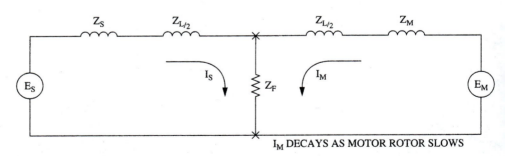

**FIGURE 10.12**  Motor delivering current to a fault

The equivalent circuit for a synchronous motor is the same as that of a synchronous generator for fault current calculations. The positive and negative sequence networks for an induction motor are shown in Figure 10.13. The actual values of

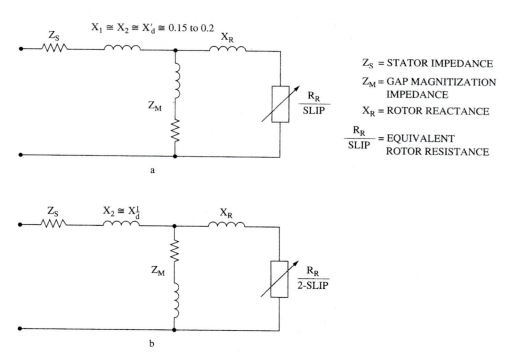

**FIGURE 10.13** Induction motor equivalent circuit ($Z_1$ and $Z_2$ furnished by motor manufacturer) (a) Positive sequence and (b) Negative sequence

the elements must be obtained from the manufacturer or measured. The zero sequence impedance of an induction motor is infinite because induction motors are not normally grounded neutral for voltages less than 4160 V. At 4160 V and above corona in the windings may be a problem if the motor is not grounded. The zero sequence impedance of a grounded neutral induction motor is not infinite.

## 10.1.11 Phase to Ground Fault

The most commonly occurring system fault is phase to ground, as illustrated in Figure 10.14a. If we ignore $R$ and use the total sequence impedance we get

$$Z_1 = Z_2 = X''_d + X_1$$

$$Z_0 = 0.5X''_d + 3X_1 + 3Z_n$$

Solving for the sequence currents we obtain

$$I_{a1} = \frac{1}{3}(I_a + aI_b + a^2I_c) = \frac{I_a}{3}$$

$$I_{a2} = \frac{1}{3}(I_a + a^2I_b + aI_c) = \frac{I_a}{3}$$

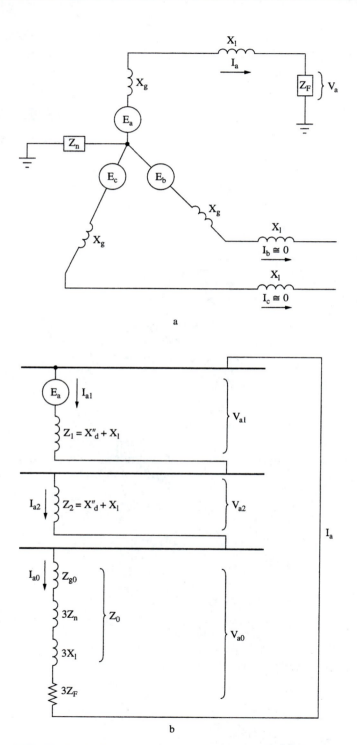

**FIGURE 10.14**  Phase to ground fault on line (a) Equivalent circuit and (b) Sequence network

$$I_{a0} = \frac{1}{3}(I_a + I_b + I_c) = \frac{I_a}{3}$$

$$I_{a1} = I_{a2} = I_{a0} = \frac{I_a}{3}$$

Therefore $I_{a1} = I_{a2} = I_{a3}$

Referring to the sequence networks of Figure 10.14b we see that the voltage across the fault impedance, $V_a = I_a Z_f$ and

$$V_a = V_{a1} + V_{a2} + V_{a0}$$

substituting from the sequence network

$$I_a Z_f = E_a - I_{a1} Z_1 - I_{a2} Z_2 - I_{a0} Z_0$$

But from above we see $I_a = 3I_{a1}$, and $I_{a1} = I_{a2} = I_{a0}$, so

$$3I_{a1} Z_f = E_a - I_{a1} Z_1 - I_{a1} Z_2 - I_{a1} Z_0$$

from which

$$I_{a1} = \frac{E_a}{Z_1 + Z_2 + Z_0 + Z_f \cdot 3}$$

Recall that the fault current $I_a = I_f = 3I_{a1}$ so

$$I_f = \frac{3E_a}{Z_1 + Z_2 + Z_0 + Z_f \cdot 3} \tag{10.18}$$

The sequence network connection is shown in Figure 10.15a.

Many faults greatly reduce the line current in the remaining two lines because the generator terminal voltage is reduced.

The maximum fault current occurs when $Z_f = 0$ and the fault is close to the source. The minimum fault current occurs when the $Z_f$ is maximum and the fault is at the far end of the line. The intervening line impedance lowers the fault current. The fault impedance varies with the resistivity of the soil in a ground fault and the arc resistance. The commonly used fault impedance by most electric utilities is $Z_f = 20\ \Omega$ resistive, but the REA specifies $Z_f = 40\ \Omega$. Once a faulted conductor is in good contact with the ground the fault impedance is very low, almost zero.

The major problem associated with the use of symmetrical components to calculate fault currents is the simplification of the networks into sequence networks. The methods used are the same as those used to simplify any network into a pu equivalent circuit. The system network must be simplified into three sequence networks instead of only one as for steady state calculations.

## 10.1.12 Summary of Sequence Network Configurations

The fault current equations for the other common fault conditions are found by using the appropriate sequence networks and solving for the current as is done in

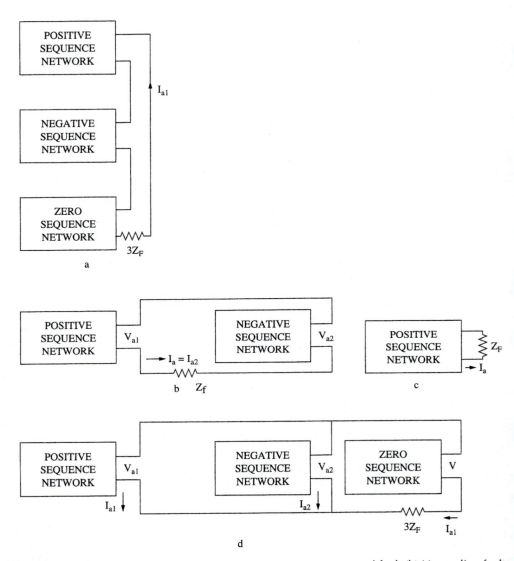

**FIGURE 10.15** Sequence network configurations (a) Line to ground fault (b) Line to line fault (c) Three-phase fault and (d) Double line to ground fault

the preceding section. We will present the final equations in this section. In the following equations $Z_1$, $Z_2$, and $Z_0$ are the sequence impedances viewed from the fault. $Z_f$ is the fault impedance and $V_f$ is the positive sequence voltage, usually the equivalent system supply voltage. Figure 10.15 shows the sequence network connections for each major fault current.

*Three-phase fault:* A balanced three-phase fault has no zero sequence impedance because the neutral current is zero, and because the system is still in balance there is no negative sequence impedance. The fault current is calculated using only the positive sequence network, as shown in Figure 10.15c. The equation is

$$I_f = \frac{V_f}{Z_1 + Z_f} \tag{10.19}$$

*Line to line fault:* A line to line fault causes no fault current in the system neutral, so the zero sequence impedance is not used to calculate fault current. The system is unbalanced by a line to line fault so the negative sequence impedance must be used in the fault current calculations. The sequence network is shown in Figure 10-15b, and the equation is

$$I_f = \pm j \frac{\sqrt{3}\, V_f}{Z_1 + Z_2 + Z_f} \tag{10.20}$$

*Double line to ground fault:* A double line to ground fault causes both system neutral current and system unbalance so all three sequence networks must be used. The positive and negative sequence networks appear as in a line to line fault except both are supplying current to the fault. Thus the networks appear as shown in Figure 10.15d looking back from the fault. The equation for the phase one and two fault currents are:

Phase one

$$I_f = -j\sqrt{3}V_f \frac{Z_0 + 3Z_f - aZ_2}{Z_1 Z_2 + (Z_1 + Z_2)(Z_0 + 3Z_f)} \tag{10.21a}$$

Phase two

$$I_f = +j\sqrt{3}V_f \frac{Z_0 + 3Z_f - a^2 Z_2}{Z_1 Z_2 + (Z_1 + Z_2)(Z_0 + 3Z_f)} \tag{10.21b}$$

All of the quantities in the sequence networks can be complex.

## 10.1.13 Symmetrical Component Example

The following example is to illustrate the use of symmetrical components to calculate a fault current. The network in the example is simple. The calculations for a complex network use the same equation, but the sequence networks have more elements.

### Example 10.2:

Calculate the minimum and maximum fault current for a phase to ground fault at point A for the circuit shown in the one line diagram of Figure 10.16. Use $Z_f = 20\ \Omega$ for minimum fault current and $Z_f = 0\ \Omega$ for maximum fault current. The line is a 138 kV with spacing as shown in Figure 9.31. It is constructed with Linnet conductors. Include the transmission line resistance.

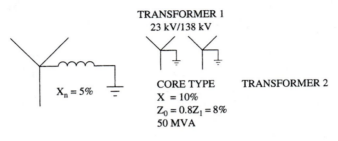

TRANSFORMER 1
23 kV/138 kV

CORE TYPE       TRANSFORMER 2
X = 10%
$Z_0 = 0.8Z_1 = 8\%$
50 MVA

$X_n = 5\%$

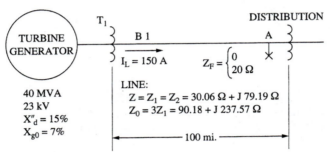

T$_1$                              DISTRIBUTION

TURBINE GENERATOR        B 1                        A

$I_L = 150$ A        $Z_F = \begin{cases} 0 \\ 20\ \Omega \end{cases}$

LINE:
$Z = Z_1 = Z_2 = 30.06\ \Omega + J\ 79.19\ \Omega$
$Z_0 = 3Z_1 = 90.18 + J\ 237.57\ \Omega$

40 MVA
23 kV
$X''_d = 15\%$
$X_{g0} = 7\%$

— 100 mi. —

**FIGURE 10.16**    Circuit for Example 10.2

**Solution:**

The system impedances are given in Figure 10.16. First we must select a system base and express all system quantities in that base. Let the base $S = 50$ MVA, The base $V = 138$ kV. Using the change of base equation

$$Z_{pu2} = Z_{pu1} \frac{kV^2_{B1}}{kV^2_{B2}} \frac{MVA_{B2}}{MVA_{B1}}$$

we calculate

$$X''_d = 0.15(0.02778)(1.25) = 0.00521\text{pu}$$

$$X_{g0} = 0.07(0.02778)(1.25) = 0.00243\text{pu}$$

$$X_n = 0.05(0.02778)(1.25) = 0.00174\text{pu}$$

The transformer is already expressed in base pu. Next we calculate the pu impedance of 100 miles of the transmission line.

$$Z_{BASE} = \frac{(BASE\ kV)^2}{BASE\ MVA} = 380.88\Omega$$

thus

$$Z_{pu\ TL} = \frac{(30.06 + j79.19)\Omega}{380.88\Omega} = 0.0789 + j0.207\text{pu}$$

and the fault impedance is

$$Z_f = \frac{20\Omega}{380.88\Omega} = 0.052\text{pu}$$

The sequence networks are shown in Figure 10.17.

Using Equation 10.18 to calculate the minimum fault current

$$I_f = \frac{3(1\text{pu}V)}{[(0.0789 + j0.312) + (0.0789 + j0.312) + (0.393 + j0.869)]}$$

$$= \frac{3\text{pu}}{(0.551 + j1.493)\text{pu}} = \frac{3\text{pu}}{1.59\text{pu}\angle 70°} = 1.86\text{pu}\angle -51°$$

$$I_f = I_{pu}I_B = 1.886\angle -70°(362.3A) = 683.2A\angle 70°$$

We use the same equation without $Z_f$ to calculate the maximum fault current at the end of the line.

$$I_f = \frac{3\text{pu}V}{[(0.0789 + j0.312) + (0.0789 + j0.312) + (0.234 + j0.869)]\text{pu}Z}$$

$$= \frac{3\text{pu}V}{(0.394 + j1.493)\text{pu}Z} = \frac{3\text{pu}V}{1.554\text{pu}Z\angle 75.2°} = 1.943\text{pu}\angle -75.2°$$

$$= (1.943\text{pu}\angle -75.2°)362.8A = 703A\angle -75.2°$$

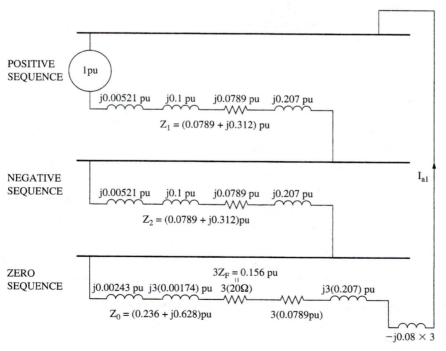

**FIGURE 10.17**  Sequence networks for Example 10.2

The maximum fault current the protective system has to interrupt occurs when a fault is close to the supplying substation, point B in Figure 10.16. At this point the transmission line impedance is zero. The interrupt capacity needed by the station circuit breaker is greater than the maximum close in fault current. First we calculate the sequence impedances without the line impedance.

$$Z_1 = Z_2 = j.000521 + j0.1 = j0.10521\text{pu}$$

For $I_{fmax}$ we let $Z_f = 0$

$$Z_0 = j0.00243 + j3(0.00174) + 3(j0.08) = j0.248\text{puZ}$$

and

$$I_f = \frac{3\text{pu}V}{(j0.10521 + j0.10521 + j0.248)\text{puZ}}$$

$$= \frac{3\text{pu}V}{0.459\text{puZ}} = 6.543\text{pu}I$$

Thus the maximum close in fault current is

$$I_f = 6.543\text{pu}(362.3\text{A}) = 2{,}367 \text{ A}$$

The protective relays must detect a fault current of 674 A and open the circuit breakers. The relays must also detect 4,984 A and open the breakers very fast to prevent equipment damage.

## 10.2

## TRANSMISSION LINE PROTECTION

The subject of transmission line protection contains enough detail to fill a book. Our purpose in this section is to introduce the major concepts of transmission line protection and some commonly used relays. The reader may wish to refer again to Chapter 5 to review basic relay mechanisms.

    High speed fault clearing provides the following three advantages for a transmission system: (1) Limits conductor damage; (2) Improves system stability; and (3) Permits high speed reclosing that prevents extended service outages from transient faults, and also improves system stability. The primary function of the protection for a transmission system is to preserve the integrity of the transmission system by clearing only the faulted section. The relays must detect the fault, determine which line is faulted, and clear only the faulted line. The idea is to trip the line but not the load. Ideally the load will continue to be served by an auxiliary line.

    In this section we will discuss pilot line and pilot carrier protection, distance relays, directional ground relays, auxiliary carrier relays, and back-up protection.

    Microcomputer controlled relays continue to be developed that perform the functions of electromechanical relays more precicely at a cost per function that is below that of electromechanical relays. As with the relays discussed in Chapter 5 they consist of data acquisition systems to acquire the line voltage and current, and sometimes conductor and ambient temperature along with other desirable data. From the relative values and phase of the line parameters the microcomputer program decides what action must be taken to assure faults are interrupted reliably.

## 10.2.1 Pilot Wire Carrier

A pilot wire is a metallic conductor that carries an ac signal from one end of a transmission line segment to another to prevent a relay from tripping (provide a blocking signal). The conductor is usually a shielded twisted pair. The blocking signal is sent to keep relays from tripping from a fault detected outside of the transmission line signal. Pilot wire carrier is used for fairly short transmission line segments because the pilot line resistance is limited to about 2 k$\Omega$, about 10 to 20 miles. The pilot line signal is about 20 V at 60 Hz. The pilot line is protected from over voltage by a gas tube and two drainage reactors, which prevent the gas tube from shorting out the pilot line when the gas tube fires. The pilot wire is normally connected to ground through lightning arresters at either end to prevent excessive voltage on the pilot relays during a transient on the line. The lightning arrester break over voltage is usually about 3 kV. The pilot line protection is shown in Figure 10.18.

The most frequently used pilot relays are the HCB and HCB-1. Figure 10.18a shows a simplified schematic of these relays, which differ only in the design of their sequence filters. The sequence filter combines the three phase currents into a single phase voltage of about 20 V rms. The HCB relay cannot detect negative sequence current, and therefore cannot detect phase to phase faults. The HCB-1 can detect both negative and positive sequence currents, so it detects both phase to phase and three-phase faults. Both relays can detect ground faults.

The ac voltage from the sequence filter is transferred to the relays by a tapped saturating transformer. The saturating transformer, along with the zener diodes, limits the voltage on the relays in the event of a large, close in fault. The saturating transformer tap adjusts overall relay sensitivity. The relay taps should be set as per the instructions in the operating manual. The ac signal from the saturating transformer is rectified and applied to the restraint coil, which is tapped so the restraint sensitivity can be adjusted. The restraint coil is effectively in series with the operate coil, which also operates on rectified ac current. The operate and restraint coils are both wound on the center leg of a dc polar unit. A polar unit is a relay core with a magnetic shunt that causes the armature to move preferentially in only one direction when the coil is energized. A permanent magnet attached to the core structure causes the relay to respond to current of one polarity only. Two additional taps in the sequence filter allow the relay sensitivity to phase and ground faults to be adjusted.

The normal connection of the relays in the system is shown in Figure 10.18b. Under normal operating conditions the phase of the pilot line ac is such that the restraint coils receive most of the relay power while the operate coils receive very little. The relay coils operate on the same principle as a differential relay. If a fault

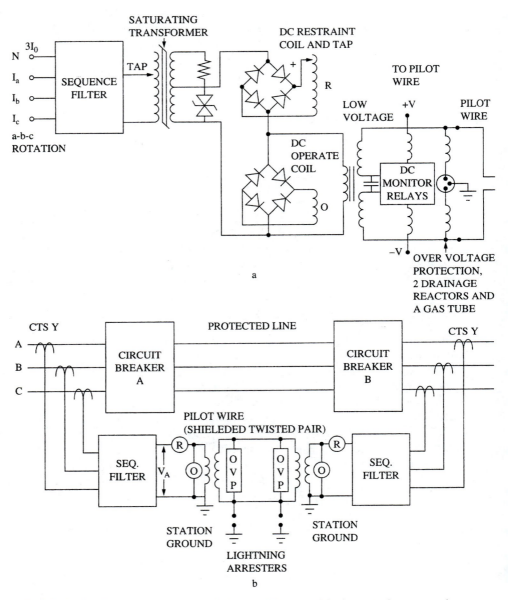

**FIGURE 10.18** Pilot wire carrier (a) HCB and HCB-1 simplified circuit diagram and connections (b) System configuration

occurs outside the protected line section the current through the restraint coils becomes stronger. If a fault occurs inside the protected area, the voltage from the two sequence filters shift in phase relative to each other causing the torque from the operate coil to strengthen as the restraint coil torque weakens, once again like a differential relay. The signal from one relay is strong enough to operate the relay coils at both ends of the line in the event the sequence filter on one end fails. The sequence phase shift in one filter is enough to cause the restraining coil torque to weaken and the operate coil strengthen.

A circuit is connected to inject about 20 V dc across the capacitor into the pilot line, as shown in Figure 10.18a. The circuit has sensitive dc monitor relays that are held closed by the current from the injected dc current. The dc monitor relays open from a current shortage if the pilot line becomes shorted across, shorted to ground, open, or is connected in reverse phase. The dc monitor relays send an alarm signal to the appropriate center if the pilot line fails.

Recently fiber optic cables have been used to provide pilot line protection. Fiber optic cables are light, and they are immune to electromagnetic interference, which is a problem on shielded twisted pairs. The pilot signal is a code sent by pulses of light on the fiber optic cable. Electronic circuitry generates and interprets the pilot signal on the fiber optic cable.

## 10.2.2 Pilot Carrier Relaying

Pilot carriers, like pilot wires, are normally used to block, or prevent, a relay from operating in response to a fault outside of the protected zone. Two major versions of carrier are used: power line carrier and microwave carrier.

Power line carrier, in which a radio frequency signal, usually between 30 kHz and 300 kHz, is injected on one phase of the power circuit. Power line carrier is economical and reliable. If the phase that has the rf carrier is faulted to ground a blocking signal is not desired so no relaying reliability is lost. Additionally the power line carrier can carry voice information if more than one frequency is used. With more frequencies supervisory control and data acquisition (SCADA) data channels can be transferred on the carrier. Sometimes the carrier is injected on the static wire. In this case the static wire must be insulated from the support structure yet still conduct transients to ground. Pilot voltages are usually less than 120 V so insulators or lightning arresters with breakdown voltages of about 300 V are used to support the static wire. This voltage is negligible in comparison to a switching or lightning induced transient. Phase conductors are preferred for the carrier. Power line carrier is useful to over 100 miles.

Microwave carrier is used when the distance between substations is too great for power line carrier because of signal radiation from the power line. The carrier signal is at a much higher frequency (in FCC allocated channels around 2, 6, and 12 GHz) and can carry many channels of communication of various types. Microwaves can be used for carrier signaling over many hundreds of miles, but the antennas must be in line of sight from each other. Relaying receiver/transmitter (R/T) stations must be used for irregular terrain. The microwave signal processing equipment is expensive and bad weather can cause carrier signal fading, but the additional SCADA channels are often a deciding factor in microwaves over power line carrier.

We will discuss only with power line carrier here. Even though the carrier transmission medium and frequency are different, the relaying action of the carrier is the same.

Figure 10.19 shows a basic power line carrier system. When a fault outside the protected line is detected by a relay the carrier is keyed (turned on) and the relays protecting the line between CB2 and CB3 are prevented from generating a trip signal. If the fault external to the protected line segment continues for long enough that it is probable the protective system of the faulted line segment has failed to operate, the relaying for the line segment between CB2 and CB3 ignores the carrier blocking signal and allows the breaker closest to the external fault to trip as back-up protection. Relay time delays are set for this function.

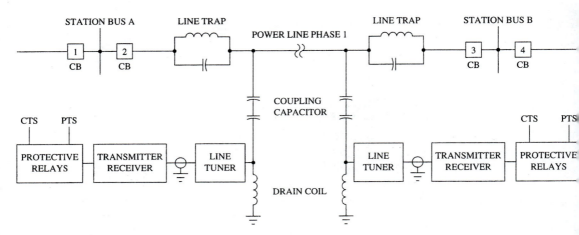

**FIGURE 10.19**   Power line carrier

If several channels of information are carried on the carrier, only one frequency is used for relay blocking. The line tuner separates the blocking signal from the other signals to activate the relays. The communication and SCADA channel frequencies may be present all of the time but the carrier frequency rf is only present when it is needed.

The rf carrier must be coupled to the line tuner and R/T units where the carrier signal is turned into a dc signal for the relays to use. A capacitive voltage divider called a coupling capacitor (which it is) is used to transfer the rf to the signal conditioning units. The line tuner is a band pass filter that only passes the desired frequency. A coupling capacitor and single frequency line tuner are illustrated in Figure 10.20a.

The rf carrier is kept out of the substation by a parallel tank circuit known as a line, or wave, filter, shown in Figure 10.20b. The filter coil must be constructed of conductor heavy enough to carry the entire line current without excessive loss. The line filter capacitors are usually placed inside of the coil.

Carriers are not used alone. They must be used with auxiliary carrier relays that can block other relay trip signals when a carrier is received. A popular carrier auxiliary relay is the *KA-4* (type 85). The trip circuit of the KA-4 is shown in Figure 10.21, and the current sensing coil is shown in Figure 10.21. The relay has a receiver unit that consists of the RRH and RRT coils wound on a core with an

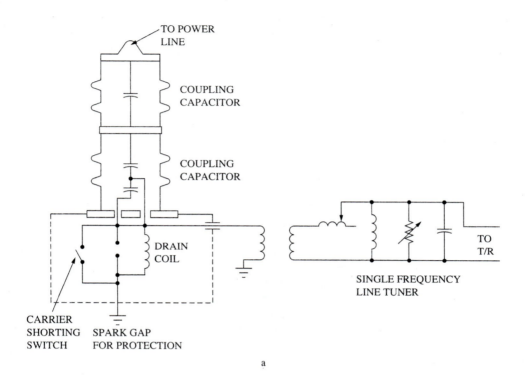

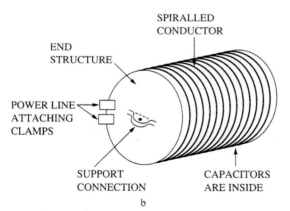

**FIGURE 10.20** Power line carrier high voltage components include (a) Coupling capacitor and (b) Line (or wave) trap. Uses large conductors for the coil to carry line current.

armature that has two normally open contacts, RRP and RRG. The RRP is a polar restraining coil that does not allow the RRT operating coil (T for trip) to close the contacts when the carrier signal is present. The RRT is supplied with a continuous dc so it holds the RRP and RRG contacts closed as long as the circuit breaker is closed, except when a carrier signal is received. The carrier signal indicates a fault

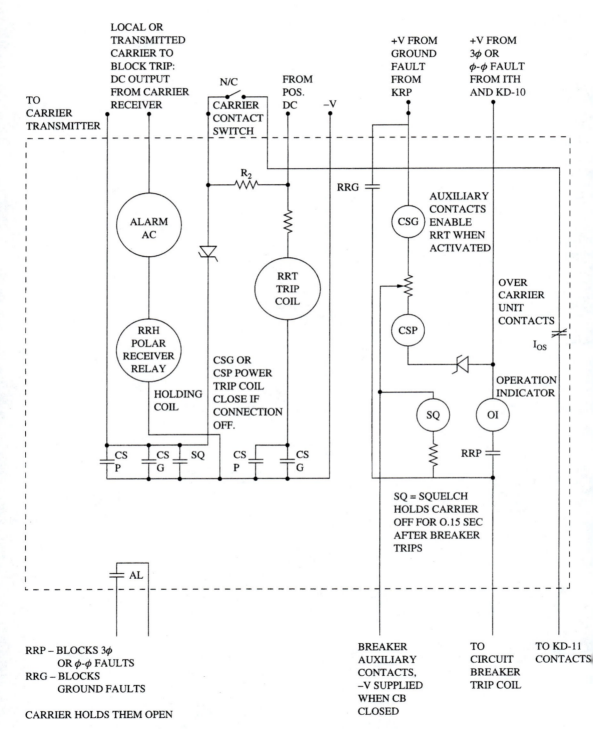

**FIGURE 10.21**   KA-4 carrier auxiliary relay (current sensing element shown in Figure 10.28)

outside of the protected zone. The CSG and CSP are coils that are energized from the negative dc station voltage when the CB is closed and when a phase to phase fault is indicated by +V from a KD-10 relay or a phase to ground fault is indicated by +V from a KRP relay. (We will discus the KD-10 and KRP relay in a later section.) The RRG contacts pass along a phase to ground trip signal when the CB is closed and the carrier is not blocking, and the RRP contacts pass along a phase to phase or three-phase trip signal under the same conditions. Summarizing the essential operation: if the circuit breaker is closed and a phase to phase or three-phase fault is detected by an ITH or a KD-10 relay the CSP coil operates, closing the CSP contacts allowing the RRT coil to operate and close the RRP contacts, tripping the circuit breaker unless the carrier restrains the RRT coil by energizing the RRH coil; and if the circuit breaker is closed and a phase to ground fault is detected by the KRP relay the CSG coil energizes, operating the CSG contacts, which energizes the RRT coil, which then closes the RRG contacts tripping the circuit breaker, unless a carrier signal is applied to the RRH coil blocking the trip.

An alarm (AL) coil in the KA-4 relay will not close if the carrier strength is below acceptable limits and cause an alarm. The alarm unit is less sensitive than the receiver unit so the carrier will still be strong enough to operate the holding coil. A receiver alarm indicates a loss of carrier. The squelch unit (SQ) holds the carrier off for 150 ms after the circuit breaker status contacts (and therefore the circuit breaker) open to allow all of the faulted line circuit breakers to trip before a carrier is transmitted. OI is an operation indicator. The overcurrent contacts allow +V to be applied to the carrier transmitter when they open. The induction cylinder overcurrent unit opens these contacts when it detects an overcurrent. This starts the carrier on a phase to ground fault. The KA-4 overcurrent unit starts the carrier because it is not directional, so a ground fault it detects could be outside the protected zone. If the fault is within the protected zone another relay trips the circuit breaker.

## 10.2.5 Directional Ground Overcurrent Relay (Type 67)

Two commonly used electromechanical directional overcurrent ground relays are the KRP and IRP, shown in Figure 10.22.

The KR type relays are directional and can be blocked by a carrier. The relays come in voltage polarized (KRP), current polarized (IRP), and dual current and phase polarized (KRD) models. They differ in the reference source used for the directional unit. We will discuss the KRP. The KRP directional unit is an induction cylinder type. The polarizing (reference) flux is from two different coils placed opposite one another with respect to the induction cylinder. The operate coils are set to provide maximum torque when the voltage leads the current by 60°, an angle that only occurs under fault conditions. The directional unit responds to a fault in one direction only.

The overcurrent unit of a KRP is a typical induction cylinder unit. While it does have an inverse time characteristic it is very fast and is considered instantaneous. The overcurrent unit is fed by a saturating transformer, with setting taps, which reduces the burden at high fault currents, and protected from excessive voltage by

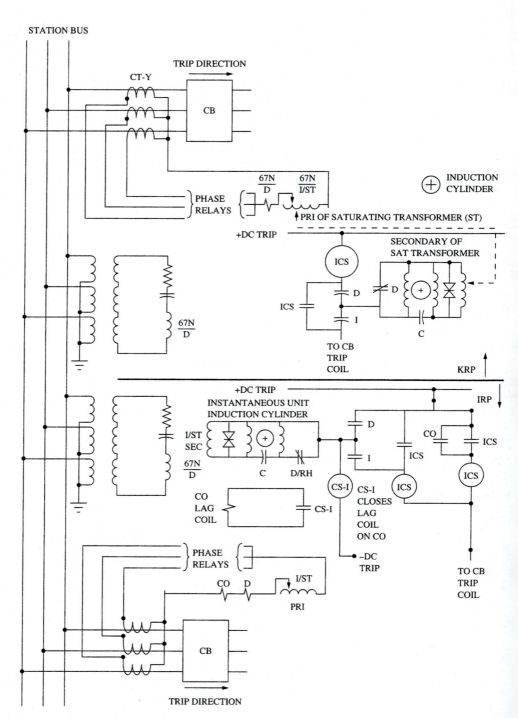

**FIGURE 10.22** KRP and IRP directional ground fault relays and connections

a varister type device. The relay also has an ICS unit to indicate trip and carry circuit breaker trip current.

The IR series is similar to the KR series except that they are not designed for carrier blocking, and they have with inverse time CO type overcurrent units. They are available with voltage (IRP), current (IRC), and dual phase and current (IRD) directional polarization. The directional unit is almost identical to that used in the KR series, and must indicate a fault in the trip direction before the overcurrent unit lag coil is enabled (by CS1). The IRP overcurrent unit is similar to the CO induction disk overcurrent relays (type 51) and has very similar tap settings. The relays are purchased with the inverse time characteristic desired. As in the type 51 relays the amount of inverse time characteristic is specified by a number, so one purchases an IRP-X, where X is the inverse characteristic number from 2, slightly inverse, to 11, extremely inverse. The overcurrent unit is fed by a saturating transformer. The IRP relay also has an instantaneous overcurrent unit to detect very high current, close in faults, and two ICS units that indicate if the relay was tripped by the inverse time or the instantaneous unit.

Both the KR and IR relays are in the K-DAR family of directional, carrier blocking relays. The K-DAR family of relays also includes directional phase tripping relays (KD-4), phase start relays (KD-41), and distance relays, to be covered next.

## 10.2.6 Distance (Impedance) Relays (Type 21)

A distance relay compares the current and voltage of a power system at the relay installation site to determine if a fault is within the length of line the relay is set to protect, or beyond. If the distance to the fault is within the protected distance the relay trips. If the distance to the fault is farther than the distance the relay is set to protect the relay does not trip. The voltage, current, and phase sensed by the relay are proportional to the line impedance to the fault, thus the relays are often called impedance relays. The total impedance of a line to a fault varies with the distance to a fault, so the relays are most often called distance relays.

The impedance, therefore the distance, is set by taps inside the relay. The tap setting is the total impedance of the line to the maximum protection distance (called the balance point).

$$Z_{TAP} = Zd \qquad (10.22)$$

where

$Z$ = impedance of the line per unit length

$d$ = number of unit lengths (usually miles) protected

To obtain an intuitive idea of how electromechanical distance relays work refer to Figure 10.23a. A balanced beam holds a moving contact. A voltage coil acts as a restraining coil, and a current coil acts as an operating coil. Under fault conditions the current rises, and the voltage drops, causing the effective impedance of the system to drop. When the impedance is as low as the faulted line impedance the beam tips toward the current coil, closing the contact. The beam is balanced when

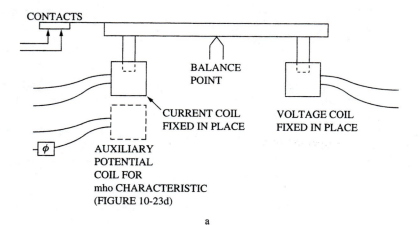

a

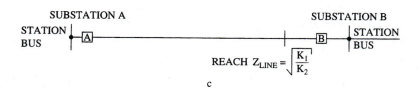

b

c

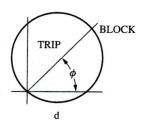

d

**FIGURE 10.23** Impedance (distance) relay (a) Balanced beam anology (b) Characteristic of impedance relay (c) Reach of distance relay and (d) Mho characteristic

the torques from the two coils are equal. This is called the balance point and the impedance the voltage and current represent at that point is the maximum range of the relay.

In a commercial electromechanical relay the coils provide torque on an induction disk or cylinder, which holds the moving contact. The contacts close when

$$K_1 I^2 < K_2 V^2$$

$$\frac{V^2}{I^2} < \frac{K_1}{K_2}$$

so the reach to the balance point is

$$\frac{V}{I} = Z_{TAP} = \sqrt{\frac{K_1}{K_2}}$$

The reach to the balance point is called the relay's *zone of protection*.

The characteristic of the balance beam, or induction disk equivalent, is shown in Figure 10.23b. The characteristic is nondirectional. A directional unit can be added to obtain the half circle characteristic indicated by lining in Figure 10.23b. The zone of protection is shown in Figure 10.23c.

It is usually desirable for a distance relay to be sensitive only in the quadrant in which the fault impedance is likely to lie. This is of course the first quadrant for down line faults. The mho type relays accomplish this by modifying either the restraint or operating quantities. Adding a potential coil to the operate side as shown in the dotted part of Figure 10.23a to close, if $K_3$ is a torque constant proportional to operate coil watts, when

$$K_3 VI\cos(\theta - \phi) > K_2 V^2$$

$$\frac{V}{I} = Z_{TAP} = \frac{K_3}{K_2} \cos(\theta - \phi)$$

where

$\theta$ = the angle of the line impedance

$\phi$ = the phase shift between the $V$ and $I$ coils of the coils on the operate side caused by a phase shift network

Figure 10.23d shows the mho characteristic, which is inherently directional. The characteristics of a distance relay can be tailored to just about any desired amount of directionality.

Two common electromechanical distance relays are the K-DAR family KD-10 and KD-11, both with mho characteristics. The KD-10 is a three-phase relay that provides a single zone of protection for three-phase, two phase to ground, and phase to phase faults. It provides instantaneous trip for those faults. The KD-11 is similar to the KD-10 except that the origin is just within its characteristic (meaning it can detect faults where it is) instead of on the circumference as in the KD-10. The KD-11 can be used with a KA-4 relay to initiate carrier blocking, and with a time delay.

Both relays, see Figures 10.28 (the sensing elements) and 10.29 (the trip circuits), have separate three-phase and phase to phase distance units. Both relays also have ICS units to indicate a trip. The distance units have taps to vary the distance of the balance point. The phase to phase units have three tap ranges, with seven tap settings in each range, from 0.2 $\Omega$ to 4.5 $\Omega$, 0.75 $\Omega$ to 21.2 $\Omega$, and 1.27 $\Omega$ to 36.7 $\Omega$, as well as multipliers to extend the basic range. The three-phase unit has taps from 0.87 $\Omega$ to 5.8 $\Omega$ with multiplier taps to extend the range. The taps should be set with the relay operating manual in hand until one is very experienced.

Solid state microcomputer controlled distance relays are becoming popular. They use voltage, current, and phase measuring circuits to obtain trip information instead of electromechanical components. They are more flexible than the electromechanical distance relays.

## 10.2.7 Zones of Protection and Carrier Blocking

The zone of protection is that distance of line that is protected. A forward zone of protection is one in the direction of the protected line, and a reverse zone is in the opposite direction of the protected line. More than one zone of distance relaying is normally used, with a different distance relay needed for each zone. The three zones of distance relay protection are referred to by number.

*Zone 1* is a forward zone with the balance point set to 80% to 90% of the length of the protected line, as shown in Figure 10.24a. Note that the zone 1s from each end of the line overlap. This is normally a KD-10 relay.

*Zone 2* is a forward zone set to look about 50% to 80% of the distance between station B and the next station beyond it (125% to 180% if the station distances are equal). This zone is to detect faults beyond the next station and key carrier blocking in zone 1, and with time delay as a back-up for the zone 1 relaying and the forward relaying of station B. Zone 2 relays set near 125% are frequently used with time delay alone for backup, while Zone 2 relays that reach further are used to key carrier blocking also. Figure 10.24b illustrates zone 2, which is normally provided by a KD-11 or a KD-10 with an auxiliary time delay relay.

*Zone 3* is either forward or reverse looking depending on the philosophy of the line protection designer. Zone 3 relays are also timed relays with time delays greater than zone 2 relays and are also carrier blocked. Forward looking zone 3 relays look beyond the receiving substation to at least the one beyond it and perhaps even further (200 to 300% for equidistant substations). Zone 3 is often set to look in reverse and key carrier blocking because a fault detected with a zone 3 relay is outside of line section that relay system is protecting. The zone 3 relay operates after a time delay of sufficient duration (normally about 90 seconds) to know the proper relay set for the line segment in reverse from station A has failed to operate, acting as a delayed back-up for that section. The zone 3 reach is normally set for 200% to 300% of the distance between stations A and B. Zone 3 may not reach or may reach beyond the next station. Reverse looking zone 3 is illustrated in Figure 10.24c. Again, zone three may also look forward beyond the next station

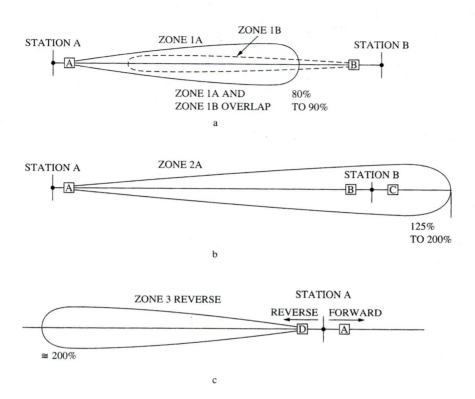

**FIGURE 10.24** Distance relay zones (a) Zone 1, typically 80% to 90% of the protected line length (distance from substation A to substation B) (b) Zone 2, typically 125% to 200% of protected line length (c) Zone 3, typically 200% to 300% of protected line length, may or may not reach next station, or may reach beyond

beyond station B. It functions then as back-up for zone 2 relaying. Zone 3 relays are usually KD-11 relays.

Figure 10.25 illustrates how the time delays of the zone relays provide back-up protection for each other. Figure 10.26 summarizes the relationship between the zones of protection and the carrier blocking. The carrier can be either power line, fiber optic, or microwave; the operation is the same. In every case the carrier is keyed to prevent the protected line section circuit breakers from tripping in response to a fault outside the protected section. This allows instantaneous tripping to be used on internal faults for quicker fault clearing.

Note that the KD-10 and KD-11 relays have phase to phase and three-phase direction units. They can detect double line to ground faults but not ground faults. The relay system must have ground fault relays to provide ground fault protection.

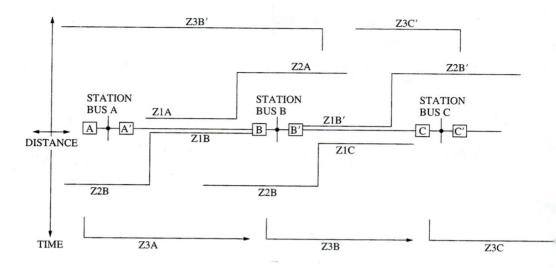

**FIGURE 10.25**   Time delays for zone coordination

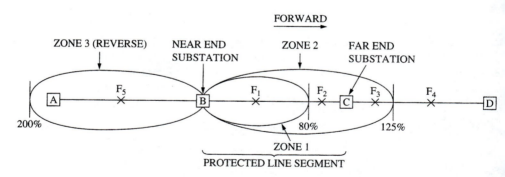

F1 - ZONE 1 TRIPS BREAKER B INSTANTANEOUSLY

F2 - ZONE 2 TRIPS BREAKER B INSTANTANEOUSLY (WILL NOT BE BLOCKED BY CARRIER)

F3 - ZONE 2 BACKS UP RELAYS AT C CARRIER BLOCKING AND TIME DELAY PREVENTS B FROM
TRIPPING UNLESS C RELAYS FAIL TO

F4 - NO ACTION

F5 - ZONE 3 (REVERSED) PUTS BLOCKING CARRIER ON LINE FROM B TO C TO PREVENT TRIP
AT B AND C. KA-4 OR KD-11 MAY KEY CARRIER DEPENDING ON FAULT TYPE
ZONE 3 MAY TRIP AFTER A TIME DELAY IF FAULT IS NOT CLEARED NORMALLY

**FIGURE 10.26**   Distance relay zones of protection and carrier coordination

## 10.2.8 Transmission Line Relaying Example

Figure 10.27 shows a fairly typical differential scheme for transmission line protection. This is for a line fed by a ring or similar bus, or that can be fed by more than one source. The protection is split into main and back-up systems, where each system is fed by its own instrument transformers. The actual relay sensing system is shown in Figure 10.28. The trip circuit for the main system is shown in Figure 10.29, and the trip circuit for the back-up relays is shown in Figure 10.30

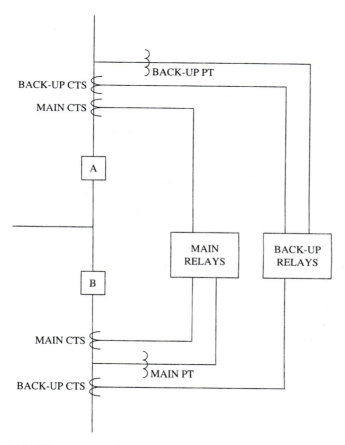

**FIGURE 10.27** Back-up protection, differential connection for line with two breakers. Each relay set is fed by separate instrument transformers.

The main relay circuit of Figure 10.28 starts with an ITH instantaneous overcurrent relay set to detect very high current close in faults. Next is a KD-10 relay set forward for zone 2, and a KD-11 relay set reverse for zone 3. Some schemes use a KD- 10 zone 1 relay in the main relay system. What is important is that all of the protection zones are not on the same instrument transformers. The KRP directional ground fault relay is next to detect ground faults in the forward direction.

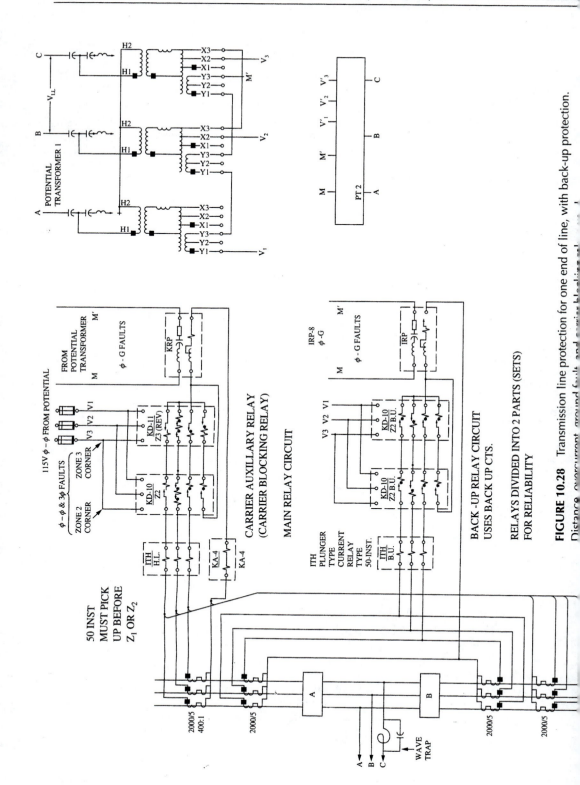

**FIGURE 10.28** Transmission line protection for one end of line, with back-up protection.

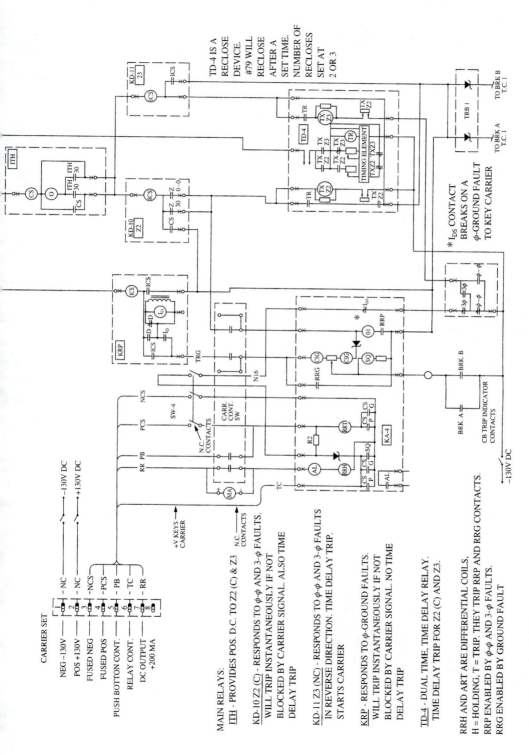

**FIGURE 10.29**  Trip circuit of main relaying in Figure 10.28

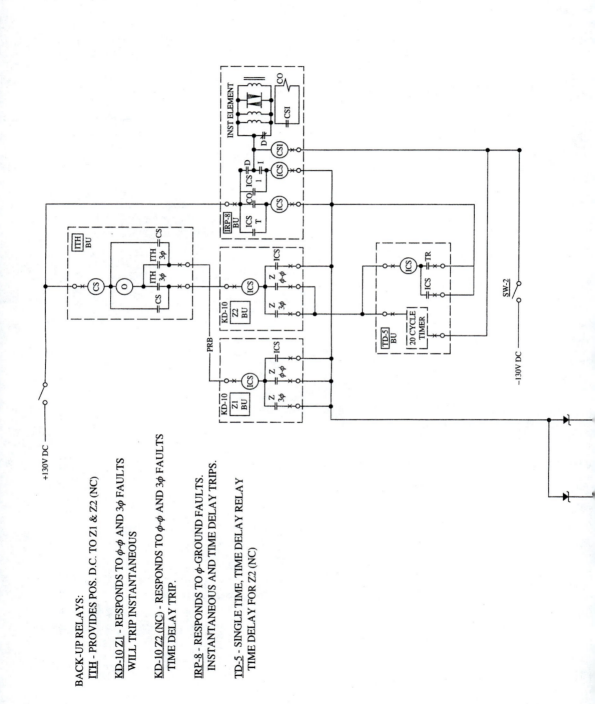

BACK-UP RELAYS:

ITH - PROVIDES POS. D.C. TO Z1 & Z2 (NC)

KD-10 Z1 - RESPONDS TO φ-φ AND 3φ FAULTS
    WILL TRIP INSTANTANEOUS

KD-10 Z2 (NC) - RESPONDS TO φ-φ AND 3φ FAULTS
    TIME DELAY TRIP.

IRP-8 - RESPONDS TO φ-GROUND FAULTS.
    INSTANTANEOUS AND TIME DELAY TRIPS.

TD-5 - SINGLE TIME, TIME DELAY RELAY
    TIME DELAY FOR Z2 (NC)

The KA-4 relay comes next. Its overcurrent function is to detect ground faults in the reverse direction and key the carrier.

The trip circuit for the main relays is shown in Figure 10.29. The relationship of the relay contacts and operation was detailed in the KA-4 relay section earlier. Notice that the KD-11 zone 3 relay is set to key the carrier by opening contacts in series with the KA-4 IOS contacts. The recloser is set to reclose the circuit breakers a short time (usually one or two seconds) after a fault is sensed by the KD-10, KD-11, or ITH relays. If the trip signal has been removed when the circuit breaker recloses, it stays closed. If the fault was not transient the breaker trips again. The TX coils close the contacts that activate the recloser timer. The recloser is normally set so that after it tries to reclose and trips two or three times it locks the circuit breaker open.

The back-up relay system of Figure 10.28 has an ITH relay, a KD-10 zone 1 relay, and a KD-10 zone 2 relay. The zone 2 relay is a back-up for the zone 2 relay of the main relay system, but it has a time delay only and does not key the carrier. The zone 1 relay is the only zone 1 relay and the main zone 1 protection for every fault type except ground faults. The philosophy is to have two zone relays in each relay set. Usually both the main and back-up relay systems both contain zone 2 relays, but only one zone 1 and zone 3 relay is used, one in the main system and the other in the back-up. The inverse time IRP directional ground fault relay is used in this back-up relay system. The trip circuit of the back-up relay system is shown in Figure 10.30. Notice that an auxiliary timer, a TD-5, is used to provide a time delay before tripping from the zone 2 KD-10 or IRP-8 relays. The ITH and the zone 1 KD-10 are set to trip at the same current in this particular system. The ITH is not directional so it trips on a fault from either direction. The zone 1 KD-10 relay, which is directional, passes on the trip signal only if the fault is in zone 1 forward.

**0.3**

## BULK POWER SUBSTATIONS

A bulk power substation changes voltage to or from transmission level voltages and operates circuit breakers in response to transmission line or substation faults. In addition a bulk power substation has the following functions: controlling power to an area, housing protective relays and instrument transformers, and housing switching arrangements that allow maintenance of any substation equipment without disrupting the power to any area served by the substation. Basically it performs all of the functions of a critical distribution substation at much higher voltage and power levels. Most of the equipment is the same, and operates the same, as the equipment at a distribution substation except that it is scaled up in size and capacity. Thus the reader is referred back to Chapter 4 for power and instrument transformer information, and to Chapter 5 for information about circuit breakers, lightning arresters, switches, and substation protection. Substation switching arrangements are shown in Chapter 6, and are useful for high or low voltage. Major differences between bulk power and distribution substations are that spacings are greater, autotransformers are often used, reactors are sometimes housed, and grounding is more critical in bulk power substations. In this chapter we will discuss the bulk power substation attributes that are different from distribution substations.

### 10.3.1 Compensating Reactors

Reactors are used to compensate transmission line capacitance to avoid excessive line voltage. The reactors may be switched, similar to capacitors, to compensate for loading changes. Reactors switched by SCRs are being used to respond rapidly to load changes. At heavy loads some inductive reactance is switched out to improve line voltage regulation, and at light loads more reactance is switched in to avoid excessive line voltage from the line capacitance. Normally half of the reactance is located at the substation from which the transmission line originates, and half at the substation at which it terminates.

### 10.3.2 Autotransformers in Bulk Power Substations

We noted in Chapter 4 that an autotransformer is never used as a distribution transformer because the lack of isolation can cause dangerously high voltages in a customer's premises if the neutral opens. In a bulk power substation, where voltages are transformed from one high voltage to another high voltage or from a transmission voltage to a subtransmission voltage, and which are accessible only by well equipped and trained workmen, autotransformers are often used. The first transformer after the generator is almost always a two winding transformer for dc isolation, and the last transformer in the substation feeding a customer is always a two winding transformer for safety.

Recall from Chapter 4.6.2 that the advantage of an autotransformer is that it can conduct some of the volt-amperes it transforms instead of transforming it all by induction. Thus the core loss is smaller in an autotransformer, and less magnetic core material is needed for magnetic induction. This allows autotransformers to be lower in cost, size, and cooling requirements than two winding transformers of the same MVA rating. The major disadvantage of autotransformers in applications for which they are suited is that their impedance is low so fault currents are high in comparison to two winding transformers.

### 10.3.3 Types of Grounding

Most transmission and subtransmission systems are solidly grounded to provide a solid reference voltage for relaying equipment, provide ample fault current for relays to detect in the event of a ground fault, and reduce voltage stress on the equipment. Transmission systems are usually grounded wye, but subtransmission systems are often delta connected to raise the zero sequence impedance in the event of a ground fault. To avoid voltage stress phase 1 of a delta system is often grounded. Delta systems may also be grounded through grounding transformers.

Grounding transformers, shown in Figure 10.31, are normally wye–delta or interconnected wye, called zig-zag. These transformers are designed to provide neutral potential equivalent to that of a wye system with the same line voltage while using very little power themselves. The windings are connected to limit the current from a ground fault to the rated neutral current by preventing core saturation. The transformers are used only for grounding and carry little current except during a ground fault, so they are fairly small. They are usually rated with the stipulation that they carry current for no more than 5 minutes. The relays should operate long before that. They are connected to the station ground.

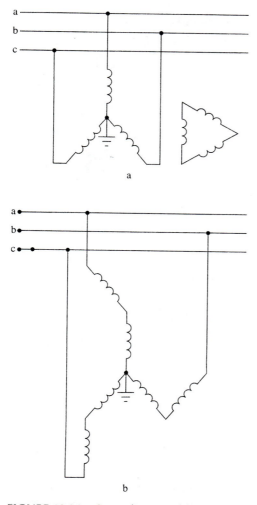

**FIGURE 10.31**   Grounding transformers (a) Wye-delta and (b) Zig-zag. Windings drawn parallel are wound on the same core.

Wye system grounding may be solid, through a conductor, reactance (imped-ance), through a reactor, resistance, through a resistor, or resonant, through both a reactor and a capacitor. These are shown in Figure 10.32. Reactance and resistance grounding are used to reduce zero sequence ground fault current while still pro-viding an adequate ground for protective relaying. Reactance grounding reduces zero sequence current, but normally resistance grounding reduces it more. In some instances resistance grounding provides enough power loss to reduce gen-erator power angle advance during a fault thereby improving stability. Resonant grounding uses a capacitor to cancel the grounding reactance during a fault, thereby providing a high impedance to the ground fault. The intent is for the par-allel resonant circuit impedance to be so high that a ground fault arc cannot main-tain itself. Resonant grounding is the highest cost and may cause problems when systems are interconnected.

A solidly grounded system allows the use of the most effective and economical lightning protection. It is the most common transmission system, but reactance and resistance grounding are used in many systems.

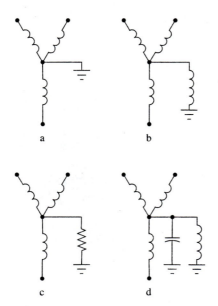

**FIGURE 10.32**   Grounding methods include: (a) Solid (b) Reactance (c) Resistance and (d) Resonant.

## 10.3.4  Bulk Substation Grounding

Equipment protection is only part of the reason that substations are so well grounded. Personnel protection is a major consideration. A continuous current through the trunk part of the body of about 0.15 A is almost always fatal. Electric shock death usually occurs from ventricular fibrillation, where the electrical signals that drive the heart contractions go wild causing very rapid but ineffectual heart contractions to occur. The trunk of the body can withstand higher currents, but only for short times, and any current in the trunk over 0.1 A is dangerous. The Institute of Electrical and Electronic Engineers (IEEE) has set a standard for nonfatal body current at 201 mA for 0.33 s for a 50 kG (110 lb) person for the general public.

$$I_{MAX\ PUBLIC} = 0.116/t$$

For workmen inside a substation the limit is 272 mA for 0.33 s for a 70 kG (110 lb) person.

$$I_{MAX\ WORKMEN} = 0.157/t$$

These currents would be reached in a person only under unusually poor circumstances, such as when wet from a rain and standing on wet ground with wet shoes.

The body current is limited around a substation, or any other electrical power structure such as a transmission tower, by reducing the ground impedance to such a degree that any easily touchable point will not develop enough voltage to cause a fatality during a ground fault. The assumed body resistance for such calculations is 1000 Ω, much less than normal body resistance. Two safety related voltages are defined, touch voltage and step voltage.

*Touch voltage* is the maximum voltage allowed on any structure from any point within reach from the ground, standing 3 feet from the structure, because of ground fault current. The maximum touch voltage is 653 V inside a substation, where only competent workers in proper clothing should be allowed, and 207 V on the substation fence. The magnitude of fault current that would produce the maximum touch voltage would be very large. Figure 10.33a illustrates the touch voltage. To minimize the touch voltage, which would be dropped along the structure, a ground conductor is buried 18 inches below the surface 3 feet from the structure as shown in Figures 10.33a and b. The buried copper cable is attached to the structure by conductor, and also connected to 58 inch copper clad rods driven 10 feet into the ground. The principle is to give the ground fault current a better path to ground than through a person hapless enough to be touching the structure at the instant of a ground fault. The buried cable also prevents a voltage from becoming too high because of fault current flowing through the soil near the structure. The 10 foot rod is needed to assure good connection of the grounding conductors to the low resistance subsoil. The resistance of topsoil is much higher than that of subsoil.

Step voltage is the maximum voltage that can flow from foot to foot as one is walking during a fault condition, as shown in Figure 10.34a. The maximum step voltage allowed is 2010 V in a substation, and 225 V at the substation fence. A high step voltage normally occurs only during a phase to ground fault. If a phase should fall to ground between two towers on a right of way, as shown in Figure 10.34b, the phase voltage will be distributed evenly along the Earth between the conductor ground point and the ground return points. If the line potential is 100 kV and the line is faulted 1000 feet from the tower on the left the ground potential per foot is

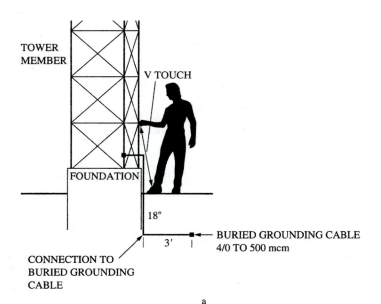

a

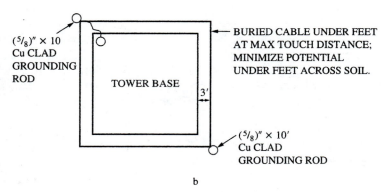

b

**FIGURE 10.33** Touch voltage (a) Definition and (b) Minimized by grounding

$$V_{ft} = \frac{V}{d} = \frac{100 \ kV}{1000 \ ft} = 100 \text{ volts per foot}$$

If the foot to foot body resistance of a person under the line walking with 3 feet strides is 1000 $\Omega$, the step voltage is 300 V and the foot to foot current is

$$I_{\text{foot to foot}} = \frac{300V}{1000\Omega} = 0.3 \text{ A}$$

which is fatal. Of course foot to foot resistance is normally much higher because people normally wear shoes. Additionally the protective relays open the circuit breaker within five cycles, and the voltage at the ground point of the faulted phase

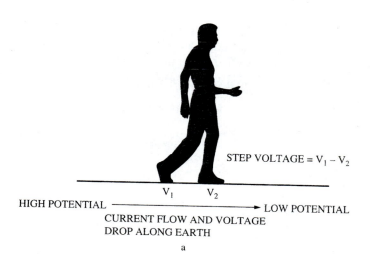

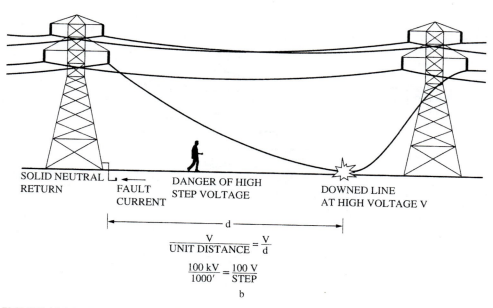

**FIGURE 10.34** Step voltage (a) Definition and (b) How a high step voltage can occur

is reduced by the voltage drop along the other impedances in the circuit. Within a bulk power substation the ground resistance is so low that the step voltage is less than the maximum except during the most severe ground faults.

All substations have a ground mat. Every piece of equipment and every support structure in the substation is connected to the ground mat, as is the substation fence. In generation substation s the generator is also grounded. The purpose of the ground mat is to ensure personnel safety by keeping both the touch and step voltages low during ground faults, and to prevent induced voltages on equipment and structures during normal operation. The same equipment and support structure grounding system is used even with delta and ungrounded wye systems that have no neutral return. A small bulk power substation ground mat is shown in Figure 10.35.

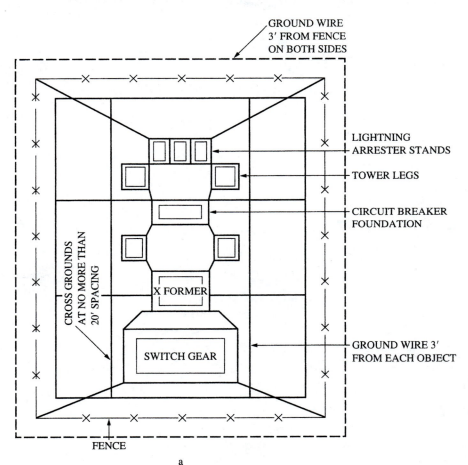

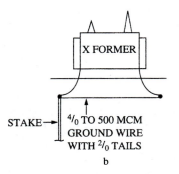

**FIGURE 10.35**   Small substation ground mat (a) Overview and (b) Transformer ground cross section

The bulk of the resistance associated with a grounding system is in the contact between the Earth and the ground conductors. If a good contact is established between the conductor and the deep subsoil the Earth resistance is very small. In fact the Earth resistance is so low that it can be used as the return line for a dc system in which the positive or negative line is out of service so the system can operate at about half capacity. Such returns may be over a thousand miles long. The ground contact consists of deep buried cable (8 feet or more) with special soil treatment, such as enclosing the 500 MCM ground cable inside compacted coke or charcoal. The buried ground cable at each end of the system may be over 3 miles long. The low resistance of the soil once good contact is established is why a well-established fault has a negligible resistance.

The resistance of the grounding conductors are in the low milliohm range in a substation. The #6 conductor that is used to ground a wood pole has less than an ohm of resistance on a 50 foot pole. Because the bulk of the resistance of grounding systems is in the conductor to soil interface, to measure the effective ground system resistance the soil to ground resistance must be included. A popular method of measuring ground system resistance is the fall of potential method, shown in Figure 10.36a. A current in the low ampere range is injected into the grounding system, usually through one of the two transformer grounding connections. The current return is placed 162 feet away from the instrument, and a voltmeter measures the potential between the ground mat and the voltage probe located 100 feet from the instrument. The resistance is the ratio of the measured voltage and the injected current. The remote probes are copper rods driven 18 inches into the subsoil. Another common probe connection is shown in Figure 10.36b.

The specified ground resistance, including the ground conductor to soil interface, for some common electrical power installations are: bulk power substations—$R_g < 0.005\ \Omega$, distribution substations—$R_g < 5\ \Omega$, transformers (other than bulk power substation) and lightning arresters—$R_g < 5\ \Omega$, transmission tower grounds—$R_g < 10\ \Omega$, and distribution poles—$R_g < 25\ \Omega$. The ground systems of electrical power installations are complex, efficient, and expensive. They are very important for safety as well as a protective relaying reference and a low impedance path to ground for transient current, such as that from a lightning stroke.

## 0.4

## SUMMARY

Transmission line fault current calculations for balanced three–phase systems can be done on a per phase basis using either actual values or the pu method. Asymmetrical system fault current can be calculated using the method of symmetrical components. In this method a three–phase system is resolved into three sets of balanced phasors. They are:

1. *Positive sequence components* are three phasors of equal magnitude and offset from each other by 120° that rotate in the same direction as the original phasors.

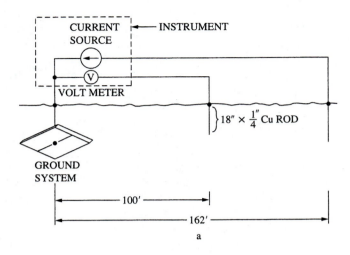

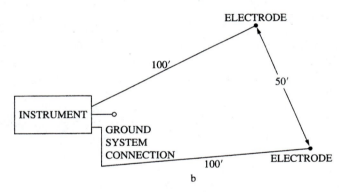

**FIGURE 10.36** Fall-of-potential method of ground resistance measurement
(a) Most common connection and (b) Alternate connection

2. *Negative sequence components* are three phasors of equal magnitude and offset from each other by 120° that rotate in the opposite direction as the original phasors. Normally they are considered to rotate in the same direction as the original with phases b and c reversed.

3. *Zero sequence components* are three phasors of equal magnitude with no offset in phase between them.

Sequence impedance is obtained by providing unit current to the three phases of the apparatus under question and writing equations for the voltage drop. The sequence impedance of the system is found by properly including the sequence impedance of each apparatus in the system. Fault current is then found by dividing the system voltage by the system sequence impedance.

Transmission lines and substation represent a substantial investment and must be protected well. Transmission lines are protected by time and instantaneous overcurrent relays and ground current relays, often directional. Differential schemes in which the current entering and leaving a section of line are compared, with the sending and receiving information transmitted by a variety of media and encoding schemes depending on the length of the line.

Specialized impedance relays called distance relays are often used for transmission line protection. The distance to a fault is proportional to the line impedance to the fault, so distance relays allow good definition of zones of protection. Normally three zones are used with distance relays to provide both main and back-up protection. They are:

1. *Zone 1* is a forward looking zone with the impedance of the distance relays set to about 80% to 90% of the impedance of the line segment impedance between substations. Zone 1 is normally set with no time delay.

2. *Zone 2* is a forward looking zone in which the relay is set to look between the receiving substation and the one beyond it. If all substations were the same distance apart zone 2 would be set to about 125% to 180% of the distance between substations for which the zone 1 relay is protecting the transmission line. Zone 2 is time delayed to operate only if the zone 1 of the receiving substation does not operate for faults in the overlap. Zone 2 relays are blocked from operating by a carrier signal if zone 1 relays operate anywhere in the distance they protect.

3. *Zone 3* is either forward or reverse looking depending on the philosophy of the line protection designer. Zone 3 relays are also timed relays with time delays greater than zone 2 relays and are also carrier blocked. Forward looking zone 3 relays look beyond the receiving substation to at least the one beyond it and perhaps even further (200% to 300% for equidistant substations). Reverse looking zone 3 relays look backward past the nearest substation and often to the one beyond it, again 200% to 300% for equidistant substations. Zone 3 relays function as back-up to the back up relays.

Transmission substations never feed customers directly and thus autotransformers are often used in them. The high value of fault current available in a transmission substation makes proper grounding of a substation essential.

0.5

## QUESTIONS

1. List the most common short circuit faults and their approximate percentage of occurrence.
2. Define the term fault.
3. State the basic principle of symmetrical components.
4. List the symmetrical components of a three-phase system.
5. The currents in Figure 10.4 are changed to $I_a = 12 \, A\angle 60°$, $I_b = 12 \, A\angle -60°$, and $I_c = 0$. Calculate the positive, negative, and zero sequence phasors, then check your work by calculating the vector sum of the symmetrical components.
6. Why are electrical machine transient and subtransient impedances used for most symmetrical component problems in electrical power networks?

7. Define a sequence network and list the major components of the three-phase sequence networks.

8. A Δ-Y transformer with high side (primary) phase currents of 300 A∠0°, 300 A∠-120°, and 300 A∠120°, and a turns ratio of 0.1 is connected by the United States standard connection method. Calculate the low side (secondary) currents.

9. Why is the zero sequence impedance infinite in an ungrounded system?

10. The system of Figure 10.16 has the following specifications:

    Generator: two pole, conductor cooled, turbine driven, 100 MVA at 23 kV, $X_d'' = 20\%$, $X_2 = X_d''$, $X_0 = 12\%$.
    Transformer 1: core type, Y-Y, grounded neutral, 23 kV/138 kV, 150 MVA, $X = 10\%$, $X_0 = 7\%$.
    Line: three-phase, Drake conductor, spacing of 14 feet, 14 feet, 28 feet, $Z_1 = Z_2$, $Z_0 = 3Z_1$.
    Fault impedance: max = 20 Ω, min = 0 Ω.

    a. Calculate the maximum and minimum fault current for a line to ground 100 miles from transformer 1.

    b. Calculate the maximum and minimum fault current for a three-phase fault 50 miles from transformer 1.

    c. Calculate the maximum line to line fault current at a distance of 25 miles from transformer 1.

    d. Calculate the maximum double line to ground fault current at the secondary of transformer 1.

11. State the basic philosophy of transmission line protection.

12. List three advantages of clearing a fault quickly.

13. Briefly state the function of a pilot wire carrier.

14. State the function of the sequence filter in the HCB and HCB-1 relays.

15. Why is a low current dc injected on a pilot wire?

16. What is power line carrier relaying?

17. State the function of the carrier in relaying.

18. State the function of the wave trap and coupling capacitor in power line carrier relaying.

19. Briefly state the function of the following relays: A. KA-4, B. KRP, C. IRP.

20. Why must directional relays be used in transmission line protection?

21. What is meant by zones of protection of distance relays?

22. What is the most common coverage of the three most commonly used distance relay zones.

23. How are the distance relays coordinated?

24. What property does the KD-11 distance relay have that makes it useful for zone 2 and 3 use?

25. State the principle of back-up protection for transmission line relays.

26. In what ways do bulk power substations differ from distribution substations?

27. Define touch and step voltages.

28. List the four common types of grounding and one advantage of each.

29. What is the source of most of the resistance of electrical transmission and distribution equipment grounding systems?

# CHAPTER 11

# Transmission Line Construction

This chapter discusses the steps involved in constructing a transmission line.

## INTRODUCTION

Transmission line design has become a classic example of the adaptation of available standard designs to best fit the needs of a particular job after consideration of power loss, voltage drop, reactance, and thermal overload conditions. To make proper design adaptations, one must understand all factors influencing that design decision.

Complete transmission design is complex, but once a good design is developed, it can be used repeatedly. Good design is changed only when experience with that design indicates it will not handle a particular transmission condition efficiently.

An example of this is the Texas Inter-connected Network, the network that connects Houston Lighting and Power (HL&P) and Texas Utilities of Dallas. Although Texas Power and Light had originally designed double circuit 345 kV towers, the design was inadequate. As a result, before the Dallas/Houston intertie was built, a new tower design was developed and approved by both companies for their new 345 kV line. Since then, the majority of all 345 kV double circuit lines in Texas have been built using that new tower design system. The towers were designed, (dead ends, swing angle, and tangent towers) so that on any 345 kV line, the layout could be made using those towers. Now, the designer has only to adapt the towers to the job at hand, select the conductors, the tower spanning, the foundations, and the auxiliary equipment. The details of the towers, the hardware, and the fittings are already accomplished and decided as policy. In fact, about 75% of the total design has been previously established. By selecting tower materials already designed and in use, the cost is significantly lowered. For their own use, HL&P has developed a tall, double circuit 345 kV line with vertical phases for use in narrow right-of-ways (ROW) (those that are seen within Houston city limits), but this more expensive tower is never used in rural areas where wider ROW are available.

With computers and standard line designs, a new line can be designed quickly. Companies only change a standard structure after much consideration, and after allowing for testing and production of the required materials.

Each power company has developed its own standards, and insist on their use, but because of the common designs, only a few systems have been developed. Originally, three engineering companies were responsible for all of the design work for approximately 80% of the power companies in the United States. Engineers went from one company to the other and the best of all available designs were adopted by each engineer and recommended to their utility clients. Conductors, fittings, and hardware were standardized by the Institute of Electrical and Electronics Engineers (IEEE) and the National Electrical Manufacturing Association (NEMA), and towers were standardized by major steel tower manufacturers. Now, common understanding makes all lines similar.

As Texas Interconnected Network developed a standard 345 kV line, the Southern Network (Georgia, Alabama, Mississippi, Louisiana, Oklahoma, Arkansas, and Tennessee), also developed a 500 kV single circuit with a bundle of three 1033 MCM ACSR per phase as their standard. Commonwealth Associates of Jackson, Michigan was selected to design and oversee this work. Now, the 500 kV lines in those states are alike. The same process has occurred in every regional power pool in the United States.

The federal government is in the electrical power business with the Tennessee Valley Authority (TVA), Bonneville, Bureau of Water Power, and the Rural Electrification Association (REA). The most readily available design books are those published by the REA, and are mandatory design guides for all borrowers of federal funds through the REA program. The REA books are available from the Bureau of Printing, and are public property.

## 11.2

## TRANSMISSION LINE SUPPORT STRUCTURES

The transmission line support structure function is to support the conductor. The conductor is the line. The structure must support the line, but it must conform to aesthetic requirements for an area, and must fit within the available ROW. Some available structure configurations are shown in Figure 11.1.

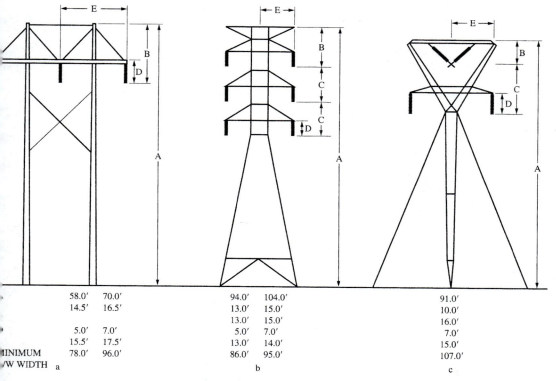

| | 58.0′ | 70.0′ | | 94.0′ | 104.0′ | | 91.0′ |
|---|---|---|---|---|---|---|---|
| | 14.5′ | 16.5′ | | 13.0′ | 15.0′ | | 10.0′ |
| | | | | 13.0′ | 15.0′ | | 16.0′ |
| | 5.0′ | 7.0′ | | 5.0′ | 7.0′ | | 7.0′ |
| | 15.5′ | 17.5′ | | 13.0′ | 14.0′ | | 15.0′ |
| MINIMUM | 78.0′ | 96.0′ | | 86.0′ | 95.0′ | | 107.0′ |
| ROW WIDTH | a | | | b | | | c |

**FIGURE 11.1** Common transmission line support structures include: (a) 138 and 230 kV single circuit wood H-frame (b) 138 and 230 kV double circuit lattice steel (c) 230 kV single circuit guyed aluminum

We have already discussed wood poles. In transmission sizes, wood poles are southern yellow pine in lengths to about 75 feet. Poles from 80 feet to 130 feet in length are usually Douglas Fir.

Wood poles are superior for constructing transmission structures because they are readily available, inexpensive, require simple foundations, and are easily erected. Additionally, the fiber in wood poles will deflect under excessive strain without losing its integrity when strain is removed. Also, because soil foundations give under excessive strain, the wood pole will lean rather than break, and can then be inexpensively straightened.

Depending on the need, wood poles can be used in singles or in "H" structures. The advantage of the single pole structure is that the required ROW width is minimal. Limiting factors are the maximum voltage that can be placed on a single pole, conductor separation, and weight of conductors.

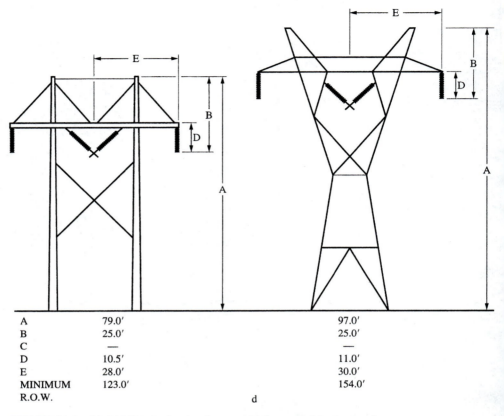

| | | |
|---|---|---|
| A | 79.0' | 97.0' |
| B | 25.0' | 25.0' |
| C | — | — |
| D | 10.5' | 11.0' |
| E | 28.0' | 30.0' |
| MINIMUM R.O.W. | 123.0' | 154.0' |

d

**FIGURE 11.1**    (d) 345 kV single circuit wood H-frame and lattice steel

The advantages of the "H" fixture are that it is a strong rigid structure that can support the weight of a long span of large conductors, and can be used at higher voltages because of the capacity for increased conductor spacing. The disadvantage is that the ROW must be wide.

Steel poles have many of the same advantages and disadvantages that wood poles do. Steel poles can be directly imbedded in the ground just as wood poles can be. Directly imbedded poles, wood or steel, can be set with a modified foundation consisting of stabilized sand, shell, or gravel with a mixture of dry cement. While this method is preferable to natural backfill, it is not as stable as concrete. Steel poles are usually set on a reinforced concrete foundation and attached to a base plate or anchor bolts. This is a strong, but expensive, foundation. We will discuss foundations in greater detail later in the chapter.

Steel poles can be shipped in 20 to 40 foot sections, which can be shipped on trucks or a single railcar, avoiding the problem of shipping longer poles on multiple flat cars. Steel poles are always more expensive than wood, but are not limited to tree heights.

The most viable structure for strength and height capabilities greater than poles can provide is the lattice steel tower structure. The lattice steel tower provides the greatest strength-to-weight ratio available. The steel can be shipped in convenient packages, and can be easily assembled and erected. Steel towers are durable and reasonable in cost. The steel tower requires a special foundation that is not excessively expensive or difficult to construct.

## 11.3

### STRUCTURE CONSTRUCTION

Erection of structures is an important part of transmission line construction. The method chosen may vary widely depending on: (1) terrain, (2) access roads, (3) work space, (4) experience and ability of workmen, (5) availability of equipment, and (6) allowed time for the completed job.

Terrain factors include the topography and the condition of the soil of the area through which the line will pass.

### 11.3.1 Factors in Choice of Structure

Certain factors must be considered, when choosing the methods of construction.

1. What type of structures are to be erected?
2. What are their natural divisions?
3. What are the dimensions of the natural divisions? (length, width, weight)
4. What are the conditions of access to ROW?
5. What are the conditions of access along the ROW?

The answer to questions 1 through 5 determine whether to use maximum equipment and minimum manpower, or minimum equipment and maximum manpower to construct the line.

If access is available for heavy equipment to and along the ROW, and if the terrain allows heavy equipment, then maximum equipment and minimum manpower is normally chosen. Whether swamp, mountains, or plains, equipment must be selected that operates in the requirements of the terrain. Heavy equipment is the preferred method of construction, and companies go to great effort to construct bridges, roads, and ferry crossings to facilitate the use of heavy equipment. When planning access, public roads are considered first; the balance of the access problems are then considered in terms of permission to use access and the cost of constructing access roads and bridges. We usually refer to these as temporary roads and bridges, because all private property is ordinarily restored to its original state.

Please follow the pattern of thinking for a particular ROW problem.

The ROW for a 138kV line is across relatively flat farm land with country roads on a near 1-mile grid. The ROW travels in an approximate diagonal across the grid. Since the farmers will continue to use their land except during times of construction or emergency repairs, the width of the ROW is not a major factor.

Proper horizontal clearances can be obtained for voltages below 115kV by use of a single pole of wood or steel. At times, getting the poles to the site is a major financial consideration, and transportation dictates the choice of structure. Economic considerations normally dictate wood poles across farm land. Since wood poles must be handled in a single piece, poles over 75 feet must be hauled on two flat cars, and poles over 130 feet must be hauled on three flat cars. Poles over 90 feet must come from Washington or Oregon. Poles hauled by two flat cars can travel east across the Rocky Mountains by either the northern or southern route. Poles hauled by three flat cars must follow the southern route. No special permits are required to transport long poles by rail, but special permits must be obtained to haul all poles longer than about 50 feet from each state they are transported across, and in many states require a highway patrol escort. If you cannot get the poles to the job site, you cannot use them!

Many rural transmission lines are 69kV. Most 69kV rural transmission lines are supported by wood poles from 50 to 70 feet in length. The long poles are used to level out the line and to provide clearance above roads, highways, railroads, and other lines. The designers limit the longer poles to the fewest and most practical number.

Poles of 50 to 70 feet weigh between 900 and 1200 pounds. It is possible to set poles of this size and weight by manpower alone, but it is both dangerous and difficult. It does not require a large piece of equipment to set these sizes. Since poles are a single piece, they may be hoisted from a single point (the center of gravity). The balance point on a pole is closer to the butt (bottom) than it is to the top; a 50 foot pole has a balance point about 21 feet from the butt. The equipment must have the capacity to raise the pole enough to clear the ground in order to set the pole into the prepared hole. A derrick is then placed in a position so that the lift point is in excess of 21 feet high and far enough from the equipment to allow for maneuvering the pole. If the land is flat and dry, the derrick is on a truck with an "A" frame. If the ground is muddy or rough, a crawler tractor or crane may be needed.

## 11.3.2 Structure Construction Procedures

Each special structure and each special piece of equipment must have a particular, established procedure that accounts for the capabilities of the equipment, and the requirements of the task at hand.

An "H" *fixture* is the principal structure in rural areas for voltages between 115 kV and 500 kV. We will use a 230 kV "H" structure as an example. The procedure is as follows:

1. All pole framing material must be delivered in the ROW to exact, designated positions.
2. All structures must be assembled or framed and placed so as to be set without moving the equipment.
3. All holes are dug.

4. The setting rig must come by, set the pole and hold it until the tamping or back-fill crew can secure it.

An efficient line construction job is an assembly-line-job in which the workmen and tools move past the materials and leave behind a finished product.

Materials are brought by truck to the designated positions, where they are unloaded either by power equipment or by hand. The pole holes are marked so that the structure can be assembled in a location that allows the setting equipment maneuverability. The holes are then dug by powered drills, by some special equipment, or by blasting with dynamite.

The structure is set by whatever method has been selected. Poles are often set by an athey wagon. The athey wagon was designed during World War II for handling heavy loads. It was modified to build lines following the war. An athey wagon is essentially a trailer-mounted derrick. The athey wagon may have rubber tires for use on the plains, or a crawler for use in the coastal plains areas. The wagon is pulled by a caterpillar or a tracked tractor. The athey wagon can be pulled anywhere the caterpillar can go, but lacks maneuverability.

Because of the nature of an athey wagon, it cannot maneuver a load. It can either back the load up, or pull it forward, but it cannot swing the load from side to side. For this reason, the "H" fixture must be assembled in a position so that only one move is required. In most cases, the "H" fixture is assembled along the ROW beyond the holes. With this method, the balance point of the "H" structure is in line with the center of the ROW and the two pre-dug holes. The derrick is raised and the spreader bar (the bar between the upright plates) is attached to the poles. The structure is moved forward, until centered, over the holes and raised to the set position. The structure is lowered into the holes and held firmly until the backfill crew secures the structure. The whole process then moves forward. Figure 11.2a shows an athey wagon setting a structure.

If a truck crane or a tracked crane is used, their load can be swung, allowing the structure to be framed in another position with relation to the holes. A truck crane cannot operate effectively in the mountains, or in the mud. A tracked crane is more useful than the truck crane, but is not as effective as the athey wagon. Figure 11.2b shows a free standing motor crane.

*Steel towers* (Fig. 11.3) are common because of their versatility and strength. The procedures are similar to the "H" frame:

1. Prepare foundation (types of foundations and methods will follow),
2. Deliver material to site,
3. Assemble, and
4. Erect.

Since the procedure used for assembly is dictated by the chosen erection method, both are discussed together. A lattice steel tower can be set by the athey wagon, track crane, crawler crane, or gin pole.

The most common gin pole in use today is a lightweight structure that can be moved up a tower, one level at a time, in the manner of cranes on high rise buildings. The gin pole has the potential to lift approximately 2000 pounds, and swing

a

b

FIGURE 11.2    Structure setting (a) With an Athey wagon and (b) With a motor crane

**FIGURE 11.3**   Steel tower erection (a) Materials laid out in preparation for construction

that weight to either side of the tower from its setup on each level. The crew for gin pole construction consists of a wench operator, four people on the tower, a foreman giving signals, and three people on the ground who prepare materials for the people above.

Using this method to erect towers, the tower must be paneled; meaning that sections of legs and bracing are bolted loosely together to be moved into position for final tightening.

When the final assembly crew arrives, the foundation and the first section of steel are in place and leveled. The gin pole is placed inside one corner and anchored. From this position, each leg is hoisted, one at a time, and its supporting brace is swung to the anchor position and loosely bolted. Each of the four panel sections for the second level are completed and the gin pole is hoisted to the next level. Each panel to be placed from one position of the gin pole is called a "pick." When access is difficult and manpower is cheap, this is an efficient construction method. However, if it is possible, the preferred method is truck or crawler crane over gin pole.

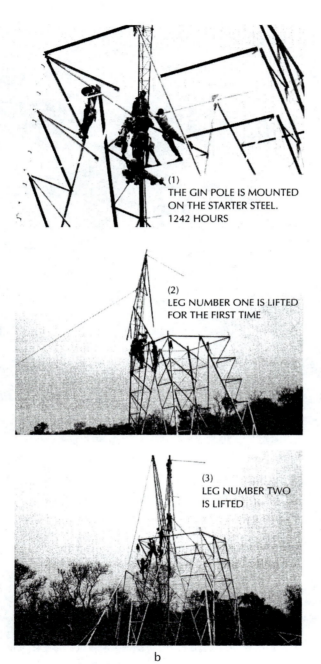

(1)
THE GIN POLE IS MOUNTED
ON THE STARTER STEEL.
1242 HOURS

(2)
LEG NUMBER ONE IS LIFTED
FOR THE FIRST TIME

(3)
LEG NUMBER TWO
IS LIFTED

b

**FIGURE 11.3**    (b) Construction steps

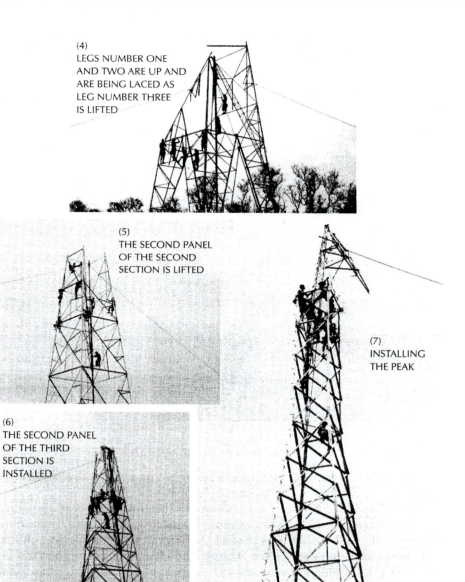

(4)
LEGS NUMBER ONE AND TWO ARE UP AND ARE BEING LACED AS LEG NUMBER THREE IS LIFTED

(5)
THE SECOND PANEL OF THE SECOND SECTION IS LIFTED

(7)
INSTALLING THE PEAK

(6)
THE SECOND PANEL OF THE THIRD SECTION IS INSTALLED

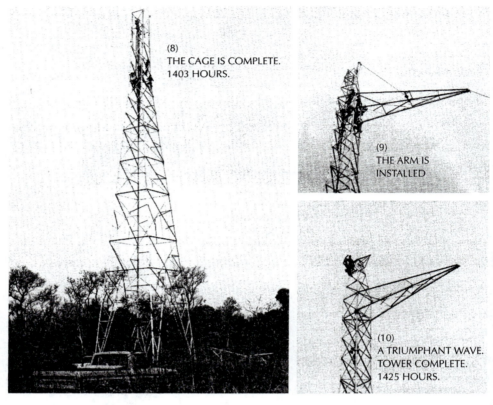

(8)
THE CAGE IS COMPLETE.
1403 HOURS.

(9)
THE ARM IS
INSTALLED

(10)
A TRIUMPHANT WAVE.
TOWER COMPLETE.
1425 HOURS.

**FIGURE 11.3**    (b) Construction steps (contd.)

The size, weight, and rigidity of a tower, along with the size of the construction equipment, dictates whether a tower is erected in pieces or as a complete unit. A tower is designed as a unit, and it requires the entire tower structure to attain its ultimate strength. No part is designed to sustain a load alone, and often does not support its own weight. It is possible to brace a flimsy tower section for erection purposes and then remove the bracing when the tower is complete.

At times, it may be necessary to assemble towers at remote and inaccessible locations. Helicopter erection is sometimes used for this purpose. Erection by helicopter demands careful planning for optimum use. For an erection planner, there are two things to consider about each helicopter: (1) Weight lifting capability and (2) Cost of operation.

Few contractors who use helicopters own them. Because of their high operating cost, it is difficult to justify ownership. A helicopter service will furnish a four passenger helicopter for use in supervision for about $400.00 per hour of flight with a minimum of 4 hours flight time per day. A helicopter with 4000 pounds lift capacity will cost about $800.00 per hour; one with 10,000 pounds lift capacity may cost $3000 per hour; while a helicopter capable of lifting 18,000 pounds may cost $6000 per hour. Considering the costs, it is essential that the helicopter never waits on anything. Every construction phase is arranged so that the helicopters work as much as they can. All other costs fade by comparison. Towers can be sectionalized, allowing the sections to be flown in, and fastened in place to form complete towers.

## 11.4

### TOWER AND POLE FOUNDATIONS

Because tower foundation is a large subject, here we introduce only tower compression and uplift forces. Foundations must overcome or accept these forces. Tower foundations are often constructed from bell-bottomed, drilled piers, filled with reinforced concrete, with a tower leg set into each pier. If the equipment and concrete is available, this method is economical and provides a good foundation.

In some cases, it is better to use hand or machine-dug holes, grillage foundations, and backfill. A grillage foundation consists of a tower leg, extended below grade and attached to heavy angle irons to which channel iron has been bolted, and then braces to make a grill-like unit. A force transmitted down the leg is transmitted into the soil, while an upward force is restrained by the weight of the soil backfill on top of the grillage.

A common foundation for poor soil is different types of piling with a "cap" as a transition piece. Poured slab types or floating foundations are sometimes required and used. Figure 11.4 shows some tower foundations.

## 11.5

### TRANSMISSION LINE CONDUCTORS

The selection of a transmission line conductor and conductor size is a critical step in transmission line design. The conductor choice effects not only the cost of constructing the line, but also the cost of transmitting electric power through the line throughout its life. The conductor choice is critical.

#### 11.5.1 Conductor Considerations

The factors that must be considered in the selection of the transmission line conductor include:

1. The required span and sag between spans
2. The tension on the conductor
3. Whether the atmosphere is corrosive
4. Whether the line will be prone to vibration
5. Power loss allowed on the line

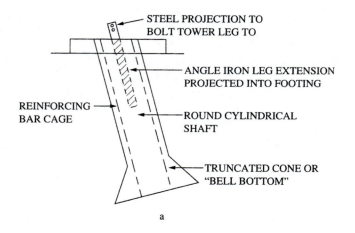

STEEL PROJECTION TO
BOLT TOWER LEG TO

ANGLE IRON LEG EXTENSION
PROJECTED INTO FOOTING

REINFORCING
BAR CAGE

ROUND CYLINDRICAL
SHAFT

TRUNCATED CONE OR
"BELL BOTTOM"

a

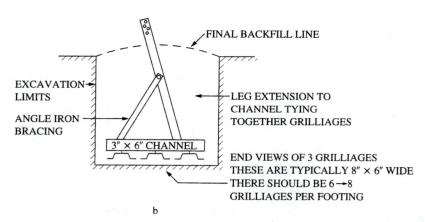

FINAL BACKFILL LINE

EXCAVATION
LIMITS

LEG EXTENSION TO
CHANNEL TYING
TOGETHER GRILLIAGES

ANGLE IRON
BRACING

3″ × 6″ CHANNEL

END VIEWS OF 3 GRILLIAGES
THESE ARE TYPICALLY 8″ × 6″ WIDE
THERE SHOULD BE 6 →8
GRILLIAGES PER FOOTING

b

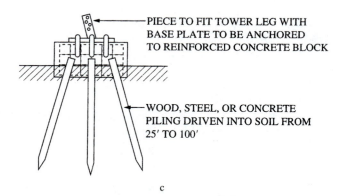

PIECE TO FIT TOWER LEG WITH
BASE PLATE TO BE ANCHORED
TO REINFORCED CONCRETE BLOCK

WOOD, STEEL, OR CONCRETE
PILING DRIVEN INTO SOIL FROM
25′ TO 100′

c

**FIGURE 11.4**   Tower foundation types include: (a) Bell bottom pier for use in
good to medium soil (b) Grillage tower footing for use in good soil and
(c) Pile and cap pier for use in very poor soil.

6. Voltage loss allowed on the line
7. Climate at the line location (for example, wind or ice loading probability and amount)

The most common types of transmission line conductor and the various considerations are discussed in the sections that follow.

## 11.5.2 Conductor Types

The most commonly used transmission conductor types are listed below:

1. ACSR (Aluminum Conductor—steel reinforced)
2. ACSR/AW (Aluminum Conductor—aluminum-clad steel reinforced)
3. ACSR-SD (Aluminum Conductor—steel reinforced/self damping)
4. ACAR (Aluminum conductor—alloy reinforced)
5. AAC-1350 (Aluminum Alloy Conductor composed of #1350 aluminum alloy)
6. AAAC-201 (Aluminum Alloy Conductor composed of 6201 alloy)

A brief discussion of each conductor follows.

*ACSR has designations (6/1, 26/7, 54/7):* The designation of 6/1 means six aluminum #1350 strands and one strand of galvanized steel for a core. This is used only for small conductors. Experience has shown that the above combinations form a strong and tightly bound conductor. ACSR is the most common conductor used today. The #1350 aluminum is a hard drawn aluminum that has good conductivity but not much strength. The core consists of galvanized steel wire. The galvanizing is a zinc coating placed on the steel by running the steel through a vat of molten zinc. The thickness of the coating is listed as "Class A" for normal thickness, "Class B" for medium, and "Class C" for heavy duty. The degree of contamination dictates the class of galvanizing for the core.

*ACSR/AW* conductor is like the ACSR above, except that the core is composed of a high-strength steel clad in an aluminum coating. This conductor is more expensive than ACSR, but it is acceptable for more corrosive atmospheres than is the ACSR with class "C" galvanizing.

*ACSR-SD* is a conductor that consists of two layers of trapezoidal-shaped strands, or two layers of trapezoidal strands and one of round strands around a conventional steel core. The strands themselves are composed of 6201 high-strength aluminum and their internal structure makes them self damping against aeolian vibration. These conductors can be strung at very high tensions and need not have auxiliary dampers installed. They have been proven to be good conductors, but the cost of conductor, splices, dead-ends, and installation labor have kept their use from becoming widespread.

*ACAR* consists of strands of 1350 aluminum and a core of 6201 high-strength aluminum. It is lighter than the ACSR and is just as strong, but higher in cost. When long spans in a corrosive atmosphere are needed, ACAR must be considered.

*AAC-1350* is used for any construction that requires good conductivity and when short spans can be utilized, AAC-1350 is a good conductor.

*AAAC-6201* conductor is composed of 6201-T81 high-strength aluminum alloy strands. It is as strong as ACSR, is much lighter, and more expensive. AAAC-6201 is good for corrosive atmospheres and long span construction.

## 11.5.3  Conductor Size

Minimum size conductors are usually dictated by either required tensions for span and sag considerations, or by the breakdown voltage of air. The factors governing tensile strength will be covered later in the section on tension. The voltage per unit surface area, for instance 50 kV per square inch, is dependent upon the voltage and the circumference of the chosen conductor.

Common minimum size conductors for given voltages are as follows:

69 kV—0000 (four ought)

138 kV—336.4 MCM

230 kV—795 MCM single conductor

345 kV—795 MCM bundle of two

500 kV—795 MCM bundle of three

750 kV—795 MCM bundle of four

Europeans have adopted a standard of 556 mm bundle of four for 500kV.

*Voltage drop considerations* require that not only must the conductor meet the minimum size requirements, but must transfer power at an acceptable loss. This is often expressed as a maximum voltage drop of 5% for a particular system, across the total series impedance. The total series impedance (z) on the line is equal to the maximum allowable voltage drop divided by the maximum load current.

$$Z = R + JX_1 = \frac{V}{I_{max}}$$

R is inversely proportional to the area of size of the conductor.

*Thermal Capacity:* When sizing a phase conductor, the thermal capacity of the conductor (ampacity) must be considered. The conductor should be able to carry the maximum expected long-term load current without overheating. Generally, a conductor is assumed to be able to withstand a temperature of 75°C (167°F) without a decrease in strength. Above that temperature, the conductor strength decreases, depending upon the amount and duration of excessive heat. A conductor's ampacity depends upon the current generated heat, the heat from the sun, and the cooling of the winds. Conductor heating is generally stated as ambient temperature plus load temperature, less cooling of the wind.

*Economic considerations* are important factors in determining conductor size. Rarely is a conductor sized to meet the minimum requirements, as stated above, the most economical. The additional cost of a larger conductor may be more than offset by the present worth of the savings resulting from the lower losses during the entire life of the conductor. A proper economic analysis should include the following factors for each of the conductor sizes considered:

1. The total cost per kilometer or mile of building the line with the particular conductor being considered.
2. The present worth of the energy losses associated with the conductor.
3. The capital cost per kilowatt of loss of the generation, substation, and transmission facilities necessary to supply the line losses.
4. Load growth.

   The results of an economic conductor analysis can often be best presented and
understood when presented in a graphical form, as shown in an REA study and in
Figure 11.5. At an initial load of approximately 200 MW, 1272 kcmil becomes more
economical than 795 kcmil. You will notice that 954 MCM is not the most economical nor the least economical at either end of the scale.

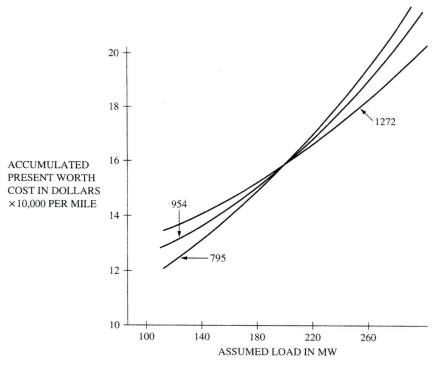

**FIGURE 11.5** Results of a typical economic conductor analysis conducted by the REA

## 11.5.4 Overhead Ground Wires

Overhead ground wires (OHGW) are the wires at near ground potential installed
above the phase wires to protect the line from lightning, to even out the ground
potential, and are sometimes used for low voltage communication. OHGW do not
conduct load current, but they rapidly conduct the heavy current of a lightning
strike to the ground, through their many grounded connections. Transmission

structures are all grounded and the OHGW is grounded at every structure. The types of OHGW are listed below.

*High Strength or Extra High Strength Galvanized Steel Wire* is used for high strength. The allowable sizes are $^3/_8$ inch and $^7/_{16}$ inch, while for extra high strength wires, the allowable sizes are $^5/_{16}$ inch, $^3/_8$ inch, and $^7/_{16}$ inch. Note the $^1/_4$ inch strand is not acceptable for use as OHGW, and neither are Siemens Martin grade wires of any size. Siemens Martin grade wire is a galvanized steel wire of low-grade steel and low stress capability, and is used only for guys to support structures.

*Aluminum-clad steel strand* conductor is covered with aluminum in place of galvanizing, and is a better conductor. This is the preferred conductor if the OHGW is to double as a communications circuit or if lightning is particularly heavy in the area for which this conductor is specified. This is a conductor consisting of seven strands of aluminum-coated steel with four different diameters of steel available, depending upon the current carrying need. The sizes are seven strands of 0.106 inch steel wire, normally designated as 7#.106, and seven strands of AWG steel shown as 7#9, 7#8, or 7#7.

### 11.5.5  Selecting Size and Type of OHGW

OHGW are sized according to current carrying demand, as stated above. The types of conductors are chosen to meet the needs of corrosion resistance. The OHGW must be capable of being installed with the same sag as the conductors, so that distance between the small OHGW and the much larger current carrying conductor remains uniform throughout the length of the span.

The specification of OHGW requires that there be no welds along their length. A weld in the metal fails in tension at a much lower temperature than the adjacent metal. A heavy current travels this wire from the point of a lightning strike to ground and if there are welds in that section, the OHGW usually fails, falling into the transmission line conductors, and putting the line out of service.

## 11.6

## CONDUCTOR TENSION

Throughout the life of a transmission line, the conductor tension may vary between 10% and 60% or more of rated conductor strength due to change in loading and temperature. Most of the time, however, the tension varies within relatively narrow limits because ice, high winds, and extreme temperature are relatively infrequent in many areas. Normal tensions may actually be more important in determining the life of the conductor than higher tensions, which are experienced infrequently.

## 11.6.1 Tension Definition

The tensions shown in Table 11.1 should be approached as closely as the special conditions will allow in order to reduce the number of structures, but rarely exceeded. There are several conditions at which maximum conductor tension limits are specified.

The initial unloaded tension refers to the state of the conductor when it is initially strung and is under no ice or wind load. After a conductor has been subject to the assumed ice and wind loads, and/or long-time creep (the inelastic elongation of a conductor that occurs with time under load), it receives a permanent or inelastic stretch. The tension of the conductor in this state, when it is again unloaded, is called the *final unloaded tension*.

The loaded tension refers to the state of a conductor when it is loaded to the assumed simultaneous ice and wind loading for the National Electrical Safety Code (NESC) loading district concerned (see Table 11.1).

The vertical load on a conductor is the weight of that span of wire with its ice loading. The horizontal load is the load due to the pressure of the wind. The total loading is the vector sum of both loads. The NESC requires that a constant be added to the vector sum to reach the standard loaded tension as follows:

|                | Heavy | Medium | Light |
|----------------|-------|--------|-------|
| Newton/meters  | 4.4   | 2.9    | 0.73  |
| Pounds/foot    | 0.3   | 0.20   | 0.05  |

Extreme wind tension refers to the state of the conductor when it has a wind blowing on it of a value not less than the 50 year mean recurrence interval wind. No ice is assumed to be on the conductor.

Extreme ice tension occurs when a conductor is loaded with what is considered to be an extreme amount of ice for an area. It is assumed there is no wind blowing when ice is on the conductor. Values of 25 to 50 mm (1 to 2 inches) of radial ice are commonly used as extreme ice loads.

## 11.6.2 Ruling Span Tension

For a given ruling span (the reference span for a particular transmission line), usually only tension limit conditions control the design of the line, while the others conditions have relatively little significance. Stringing sags should be based on Table 11.1, limits 1, 2, and 3 only, as long as tensions at conditions 4 and 5 are satisfactory. If the conductor loading under extreme ice or wind loads is greater than under the "standard load" condition, calculated sag and tension values must meet those requirements.

▆▆▆▆  **TABLE 11.1   REA Standards for Maximum Conductor and Overhead Ground Wire and Temperature Limits for ACSR & 6201 AAAC**

### A. Temperatures

1. Tension limits 1, 2, and 3 below must be met at the following temperatures:

| | | |
|---|---|---|
| Heavy loading district | −17.8°C | (0°F) |
| Medium loading district | −9.4°C | (15°F) |
| Light loading district | −1.1°C | (30°F) |

2. Limit 4 must be met at the temperature at which the extreme wind is expected.
3. Limit 5 must be met at              0°C     (32°F).

### B. Tension Limits in Percent of Conductor Rated Strength

| Tension Condition (See text for exp.) | Phase Cond. | OHGW High Strength Steel | High Strength Steel |
|---|---|---|---|
| 1. Max. initial unloaded | 33.30* | 25 | 20 |
| 2. Max. final unloaded | 25.00† | 25 | 20 |
| 3. Standard loaded (usually NESC district loading) | 50.00 | 50 | 50 |
| 4. Max. extreme wind (A) | 70.00‡ | 80 | 80 |
| 5. Max. extreme ice (A) | 70.00 | 80 | 80 |

*NOTE:* (A) These limits are for tension only. When conductor stringing sags are to be determined, limits 1, 2, and 3 should be considered as long as tensions at conditions 4 and 5 are satisfactory.

*NOTE:* Tension limits do not apply for self-damping and other special conductors.

*In areas prone to eolian vibration , a value of approximately 20 percent at the average annual minimum temperature is recommended if vibration dampers or other means of controlling vibration are not used.
†For 6201 AAAC, a value of 20 percent if recommended.
‡For ACSR only. 6201 Aluminum use 60 percent.

## 11.6.3 Overhead Ground Wire Tension

To avoid unnecessarily high mechanical stresses in the OHGW, supporting structures, and guys, the OHGW should not be strung with more tension than is necessary to coordinate its sags at different conditions with the phase conductors.

## 11.7

▆▆▆▆

## RULING SPAN

The ruling span is the assumed uniform design span that most closely resembles the variety of spans that are in any particular section of line. The purpose of the ruling span is to calculate the sags and clearances necessary for structure spotting and conductor stringing.

In a line where all spans are equal, the ruling span is equal to the average span. If the spans vary, the ruling span is between the longest span and shortest span in length.

## 11.7.1 Ruling Span Length

A satisfactory method for approximating the ruling span length is to take the average span plus two thirds of the difference between the maximum span and the average span, when the maximum span is not greatly different than the average span:

$$RS = L_{ave} + \frac{2}{3}(L_{max} - L_{ave}) \tag{11.1}$$

Where

RS = the ruling span

$L_{AVE}$ = average span

$L_{MAX}$ = maximum span

all in the same distance units. The exact equation calculating the ruling span is

$$RS = \left(\frac{L_1^3 + L_2^3 + L_3^3 + \ldots + L_n^3}{L_1 + L_2 + L_3 + \ldots + L_n}\right)^{1\backslash 2} \tag{11.2}$$

**Example 11.1:**

A line segment is 4250 feet from dead end to dead end. The individual span lengths are as follows: $L_1$ = 750 ft, $L_2$ = 800 ft, $L_3$ = 1200 ft, $L_4$ = 500 ft, and $L_5$ = 1000 ft. Calculate the ruling span by the exact and approximate methods.

**Solution:**

Using Equation 11.2 we obtain

$$RS = \left(\frac{750^3 + 800^3 + 1200^3 + 500^3 + 1000^3}{750 + 800 + 1200 + 500 + 1000}\right)^{1\backslash 2} \text{ ft} = 934.94 \text{ ft}$$

By the approximate method we obtain

$$RS = [850 + \frac{2}{3}(1200 - 850)] \text{ ft} = 1083.33 \text{ ft}$$

## 11.7.2 Limits on Lengths of Span

No span should be longer than twice the ruling span and none should be less than one half the ruling span.

By comparison to the ruling span calculated by the exact method, $L_4$ of Example 11.1 is close to one half of the ruling span. All spans pass the double ruling span test. This set of combinations of span lengths is acceptable.

By comparison to the ruling span calculated by the approximate method span four is under the minimum. The indicated solution to this problem is to even the spans up, if possible, and if not, to put dead ends on either side of the long span. Both solutions provide acceptable results.

## 11.8

### DETERMINING CONDUCTOR SAG AND TENSION

Determining conductor sags and tensions, given the restraints shown in Table 11.1 and 11.7.3 listed above, and conforming to the stress-strain curves for a compound conductor such as ACSR, are complex. The best method is to use one of several computer programs written for this purpose. The computer program must:

1. Check all limiting conditions simultaneously and follow the governing parameter
2. Account for creep
3. Use average tension values (not at the support or midway).

The source most often used prior to computer calculations were manufacturers graphical methods. A good example is:

"Graphic method for sag tension calculations for ACSR and other conductors"
—Publication #8, Aluminum Company of America, 1961.

### 11.8.1 Sag-tension Relationship

The sag is inversely proportionate to the tension in a conductor. The maximum tension limits are set by the criteria mentioned previously. The allowable sag is a quantity that is dependent upon the length of span and the height of the supporting structure. This can be shown by the following parobolic approximate relationship:

$$S = \frac{W_c L^2}{8T_h} \tag{11.3}$$

Where

$S$ = sag

$W_c$ = weight of the conductor per unit length

$T_h$ = tension used

$L$ = length of span

If we pick a limiting tension for, $T_h$, and an acceptable sag, then we can easily calculate the length of span. If, on the other hand, the length of span is critical, the sag can be determined. All of this will be brought together in the section on the use of the sag template (section 11.11.2).

### 11.8.2 Equation for Sag

When any flexible conductor is supported at the ends, the conductor assumes the shape of a catenary. The formula for sag becomes:

$$S_{cat} = \frac{T_h}{W} \cosh\left(\frac{W_c L}{2T_h} - 1\right)$$

(11.4)

Where the sag is less then 5% of the length of span, the parabolic approximation, which has a much simpler equation, is close to that of the catenary. Only on long spans such as river crossings is it necessary to use the exact equation for sag. For others we will use the equation for the parabola, Equation 11.3.

## 11.8.3 Sag Calculation Example

An example calculation of sag using drake conductor and allowable tension as 25% of ultimate strength follows.

The ultimate strength of drake is listed by the NESC as 31,500 lb, $T_h$ equals $0.25 \times 31,500 = 7875$ lb. $W_c$ is 1.094 lb per ft for drake and L = 800ft is assumed.

$$S = \frac{1.094(800)^2}{8(7875)} = 11.11 \text{ ft}$$

This sag is based upon initial tension only and ignores the modulus of elasticity and temperature coefficient of expansion. A table prepared by a computer program is shown below.

> **Input**
> Conductor 477 MCM ACSR 26/7
> Code name Flicker
> Ruling span = 701.6 ft
> Ultimate strength 17,200#
> Diameter 0.846 inches
> Tension limitations:
>   Initial 33%
>   Final 25%
>   Loaded 50%
>   Final temperature 60°F (ambient)
> Weight factors:
>   Dead weight 0.6145 lb/ft
>   1/2 inch of ice 1.4517 lb/ft
>   4 pounds wind 0.6153 lb/ft
>   Resultant weight 1.8867 lb/ft
>   Area 0.4232 in$^2$
>   Temperature coefficient of linear expansion = 0.31
>   Expansion = 0.0000108 feet per degree F
>   Modulus: Final 10.3 × 10$^6$ lbs/in$^2$
>            Initial 8.113 × 10$^6$ lbs/in$^2$

**Output**

| | Temp | Sag | Tension |
|---|---|---|---|
| Initial | Compatible with 1/2 of ice | 15.48 ft | 7500 lb. |
| Final | 0° | 12.17 | 3101 |
| | 30° | 13.87 | 2722 |
| | 60° | 15.48 | 2444 |
| | 90° | 17.04 | 2221 |
| | 120° | 18.51 | 2046 |

Calculation of initial sag by Equation 11.3 for this set of circumstances including ice and wind gives the following:

$$S = \frac{1.8867(701.6)^2}{8(7500)} = 15.48 \text{ ft}$$

This calculation shows that *if* we had a temperature that would allow for one-half inch of radial ice and a wind that produced 4 pounds per square foot wind pressure, and used a limiting factor of 7500 pounds conductor stress, a sag of 15.48 feet would be produced. This does not tell us what happens after the initial elongation of the conductor caused by applying 7500 pounds force, nor the "creep" of the conductor after time, nor does it account for changes due to temperature.

The computer program shows that if the conductors are sagged to 15.48 feet at 60°F, the tension will be 2444 pounds, and if the temperature drops to 30°F, the wire will shorten so that the tension becomes 2722 pounds, and the sag becomes 13.87 feet. Should the temperature increase to 90°F, the tension will drop to 2221 pounds and the sag will increase to 17.04 feet. It should be evident that sag calculation by computer is essential if all factors are to be considered.

## 11.9
---

## CONDUCTOR VIBRATION

Overhead transmission line conductors are subject to two different types of vibration, aeolian vibration and galloping.

### 11.9.1 Aeolian Vibration

Aeolian vibration is a high frequency, low amplitude oscillation, generated by a low velocity, steady wind blowing across the conductors. This steady wind creates air vortices on the lee side of the conductor, which break at regular intervals, creating a force on the conductor, which is alternately upward and downward. When this frequency approaches the resonant frequency of that particular span, vibration occurs. In other words, when an integral number of vortice breaks occur within a span, vibration occurs. Wind speeds of between 8 and 15 miles per hour cause aeolian vibration. Long spans with high tensioned conductors are especially prone to aeolian vibration. The wind conditions causing aeolian vibration usually occur in flat, open terrain. Aeolian vibration causes conductors to work harden, fracture, and break at the point of attachment to the suspension shoe. In other words, the conductor fails under fatigue.

The effects of aeolian vibration are mitigated by selection of a good suspension clamp and by installation of armor rods, or an armor grip suspension. Aeolian vibration is prevented by attaching dampers to the conductor, which act to destroy the effects of the steady wind and reinforcement of the vibration. The dampers oscillate at higher frequencies than the conductor, pull energy from the attempted aeolian vibration, and cause the damping.

## 11.9.2 Galloping

Galloping, or dancing, is a condition in which transmission line conductors vibrate with very large amplitudes at very low frequencies. Galloping can cause faults due to contact between phases, between a phase and a static wire, or conductor failure at a support point. The excessive stress caused by galloping has been known to cause both structure failure and excessive sag due to the conductor being stressed beyond its ultimate strength.

Galloping usually occurs when a steady, moderate, transverse wind blows over a conductor layered with ice that was deposited by freezing rain. This increases the diameter and weight of the conductor as well as the tension on the conductor.

There is one case on the Gulf Coast where a steady, moderate wind blew for 4 days before galloping was noticed. The phenomenon continued to reinforce itself until eventually several tower arms failed and the conductor lay on the ground.

On another occasion, during a hurricane at Freeport, Texas, a neoprene covered, 1000 pair, telephone cable began to gallop. Each upward swing pulled the poles from the ground a small amount. Each downward swing pushed the poles back into the ground, but not as much as they had been pulled out. This phenomenon, called "jacking," left several miles of telephone cable, with its supporting poles, on the ground.

Galloping can not be prevented but, by careful selection of span length and tension, it can be minimized.

During galloping, the conductors oscillate elliptically at frequencies under 1 Hz and amplitudes of several feet. When one loop of oscillation appears, this is called single-loop galloping, and rarely occurs in spans over 700 feet in length. Sometimes two loops appear that are superimposed on a single loop. This is called double-loop galloping. Double-loop galloping is common. In double-loop galloping, the point of maximum amplitude is at the quarter wave points. The shorter envelopes of motion make the problem of clearance required for double-loop galloping much less than that required for single loop.

Spans in excess of 700 feet are nearly immune to galloping as are most spans using bundled conductors. When galloping is found to occur, it can be negated by installation of aerodynamic drag dampers, or by use of interphase spacers. Galloping most likely occurs on a long span struck by winds that are funneled between mountains.

## 11.10

### LINE PLANNING

The steps for planning the construction of a transmission line are discussed in the following section.

### 11.10.1 Plan-profile Drawings

Plan-profile drawings are drawings that show a topographical contour map of the terrain along and near the ROW, and a sideview profile of the line, showing elevation and towers. Figure 11.6 shows a section of a plan-profile drawing.

The transmission line plan-profile drawings serve as a worksheet, and eventually an expository sheet, which shows what is to be done and the problems involved. Initially, the drawings are prepared based on a route survey showing land ownership, the locations and elevations of all natural and man-made features to be crossed or that are adjacent to the proposed line (all affect ROW), line design, and construction. The drawings are then used to complete line design work such as structure spotting. During material procurement and construction, the drawings are used to control purchase of materials and serve as construction specification drawings. After construction, the final plan-profile drawings become the permanent record of property and ROW data, which is useful in line operation and maintenance, and in planning future modifications.

Beginning with initial preparation, accuracy, clarity, and completeness of the drawings should be maintained to ensure economical design and correct construction. All revisions made subsequent to initial preparation and transmittal of the drawings should be noted in the revision block by date and with a brief description of the revision.

Drawing preparation begins with an aerial survey followed by a ground check. The proper translation of these data to the plan-profile drawings is critical. Errors that occur during this initial stage affect line design because a graphical method is used to locate the structures and conductor. The final field check of the structure site should reveal any error. Normally, plan-profile sheets are prepared using a scale of 200 feet to the inch horizontally and 20 feet to the inch vertically. On this scale, each sheet of plan-profile can conveniently accommodate about 1 mile of line with enough overlap to connect the end span on adjacent sheets.

The sample format for a plan-profile drawing is shown by Figure ll.6 with units and stations in customary United States units. Increase in station and structure numbering usually proceeds from left to right with the profile and corresponding plan view on the same sheet.

Existing features to be crossed by the transmission line, including the height and position of power and communication lines, should be shown and noted by station and description in both the plan and profile views. The magnitude and direction of all deflection angles in the line should be given and referenced by P.I. station in plan and elevation. In rough terrain, broken lines representing

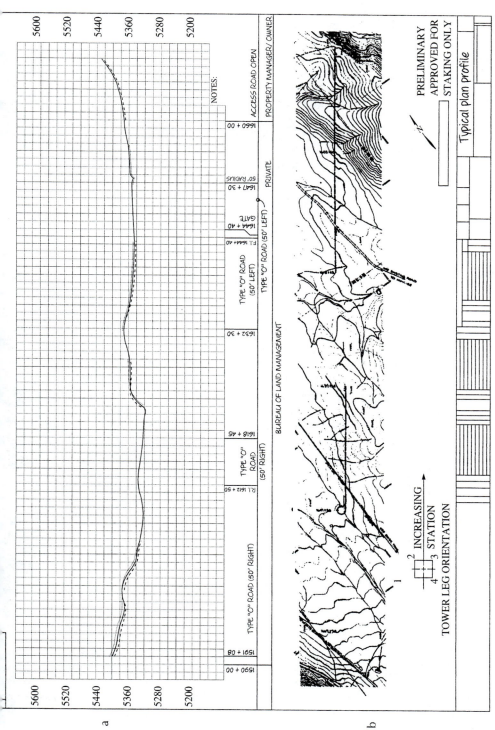

**FIGURE 11.6** Plan profile drawing (a) Profile with towers spotted and (b) Contour

side-hill profiles should be plotted to assure adequate conductor–ground clearances and pole height.

## 11.10.2 The Sag Template

The sag template is a transparent template used for graphical analysis of the best combined collection of structure locations, structure heights, and span lengths to fit a particular route with its terrain. A separate template is made for each conductor type and each tension. Each template has superimposed each of the following curves calculated with the design tension, so that all ends of curves are at the same point (The sag values for each of these conditions are available from the manufacturer.):

**A.  Hot (Maximum Sag) Curve**
The maximum operating temperature, no ice, no wind, final sag curve. Used to check for minimum vertical clearances (or if maximum sag occurs under an icing condition, this value should be used for the sag template).

**B.  Cold Curve**
The minimum temperature, no ice, no wind, initial sag curve. Used to check for uplift and insulator swing.

**C.  Normal Curve**
The 16°C (60°F), no ice, no wind, final sag curve. Used to check normal clearances and insulator swing.

A sample of the conductor sag template is shown in Figure 11.7. It is used on plan-profile drawings to determine graphically the location and height of supporting structures required to meet line design criteria for vertical clearances, insulator swing, and span limitations.

Generally, the conductor sag curves control the line design. The sag template for the OHGW is used to show its position in relationship to the conductors for special spans or change in conductor configuration. The sag curves reveal the presence of uplift if it is present. Uplift will be discussed with the use of the sag template.

The template should be made to include spans three or four times as long as the normal level ground span to allow for spotting structures on steep terrain.

The form of the template is based on the fact that at the time when the conductors are installed, the horizontal tensions must be equal in all level and inclined spans if the suspension insulators are plumb in profile. This is also approximately true at maximum temperature. To obtain values for plotting the sag curves, sag values for the ruling span are extended for spans shorter and longer than the ruling span.

## 11.10.3 Structure Spotting

Structure spotting is the design process that determines the height, location, and type of consecutive structures on the plan-profile sheets. Actual economy and safety of the transmission line depends on how well this final step in the design is performed. The structure spotting should closely conform to the design criteria established for the line. Constraints on structure locations and other physical

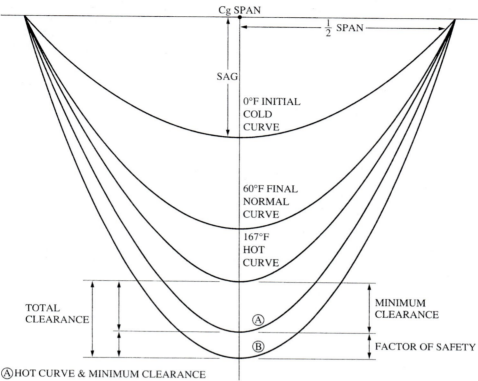

**A** HOT CURVE & MINIMUM CLEARANCE
**B** HOT CURVE & MINIMUM CLEARANCE & FACTOR OF SAFETY

HORIZONTAL SCALE 1" = 200'
VERTICAL SCALE 1" = 20'

**FIGURE 11.7**  Typical pattern for sag template

limitations encountered may prevent spotting of structures at optimum locations. The initial effort at spotting the line must be followed by a field check to determine whether or not the structures can be placed as first laid out. Corrections or alterations are then made to accommodate the field check.

Ideally, the desired properties of a well-designed and economical line layout are:

1. Spans approximately uniform in length, equal to or slightly less than the design ruling span.
2. Maximum use of the basic structure, which is the structure and height selected as the most economical for the design condition.
3. The shape of the running conductor profile, also referred to as the grading of the line, should be smooth. If the conductor attachment points at the structure lie in a smooth-flowing curve, the loadings are equalized on successive structures.

For a generally level and straight line with few constraints on structure locations, the above stated objectives can be readily achieved. Greater skill and effort are needed for lines on an abrupt or undulating ground profile and where constraints on structure location exist. Examples of these conditions are high or low points in the profile features such as line angle points, crossings over highway, railroad, water, power and communication lines, and ground with poor soil conditions. Structure locations and heights are often controlled or fixed by these special considerations. Alternate layouts between fixed locations may be required to determine the best arrangement based on factors of cost and effective design.

The principal design factors involved in structure spotting are:

1. Vertical clearances
2. Horizontal clearances
3. Uplift
4. Horizontal or vertical span limitations due to:
    a. Vertical sag—clearance requirement
    b. Conductor separation
    c. Galloping
    d. Strength of conductor.

## 11.10.4 Preparation

The following steps are required for structure spotting:

1. Plan-profile drawings of the transmission line.
2. Sag template of the same scale as the plan-profile prepared for the design temperatures, loading condition, and ruling span of the specified conductor and OHGW.
3. Table of required minimum conductor clearances over ground features and other overhead lines.
4. Insulator swing charts.
5. Horizontal and vertical span limitations due to clearance or strength requirements.

## 11.10.5 Process of Spotting

The process of spotting begins at a known or established conductor attachment point such as a substation take-off structure. For level terrain, when a sag template is held vertically and the ground clearance curve is held tangent to the profile, the edge of the template intersects the profile at the point where structures of the basic height should be set. This relation is illustrated for a level span in Figure 11.7. Curve No. 1 represents the actual position of the lowest conductor (hot curve), offset by the required total ground clearance, C.

The point where the normal curve intersects the profile determines the location of the next structure. The template should then be shifted and adjusted so that with opposite edge of the template held on the conductor attachment point previously located with the clearance curve again barely touching the profile. The process is repeated to establish the location of each succeeding structure. After all the structures are thus located, the structures and lowest conductor should be drawn in.

The above procedure can only be followed on lines that are approximately straight and that cross relatively flat terrain with the basic ground clearances. When line angles, broken terrain, and crossings are encountered, it may be necessary to try several different arrangements of structure locations and heights at increased clearances to determine the most satisfactory arrangement. Special considerations often fix or limit the structure locations, and it is advisable to examine the profile for several span lengths ahead for these conditions and adjust the structure spotting accordingly.

The relationship of the ground clearance and conductor curves is also used for spans other than level-ground spans by shifting the sag template until ground profile touches or is below the clearance curve with the previously established conductor attachment point (normally, the left) positioned on the conductor curve. The conductor curve would then indicate the required conductor height for any selected span. Structure height may be determined by scaling or use of the proper structure height template. Design limitations due to clearance or structure strength should be observed.

The vertical span is the distance between the conductor low points in spans adjacent to the structure. Horizontal span is the distance between the mid points on adjacent spans. Where conductor attachments are at different elevations on adjacent structures, the low point is not at mid-span and will shift its position as the temperature changes.

Where minimum vertical span or uplift is of concern, the cold curve should be used. The normal temperature is more critical and should be used if the vertical span is limited by a maximum value. Figure 11.8 shows examples of the relationship of conductor low points and vertical spans that may occur in a line profile.

If insulator swing is unacceptable, one of the following corrective steps, in order of preference, is recommended:

1. Relocate structures to adjust horizontal-vertical span ratio.
2. Increase structure height, or lower adjacent structures.
3. Use a different structure with greater allowable swing angle or a dead end structure.
4. Add weight at insulators to provide needed downward vertical force.

## 11.10.6 Uplift

Uplift is defined as negative vertical span and is determined by the same procedure as vertical span. On steeply inclined spans, when the cold sag curve shows the low point to be beyond the lower support structure, the conductors in the uphill span exert upward forces on the lower structure. The amount of this force at each attachment point is related to the weight of the loaded conductor from the lower support to the low point of sag. Uplift exists at a structure when the total vertical span from the ahead and back spans is negative, as shown by structure No. 4 in Figure 11.8, while no net uplift occurs at structure No. 3. Uplift must be avoided for suspension, pin-type, and post insulator construction. For structures with suspension insulators, the check for allowable insulator swing is usually the

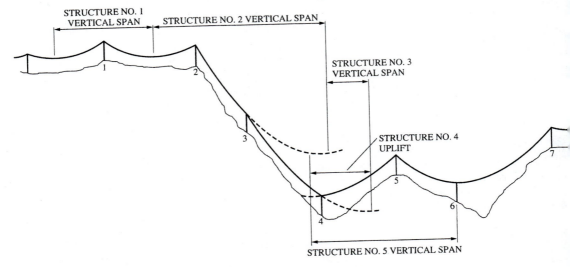

**FIGURE 11.8**   Sag low point, vertical spans, and uplift

controlling criteria on vertical spans. A rapid method to determine uplift is shown by Figure 11.8. There is no danger of uplift if the cold curve passes below the point of conductor support on a given structure with the curve on the point of conductor support at the two adjacent structures.

Designing for uplift or minimizing its effects is similar to the corrective measures listed for excessive insulator swing, except that adding excessive weights should be avoided. Double dead-ends and certain angle structures can have uplift as long as the total force of uplift does not approach the structure weight. If it does, hold-down guys are necessary. Care should be exercised to avoid locating structures resulting in poor line grading.

## 11.10.7  Final Drawings

The conductor and ground wire sizes, design tensions, ruling span, and the design loading condition should be shown on the first sheet of the plan-profile drawings. A copy of the sag template should be shown. The actual ruling spans between dead-ends should be calculated and noted on the sheets.

As conductor sags and structures are spotted on each profile sheet, the structure locations are marked on the plan view and examined to ensure that the locations are satisfactory and do not conflict with existing features or obstructions. To facilitate preparation of a structure list and the tabulation of the number of construction units, the following items, where required, should be indicated at each structure station in the profile view:

1. Structure type designation
2. Pole height and class or height of tower
3. Pole top, crossarm, or brace assemblies

4. Pole ground unit
5. Miscellaneous hardware units (vibration dampers at span locations)
6. Guying assemblies and anchors

## 11.10.8 Drawing Check and Review

The completed plan-profile drawings should be checked to ensure that the line meets the design requirements and criteria originally specified, adequate clearances and computed limitations have been maintained, and required strength capacities have been satisfied. The sheets should be checked for accuracy, completeness, and clarity.

.11

## LINE CONSTRUCTION EXAMPLE

Let us follow a line from inception to completion, using as models, companies such as Brazos Power or South Texas Electric Cooperative. These are both generation and transmission companies, and neither sells retail to individuals.

## 11.11.1 Right of Way Planning

Right of way planning begins when the need to build a new line is manifest. This need might be in relation to a new generating plant, or possibly to serve a new customer. This recently happened when Brazos Power was asked by the Texas Municipal Power Pool to provide wheeling facilities to interconnect several north and central Texas municipal power plants. The cities included Greenville, Garland, Bryan, and others.

At this juncture, the approximate amount of power to be shifted from one place to another must be determined. Maximum shifting generally occurs when a vital power plant is out of service for repairs and its load must be shared by the other municipalities.

At this time, the most suitable voltage is determined. For such a purpose, the highest approved REA design is 230 kV. Standard conductors are selected and losses checked. With this information, the most cost effective conductor size is estimated. (Computers now make these decisions easy and each company checks their transmission line designs to calculate the cost and loss per mile on each standard design they use.) Once the voltage and general class of conductors have been selected, the specific conductor and the supporting structure is chosen. If a single 795 MCM ACSR per phase is to be used, wooded "H" fixtures are the most economical.

The ROW width must now be decided. The REA dictates a standard ROW width of 100 feet unless a particular problem severely limits the available ROW. When a limiting problem exists, great effort is required to prove to REA that a narrow ROW is acceptable.

Next, a study of aerial maps determines the best transmission route. The bes route is one that has the support of the courts in condemnation proceedings. Th ROWs for transmission lines, as ROWs for highways, are determined to be in th public interest and therefore condemnation is allowed. To obtain a court order o condemnation, the following must be established:

1. Public need for the line must be demonstrated.
2. The selected ROW will cause the least damage to the people along it.
3. "Just compensation" must be offered for the property taken.
4. No more than the required land may be taken.

In general, it takes 2 to 10 years to obtain ROW for a line. When the ROW is finally made available, the line is laid out.

To "lay out the line," is a term adopted by this profession, which seems to have no logical translation into more commonly understood English, and is the mos reasonable descriptive term for its meaning. For these reasons we will continue to use the term here. To lay out the line, the location of all angles and requirements for special structures are established, and the appropriate class of structures is selected. The structure locations are decided along with the specific structure at that location. Heights of the structures are not specified until span lengths and their required clearances are determined.

The ideal span length is established by considering the proper conductor tensions, ideal structure heights, and required ground clearance. In practice, an ideal line is laid out on ideal flat terrain to obtain the ruling span on which the conductor design is based. With these considerations met, comparison to terrain features sets the required heights and locations of all structures including the tangents, as discussed more fully in the section on spotting. All clearance and strength requirements for the construction of a safe line have been defined in the National Electric Safety Code (NESC). Recall that the NESC is a consensus construction code agreed to as appropriate and safe by all organizations in the United States who have an interest in safe electrical line construction and operation. It is published under the authority of the National Bureau of Standards.

Next, all clearances are checked and structure heights are adjusted to have required clearance above all roadways, waterways, and walkways, without excessive elevation changes. An additional check at this time is made to be sure that maximum wind loading will not allow less than minimum clearance between any phase and any possible obstruction—including trees off the ROW. One additional consideration is that tall trees (called danger trees) off the ROW must be cut. Danger trees are those that if blown over by wind would interfere with the line.

The site selection for the structures (spotting) requires the most careful judgment of the designer. If the terrain is rough, there is the possibility of uplift on a structure. Downward force is required on the insulator strings to keep them in tension, thus minimizing radio interference from the loose coupling of the insulators, and to ensure that the maximum span length and conductor tension is not exceeded. The ideal tension (and therefore the sag on the conductor) is established for the ruling span. The calculation of sag for the diverse spans and ensuring that they are properly installed is the principal problem facing the field engineer.

When a line has been designed using standard structures and a selected conductor with an accessories system approved by REA, the owner can expect 40 to 50 years of good service if the construction matches the design.

## 11.11.2 Conductor Stringing

Efficient wire installation is a complicated process. Wire installation, or stringing, is an art. Those capable of the art and science of being a wire engineer, usually called a sag engineer, normally move from job to job performing only wire installation.

Using the job plan, the sag engineer organizes the job so that work can progress from one end to the other. He takes a list of the wire reels, which shows the reel number and length, and selects specific reels to go to specific points. The reels are purchased in matched lengths to avoid waste. If the wire reels are 10,000 feet long, then 10,000 feet of each conductor should be installed in one operation. If the line is to be one conductor per phase, three conductors are installed from one location, or set up in two directions. If the conductor is a bundle of three, nine conductors are installed from each set-up in two directions. While wire may be ordered in 10,000 foot reels, the wire will be shipped in lengths of 10,000 feet, plus or minus 5%, labelled as to the length of wire on the spool. Unless "matched" reels are requested, much wire may be wasted.

If, for example, we check the first part of our plan and find at 10,500 feet from the starting dead end, near midspan, a level piece of ground in the ROW of sufficient size and accessability to set up all the equipment needed to begin conductor stringing, we designate that site as "wire site #1", and so mark it on our plan. The reels are checked to find nine reels as near 10,600 feet as possible (they may vary from 10,550 to 10,610 feet). These reels are assigned to "pull #1" and delivered to that wire site. At some point, 200 to 400 feet, before dead-end structure #1, a location must be found to designate as "puller site #1," and the appropriate equipment delivered to that point.

It is most effective to use the leap-frog method, and deliver the wire for pull #2 to wire site #1 so that the same equipment can be used for pull #1, turned around, and used for pull #2. The length of each pull is set by available space near midspan and the length of the available reels. The sag engineer prepares sheets similar to the ones shown in Figure 11.9 and assigns specific reels to particular wire pulls. One "set-up" is the arrangement of equipment ready to complete one pull.

The steps of a pull are as follows:

1. Install strawline from the wire site to the pull site. The strawline is a light steel line that can be dragged on the ground (to avoid damage, the conductor must never touch the ground) and is easily handled. It is mounted on a powered reel. It is strong enough to pull in the static wire also. On a standard 500 kV line, five strawlines are needed for the pull, one for each phase bundle and one for each static wire. The strawlines are put into place by setting up the five reels of straw lines at the appropriate location on the wire site, then dragging the end along the ROW to a point just beyond the first tower. The strawlines

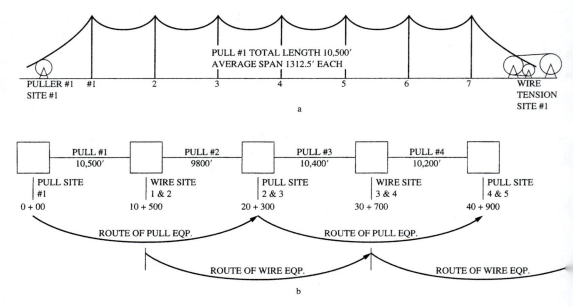

**FIGURE 11.9** Conductor stringing organization (a) Seven span pull of a conductor (b) Pull organization and equipment paths

are relaxed and disconnected from the pulling CAT. One at a time, the ends are then tied to Pea lines*, and pulled into place. Now with all five strawlines in place on the first tower, they are reattached to the CAT and dragged to a point past the second tower and the process repeated until all lines are in place for the complete path of Pull #1. This is the most time consuming task in the wire stringing operation.

2. Wiresite preparation consists of setting up the conductor tensioning machine, mounting reels on the reel stands, and preparing the tools for the wire pull. Figure 11.10 shows a wire site set-up.

3. Puller site preparation consists of setting up the wire puller and the static reels on the static wire tensioner. Figure 11.11 shows the puller site set-up.

4. The two static wires are fastened to the two strawlines that were pulled into the static positions and, one after the other, the two strawlines are pulled to the wire statics in place. This leaves one strawline in place in each conductor stringing dolly. The puller is a heavy-duty machine with one reel of pulling line, which may consist of 18,000 to 20,000 feet of $3/4$ to $1\frac{1}{8}$ inch plough steel, anti-rotational, pulling line. One strawline is fastened to the pulling line, and the pulling line is pulled into position along the route to the wire site. The pulling line is then fastened to a pulling board, which in turn is fastened to the three conductors. This pulling board is shaped so that the pulling line

---

*Pea lines are small flexible lines installed when the dolly is hung. The pea line is threaded through the proper groove with two loose ends near the ground, so that the strawline can be pulled into place in the stringing dolly without someone climbing the tower.

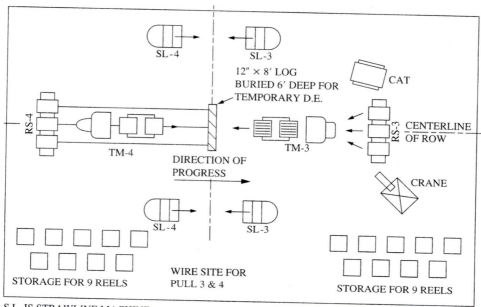

S.L. IS STRAWLINE MACHINE

SL3  STRAWLINE MACHINE IN POSITION FOR PULL #3

SL4  STRAWLINE MACHINE IN POSITION FOR PULL #4

RS3  IS 3 CONDUCTOR REEL STAND IN PLACE FOR PULL #4 TO ALLOW THREE CONDUCTORS FOR ONE PHASE TO BE PULLED SIMULTANEOUSLY

TM3  IS 3 CONDUCTOR TENSION MACHINE IN PLACE FOR PULL #3

CRANE IS 20 TON ROUGH TERRAIN CRANE TO LOAD CONDUCTOR REELS ON REEL STAND

**FIGURE 11.10**  Wire site

is pulled through the stringing dolly in its proper groove, placing each of the conductors in their proper groove in the stringing dollies (which have been previously installed on the towers at the position where the conductors are attached). When the pulling line is retrieved, the first phase of conductors are in place and temporarily dead-ended by the wire crew, the remaining two phases are done, using the same procedures. The conductors are winded (brought to approximate sag) and transferred to temporary dead-ends.

The equipment is shifted to make the next pull, and the procedure is repeated. The sagging crew arrives within 72 hours to begin sagging. Figure 11.12 through 11.14 are photographs of equipment used in conductor stringing.

Timing of critical procedures and requirements for simultaneous operations require a well-orchestrated stringing effort for efficiency. The work can be done by crews of as few as 15 people, but efficiency dictates a minimum of 35 to 53 people, depending on the required rate of progress. If five pulls per week (10 miles) are desired, 53 people will work 6 days. If three pulls (6 miles)

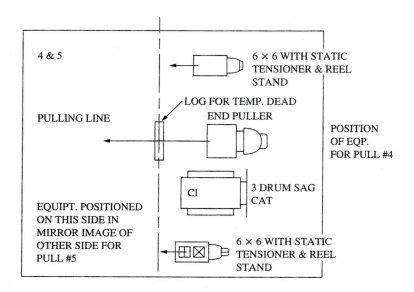

4 & 5

6 × 6 WITH STATIC
TENSIONER & REEL
STAND

LOG FOR TEMP. DEAD
END PULLER

PULLING LINE

POSITION
OF EQP.
FOR PULL #4

Cl        3 DRUM SAG
          CAT

EQUIPT. POSITIONED
ON THIS SIDE IN
MIRROR IMAGE OF
OTHER SIDE FOR
PULL #5

6 × 6 WITH STATIC
TENSIONER & REEL
STAND

**FIGURE 11.11**    Puller site

**FIGURE 11.12**    Strawline machine and reel trailer (a) Strawline machine

**FIGURE 11.12**   (b) Lowboy trailer with braked reel stands for installing bundle of four conductor

are desired, they can be done in 5 days with 35 people. The fast operation requires three foremen. One foreman is responsible for all preparatory work, one is responsible for the wire pull, and one is responsible for clean-up.

Preparatory, or front end work, before the pull, consists of the following tasks:

  a.  Roads are prepared, puller sites cleared and leveled.

  b.  Material is hauled to proper sites. This includes wire, support hardware, insulators, splicing materials, and stringing dollies. Sufficient stringing dollies for both the conductors and static wires for the three longest pulls are required. As soon as the wire has been clipped in pull #1, a crew must retrieve the stringing dollies used in pull #1 and haul them to pull #2 for installation there. These procedures are repeated in sequence throughout the job.

  c.  All insulators are washed and hung, and insulator hardware installed with the stringing dolly in position. When this procedure is complete for the length of the pull, the job is ready for the wire crew and their procedures, which have been discussed. When the wire crew completes their work, the job is passed to the clean-up crew.

The initial task for the clean-up crew is sagging. Wire is sagged between temporary or permanent dead-ends, whichever is required. The sag is started at the dead-end on one end and carried through to the other. On any sag in excess of five spans, the sag on three spans must be checked. The sagger selects a span near each end and one near the center, as close to the length of the ruling span as possible, and calculates the sags for those spans. On level

**FIGURE 11.13**   Views of a tensioner for bundle of two stringing

ground, sags are determined by mounting a target on one structure and a
transit on the other so that proper sag in the low point of the catenary of the
wire and the line between target and transit can be determined, as shown in
Figure 11.15a. The stopwatch method of conductor sagging has become the
method of choice for most sag engineers, and will be described in Appendix.

Sag for any span can be calculated by the following equation that references sag
of a span to that of the ruling span.

$$S_{REFRS} = \frac{L^2}{L_{RS}} S_{RS}$$

(10.5)

a

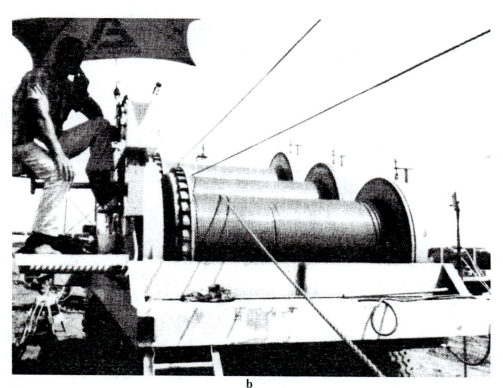

b

**Figure 11.14** Two views of a three drum wire puller for tension wire stringing

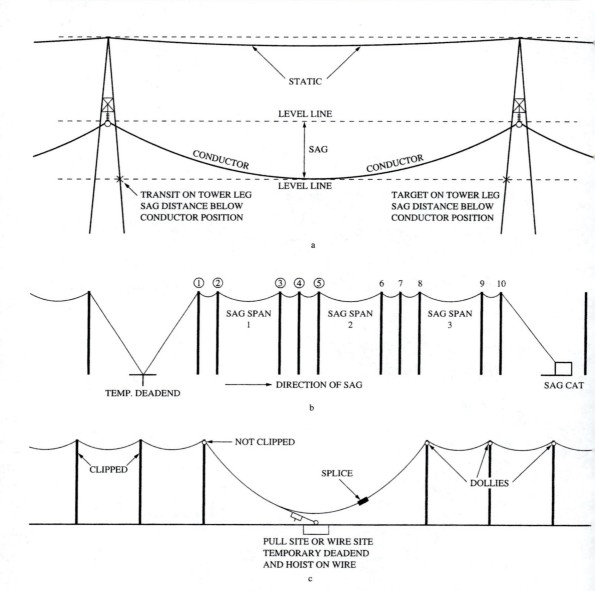

**FIGURE 11.15**  Conductor sagging (a) Sag measurement (b) Sag pull and (c) Clipping

Where

$S_{\text{REF RS}}$ = sag in any span of interest other than the ruling span

$S_{\text{RS}}$ = sag in the ruling span

L = length of span of interest

$L_{\text{RS}}$ = length of the ruling span

Sagging procedure is illustrated in Figure 11.15b and c. A caterpillar is used to pull conductor to near sag, then the wire is transferred to a temporary dead-end with hand-operated hoists, which are used to complete the sag. Figure 11.16 shows a sagging CAT.

a

b

**FIGURE 11.16** Sag cat (a) M-5 equipped with three winches for sagging bundle of three conductor (b) M-5 equipped with four winches for sagging bundle of four conductor

The span nearest to the temporary dead-end reaches its proper sag last, so both the second and third sag spans are brought to a little above their proper sag to bring span #1 into proper sag. Then, tension is relaxed to allow spans #3 and #2 to return to proper sag. The conductor is left in sag for 1 hour to allow any conductor creep to settle. The wire is then *clipped* (transferred from the pulling dolly to the permanent support) at all spans except at the last structure sagged.

The second wire set-up, which is already in place, is sagged just as the first set, except that the second pull conductor is spliced to the first pull conductor, and the tension is applied by the caterpillar. When the tension in the second pull is approximately equal to the first, the last structure conductors are clipped, the hand hoists are removed, and the sagging continues. An observer is stationed to watch the last structure clipped. When the sag is correct, the bells on that structure are perpendicular. Three people with transits verify the sag. Sagging is usually done on each set-up, but it can be delayed and done on two set-ups at a time. A delay causes lost clipping time, and requires extra stringing dollies, which are quite expensive.

The next task the clean-up crew performs is clipping the tangent structures. Clipping consists of raising the weight of the wire with a hand-operated chain hoist called a coffin hoist, removing the stringing dolly and lowering it to the ground, straightening the string of insulators to ensure that they are vertical, and transferring the conductors to the conductor support hardware previously attached to the string of insulators. Figure 11.17 shows clipping in progress. Ideally, there should be enough clipping crews to clip one set-up during the time one pull is made. This is called one turn around. If one set-up is to be pulled per day, then every other task must be done within the same time frame. In this way, each person repeats the same task daily, becomes adept at it, and maximum efficiency is obtained. A crew reaches maximum efficiency by the fourth or fifth pull.

The next task for the clean-up crew is the "dead-end" work. If the line has many dead-ends, a separate crew is organized to install all the dead-ends and jumpers of a pull during the same period that one pull can be made. Otherwise this phase of the work falls behind.

After the clean-up crew finishes the line installation, true clean-up begins. The ROW is cleaned of all debris, and the equipment sites are returned to normal. Temporary gates are removed and fences closed. Temporary bridges and roads are removed and the complete ROW restored to a condition satisfactory to the land owner.

The job is now complete.

## 11.12

### SUMMARY

Considerations for transmission line design are power loss, voltage drop, reactance, thermal overload, wheeling needs, total load, and cost. Additional considerations

**FIGURE 11.17**   Conductor clipping with a high-reach

for transmission line design are ROW, terrain, weather, skilled labor and equipment availability, and their cost. When possible a utility will develop standard support structure and line designs for commonly used voltages and terrain and will depart from them only for special situations. Experience with a particular design results in fewer construction and maintenance errors, and the stocking of fewer maintenance parts results in both convenience and savings. The REA has published successful distribution and transmission design specifications that have resulted from REA financed projects. The REA specifications are good starting points for line design. Additionally fittings, conductors, and hardware have been standardized by the Electrical Manufacturers Association.

The most used support structures are wood, which is least expensive, steel poles, which work well and can be shipped in sections but are more expensive than wood poles, and lattice steel structures, which have the greatest strength-to-weight ratio and are reasonable in cost. Wood poles are usually arranged as "H" fixtures for transmission applications, and lattice steel is used for erecting towers. Each type of structure—wood, steel, and lattice steel—has associated with it recommended foundation requirements and construction practices.

ACSR is the most commonly used conductor for transmission lines, but other conductors are used for specialized requirement. Tension and ice loading and span length as well as the electrical loading are factors in the choice of transmission line conductors. The OHGW are normally high strength galvanized steel or aluminum clad steel stranded conductors.

A typical span, called the ruling span, is used for the nominal design of a line. Adjustments are made in the design as the span must be varied from ruling to accommodate varied terrain. Typical conductor tension is calculated for the ruling span. Conductor sag is an important factor in setting the ultimate tension of the line. Too little sag results in too much tension, and too much sag in too little ground clearance. Lines must be sufficiently damped that aeolian vibration (a resonance induced from steady cross winds) does not occur, and to minimize galloping, which is a low frequency vibration induced by the lift from icing conditions and steady cross wind. Both types of vibration can cause catastrophic outages.

Line design requires the preparation of a plan profile which is an outline of the topography of the terrain along and near the ROW. The location of the towers is decided on and the proper sag calculated. The plan profile in uneven terrain is carefully checked to see that no tower experiences negative sag or uplift under any expected line tension. The final drawings are then prepared, and the construction phase can begin.

Construction consists of several steps, often occurring simultaneously at different locations along the line path. The heavy equipment needed for each phase of the construction must be arranged for and access to the ROW for the equipment must be secured. The conductor, tower parts, and fittings must be ordered for the line and transported to the correct location on the ROW at the proper time. The ROW must be cleared, foundations set, poles set or towers constructed, and line must be strung. Final inspection and connection to the substations, which must have the proper instrument transformers and protection equipment installed, precedes line energization. Finally, it must all be paid for.

## 11.13

### QUESTIONS

1. Few transmission lines are designed from scratch. Why?
2. State the advantages of standard lines.
3. List the major advantages of wood, steel, and steel lattice transmission line support structures.
4. List the factors that influence the method the transmission line support structure.
5. List the steps in transmission line structure construction.
6. Why must steel lattice towers sometimes be braced during construction?

7. State the prime consideration in helicopter tower construction.
8. List the major factors that influence the selection of the transmission conductor diameter.
9. Why must there be no welds in the OHGW?
10. What factors affect the choice of tension for the transmission line conductor?
11. Define ruling span.
12. Calculate, by both the approximate and exact equation, the ruling span for the following span lengths:
   L1 = 900 ft, L2 = 1400 ft, L3 = 1100 ft, L4 = 1000 ft, L5 = 1500 ft.
13. State the accepted limits on span length relative to ruling span.
14. Calculate the sag for a 1000 foot span of Drake conductor if the allowable tension is 30% of the ultimate conductor strength.
15. Define aeolian vibration.
16. What is galloping?
17. Why are plan-profile drawings important?
18. What is meant by the term "structure spotting"?
19. State the source and effects of uplift.
20. What are the major factors considered in ROW condemnation proceedings?
21. List the steps in conductor stringing.
22. Describe the steps performed in "clipping a conductor."

# APPENDIX A

# Three-phase Power Flow Constancy

We noted in Chapter 2 that balanced three-phase power is constant over time. Let

$$V_A = V_p \cos\omega t$$

$$I_A = I_p \cos(\omega t + \theta)$$

then

$$P_A = I_p V_p \cos(\omega t + \theta)\cos\omega t$$

noting that

$$\cos A \cos B = (1/2)\cos(A - B) + (1/2)\cos(A + B)$$

we see

$$\cos(\omega t + \theta)\cos\omega t = (1/2)\cos(\omega t + \theta - \omega t) + (1/2)\cos(\omega t + \theta + \omega t)$$

$$= (1/2)[\cos\theta + \cos(2\omega t + \theta)]$$

therefore

$$P_A = \frac{I_p V_p}{2}[\cos(2\omega t + \theta) + \cos\theta]$$

For the other two phases, B and C at $\omega t + 120°$ and $\omega t + 240°$ respectively

$$P_B = \frac{V_P I_P}{2} [\cos(2\omega t + 240° + \theta) + \cos\theta]$$

$$P_C = \frac{V_P I_P}{2} [\cos(2\omega t + 480° + \theta) + \cos\theta]$$

Note now that

$\cos(2\omega t + \theta) + \cos(2\omega t + 240° + \theta) + \cos(2\omega t + 480° + \theta) = 0$

For example let $2\omega t + \theta = \pi$

$\cos\pi = {}^-1$

$\cos(180° + 240°) = 0.5$

$\cos(180° + 480°) = 0.5$

${}^-1 + 0.5 + 0.5 = 0$

Now let $2\omega t + \theta = \frac{\pi}{2}$

$\cos\frac{\pi}{2} = 0$

$\cos(\frac{\pi}{2} + 240°) = 0.866$

$\cos(\frac{\pi}{2} + 480°) = {}^-0.866$

Thus total three-phase instantaneous power is

$P_T = P_A + P_B + P_C = (^3/_2) I_P V_P \cos\theta$

Which is a constant.

# Sample REA Specifications

REA publications are public domain. The REA provides a valuable service in developing and making available construction specifications such as these. They are used as guidelines throughout the United States.

# Specifications for Construction

1. *General*

All construction work shall be done in accordance with the staking sheets, plans and specifications, and the construction drawings.

The 1981 or latest edition of the National Electrical Safety Code (NESC), ANSI C2, shall be followed except where local regulations are more stringent, in which case local regulations shall govern.

2. *Distribution of Poles*

In distributing the poles, large, choice, dense poles shall be used at transformer, dead-end, angle, and corner locations.

3. *Pole Setting*

The minimum depth for setting poles shall be as follows:

| Length of Pole (Feet) | Setting in Soil (Feet) | Setting in All Solid Rock (Feet) |
|:---:|:---:|:---:|
| 20 | 4.0 | 3.0 |
| 25 | 5.0 | 3.5 |
| 30 | 5.5 | 3.5 |
| 35 | 6.0 | 4.0 |
| 40 | 6.0 | 4.0 |
| 45 | 6.5 | 4.5 |
| 50 | 7.0 | 4.5 |
| 55 | 7.5 | 5.0 |
| 60 | 8.0 | 5.0 |

"Setting in Soil" depths shall apply:

a. Where poles are to be set in soil.

b. Where there is a layer of soil of more than two (2) feet in depth over solid rock.

c. Where the hole in solid rock is not substantially vertical or the diameter of the hole at the surface of the rock exceeds approximately twice the diameter of the pole at the same level.

"Setting in All Solid Rock" depths shall apply where poles are to be set in solid rock and where the hole is substantially vertical, approximately uniform in diameter and large enough to permit the use of tamping bars the full depth of the hole.

Where there is a layer of soil two (2) feet or less in depth over solid rock, the depth of the hole shall be the depth of the soil in addition to the depth specified under "Setting in All Solid Rock" provided, however, that such depth shall not exceed the depth specified under "Setting in Soil."

On sloping ground, the depth of the hole shall be measured from the low side of the hole.

Poles shall be set so that alternate crossarm gains face in opposite directions, except at terminals and dead ends where the gains of the last two (2) poles shall be on the side facing the terminal or dead end. On unusually long spans, the poles shall be set so that the crossarm is located on the side of the pole away from the long span. Where pole top insulator brackets or pole top pins are used, they shall be located on the opposite side of the pole from the gain.

Poles shall be set in alignment and plumb, except at corners, terminals, angles, junctions, or other points of strain, where they shall be set and raked against the strain so that the conductors are in line.

Poles shall be raked against the conductor strain not less than 1-inch for each 10 feet of pole length nor more than 2 inches for each 10 feet of pole length after conductors are installed at the required tension.

Pole backfill shall be thoroughly tamped in full depth. Excess dirt shall be banked around the pole.

Poles which have been in storage for more than 1 year from the date of treatment shall be ground line treated when installed.

4. *Grading of Line*

When using high poles to clear obstacles such as buildings, foreign wire crossings, railroads, etc., there shall be no upstrain on pin-type or post-type insulators in grading the line each way to lower poles.

5. *Guys and Anchors*

Guys shall be placed before the conductors are strung and shall be attached to the pole as shown in the construction drawings.

All anchors and rods shall be in line with the strain and shall be installed so that approximately 6 inches of the rod remain out of the ground. In cultivated fields or other locations, as deemed necessary, the projection of the anchor rod above earth may be increased to a maximum of 12 inches to prevent burial of the rod eye. The backfill of all anchor holes must be thoroughly tamped the full depth.

After a cone anchor has been set in place, the hole shall be backfilled with coarse crushed rock for 2 feet above the anchor tamping during the filling. The remainder of the hole shall be backfilled and tamped with dirt.

6. *Locknuts*

A locknut shall be installed with each nut, eyenut or other fastener on all bolts or threaded hardware such as insulator pins and studs, upset bolts, double arming bolts, etc.

7. *Conductors*

Conductors must be handled with care. Conductors shall neither be trampled on nor run over by vehicles. Each reel shall be examined and the wire shall be inspected for cuts, kinks, or other injuries. Injured portions shall be cut out and the conductor spliced. The conductors shall be pulled over suitable rollers or stringing blocks properly mounted on the pole or crossarm if necessary to prevent binding while stringing.

The neutral conductor should be maintained on one side of the pole (preferably the road side) for tangent construction and for angles not exceeding 20°.

With pin-type or post-type insulators, the conductors shall be tied in the top groove of the insulator on tangent poles and on the side of the insulator away from the strain at angles. Pin-type and post-type insulators shall be tight on the pins and brackets, respectively, and the top groove must be in line with the conductor after tying.

For line angles of 0° to 5° in locations known to be subject to considerable conductor vibration, insulated brackets (material item da) may be substituted for the single and double upset bolts used for supporting the neutral and secondary conductors.

All conductors shall be cleaned thoroughly by wirebrushing before splicing or installing connectors or clamps. A suitable inhibitor shall be used before splicing or applying connectors over aluminum conductor.

8. *Splices and Dead Ends*

Conductors shall be spliced and dead-ended as shown on the construction drawings. There shall be not more than one splice per conductor in any span and splices shall be located at least 10 feet from the conductor support. No splices shall be located in Grade B crossing spans and preferably not in the adjacent spans. Splices shall be installed in accordance with the manufacturer's recommendations.

9. *Taps and Jumpers*

Jumpers and other leads connected to line conductors shall have sufficient slack to allow free movement of the conductors. Where slack is not shown on the construction drawings, it will be provided by at least two (2) bends in a vertical plane, or one (1) in a horizontal plane, or the equivalent. In areas where aeolian vibration occurs, special measures to minimize the effects of jumper breaks shall be used as specified.

All leads on equipment such as transformers, reclosers, etc., shall be a minimum of #6 copper conductivity. Where aluminum jumpers are used, a connection to an unplated bronze terminal shall be made by splicing a short stub of copper to the aluminum jumper using a compression connector suitable for the bimetallic connection.

10. *Hot-Line Clamps and Connectors*

Connectors and hot-line clamps suitable for the purpose shall be installed as shown on the guide drawings. On all hot-line clamp installations, the clamp and jumper shall be installed so that they are permanently bonded to the load side of the line, allowing the jumper to be de-energized when the clamp is disconnected.

11. *Surge Arrester Gap Settings*

The external gap electrodes of surge arresters, combination arrester cutout units, and transformer mounted arresters shall be adjusted to the manufacturer's recommended spacing. Care shall be taken that the adjusted gap is not disturbed when the equipment is installed.

12. *Conductor Ties*

Hand-formed ties shall be in accordance with construction drawings. Factory-formed ties shall be installed in accordance with the manufacturer's recommendations.

13. *Sagging of Conductors*

Conductors shall be sagged in accordance with the conductor manufacturer's recommendations. All conductors shall be sagged evenly. The air temperature at the time and place of sagging shall be determined by a certified thermometer.

The sag of all conductors after stringing shall be in accordance with the engineer's instructions.

14. *Secondaries and Service Drops*

Secondary conductors may be bare or covered wire or multi-conductor service cable. The conductors shall be sagged in accordance with the manufacturer's recommendations.

Conductors for secondary underbuild on primary lines will normally be bare, except in those instances where prevailing conditions may limit primary span lengths to the extent that covered wires or service cables may be used. Service drops shall be covered wire or service cable.

Secondaries and service drops shall be so installed as not to obstruct climbing space. There shall not be more than one splice per conductor in any span, and splices shall be located at least 10 feet from the conductor support. Where the same covered conductors or service cables are to be used for the secondary and service drop, they may be installed in one continuous run.

15. *Grounds*

Ground rods shall be driven full length in undisturbed earth in accordance with the construction drawings. The top shall be at least 12 inches below the surface of the earth. The ground wire shall be attached to the rod with a clamp and shall be secured to the pole with staples. The staples on the ground wire shall be spaced 2 feet apart, except for a distance of 8 feet above the ground and 8 feet down from the top of the pole where they shall be 6 inches apart.

All equipment shall have at least two (2) connections from the frame, case or tank to the multi-grounded neutral conductor.

The equipment ground, neutral wires, and surge-protection equipment shall be interconnected and attached to a common ground wire.

16. *Clearing Right-of-Way*

The right-of-way shall be prepared by removing trees, clearing underbrush, and trimming trees so that the right-of-way is cleared close to the ground and is the width specified, except that low growing shrubs which will not interfere with the operation or maintenance of the line shall be left undisturbed if so directed by the owner. Slash may be chipped and blown on the right-of-way. The landowner's written permission shall be received prior to cutting trees outside the right-of-way. Trees fronting each side of the right-of-way shall be trimmed symmetrically unless otherwise specified. Dead trees beyond the right-of-way which would strike the line in falling shall be removed. Leaning trees beyond the right-of-way, which would strike the line in falling and which would require topping if not removed, shall either be removed or topped, except that shade, fruit, or ornamental trees shall be trimmed and not removed, unless otherwise authorized.

17. *Structures Exceeding 200 Feet in Height and Structures in the Vicinity of Airports*

The Federal Aviation Administration (FAA) requires (14 CFR 77) that in cases where structures or conductors will exceed a height of 200 feet, or are within 20,000 feet of an airport, the nearest regional or area office of the FAA be contacted and FAA Form 7460-1 be filed if necessary.

Some REA drawings follow.

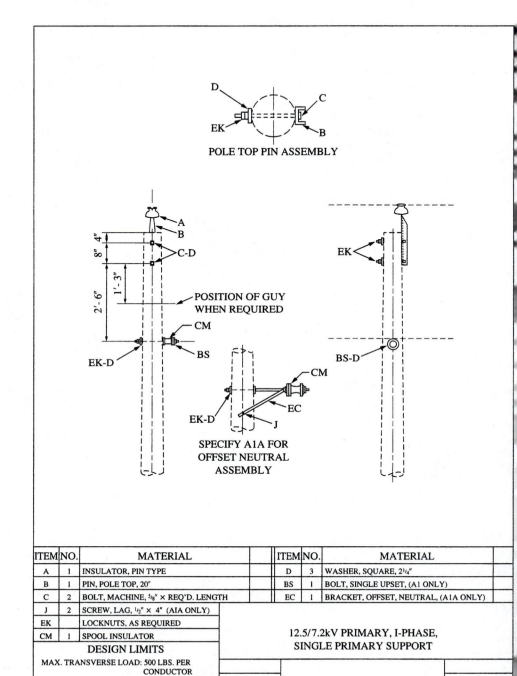

POLE TOP PIN ASSEMBLY

POSITION OF GUY
WHEN REQUIRED

SPECIFY A1A FOR
OFFSET NEUTRAL
ASSEMBLY

| ITEM | NO. | MATERIAL | | ITEM | NO. | MATERIAL | |
|------|-----|----------|---|------|-----|----------|---|
| A | 1 | INSULATOR, PIN TYPE | | D | 3 | WASHER, SQUARE, 2¼″ | |
| B | 1 | PIN, POLE TOP, 20″ | | BS | 1 | BOLT, SINGLE UPSET, (A1 ONLY) | |
| C | 2 | BOLT, MACHINE, ⅝″ × REQ'D. LENGTH | | EC | 1 | BRACKET, OFFSET, NEUTRAL, (A1A ONLY) | |
| J | 2 | SCREW, LAG, ½″ × 4″ (A1A ONLY) | | | | | |
| EK | | LOCKNUTS, AS REQUIRED | | | | | |
| CM | 1 | SPOOL INSULATOR | | | | 12.5/7.2kV PRIMARY, I-PHASE, | |

**DESIGN LIMITS**

MAX. TRANSVERSE LOAD: 500 LBS. PER
CONDUCTOR

MAX. LINE ANGLE WITHIN LOAD LIMITS: 5°

12.5/7.2kV PRIMARY, I-PHASE,
SINGLE PRIMARY SUPPORT

| APR., 1983 | | A1, A1A |

**FIGURE B.1**

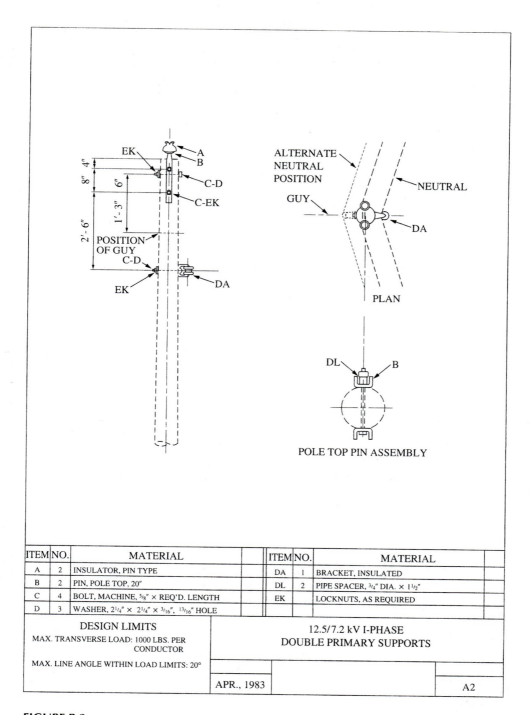

POLE TOP PIN ASSEMBLY

| ITEM | NO. | MATERIAL |  | ITEM | NO. | MATERIAL |  |
|------|-----|----------|--|------|-----|----------|--|
| A | 2 | INSULATOR, PIN TYPE |  | DA | 1 | BRACKET, INSULATED |  |
| B | 2 | PIN, POLE TOP, 20″ |  | DL | 2 | PIPE SPACER, ³/₄″ DIA. × 1¹/₂″ |  |
| C | 4 | BOLT, MACHINE, ⁵/₈″ × REQ'D. LENGTH |  | EK |  | LOCKNUTS, AS REQUIRED |  |
| D | 3 | WASHER, 2¹/₄″ × 2¹/₄″ × ³/₁₆″, ¹³/₁₆″ HOLE |  |  |  |  |  |

| DESIGN LIMITS | 12.5/7.2 kV I-PHASE |
|---|---|
| MAX. TRANSVERSE LOAD: 1000 LBS. PER CONDUCTOR | DOUBLE PRIMARY SUPPORTS |
| MAX. LINE ANGLE WITHIN LOAD LIMITS: 20° | |
| APR., 1983 | A2 |

FIGURE B.2

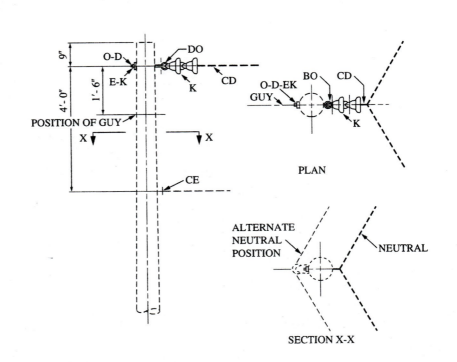

POSITION OF GUY

PLAN

ALTERNATE
NEUTRAL
POSITION

NEUTRAL

SECTION X-X

NOTE: ITEMS CD AND CE ARE SHOWN ON ASSEMBLY DRAWINGS M41-1 AND M41-10

| ITEM | NO. REQ'D | MATERIAL | | ITEM | NO. REQ'D | MATERIAL | |
|------|-----------|----------|--|------|-----------|----------|--|
| | | | | BO | 1 | SHACKLE, ANCHOR | |
| D | 2 | WASHER, 2¼″ × 2¼″ × ³/₁₆″, ³/₁₆″ HOLE | | CD | 1 | ANGLE ASSEMBLY, PRIMARY | |
| K | 4 | INSULATOR, SUSPENSION | | CE | 1 | ANGLE ASSEMBLY, NEUTRAL | |
| O | 3 | BOLT, EYE, ⅝″ × REQ'D LENGTH | | EK | | LOCKNUT, AS REQUIRED | |

| DESIGN LIMITS<br>MAX. TRANSVERSE LOAD: 4000 LBS. PER CONDUCTOR<br><br>ANGLE: 20° - 60° | | 12.5/7.2 kV PRIMARY I-PHASE | |
|---|---|---|---|
| APR., 1983 | | | A3 |

**FIGURE B.3**

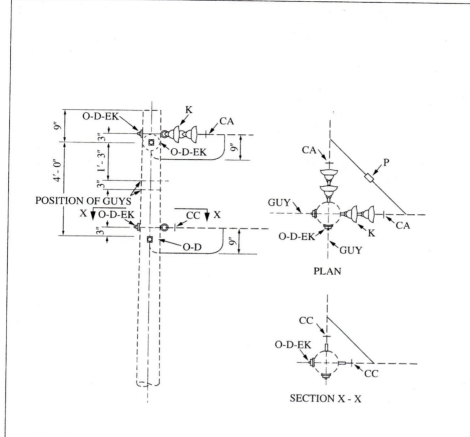

POSITION OF GUYS

PLAN

SECTION X - X

NOTE: ITEMS CA AND CC ARE SHOWN ON ASSEMBLY
DRAWINGS M42-3, M42-11, M42-13 AND M42-21

| ITEM | NO. | MATERIAL | | ITEM | NO. | MATERIAL | |
|------|-----|----------|--|------|-----|----------|--|
| D | 4 | WASHER, $2^{1/4}'' \times 2^{1/4}'' \times {}^{3/16}''$, ${}^{13/16}''$ HOLE | | CA | 2 | DEADEND ASSEMBLY, PRIMARY | |
| K | 4 | INSULATOR, SUSPENSION | | CC | 2 | DEADEND ASSEMBLY, NEUTRAL | |
| O | 4 | BOLT, EYE, ${}^{5/8}'' \times$ REQ'D LENGTH | | EK | | LOCKNUTS. AS REQUIRED | |
| P | | CONNECTORS, AS REQUIRED | | | | | |

12.5/7.2 kV PRIMARY, I-PHASE
60° TO 90° ANGLE

APR., 1983

A4

**FIGURE B.4**

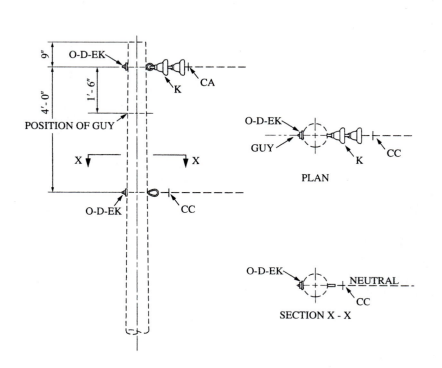

NOTE: ITEMS CA AND CC ARE SHOWN ON ASSEMBLY
DRAWINGS M42-3, M42-11, M42-13 AND M42-21

| ITEM | NO. | MATERIAL | | ITEM | NO. | MATERIAL | |
|------|-----|----------|--|------|-----|----------|--|
| D | 2 | WASHER, SQUARE, 2¼″ | | CC | 1 | DEADEND ASSEMBLY, NEUTRAL | |
| K | 2 | INSULATOR, SUSPENSION | | EK | | LOCKNUTS, AS REQUIRED | |
| O | 2 | BOLT, EYE, ⅝″ × REQ'D. LENGTH | | | | | |
| CA | 1 | DEADEND ASSEMBLY, PRIMARY | | | | | |

12.5/7.2 kV PRIMARY, I-PHASE
DEADEND (SINGLE)

APR., 1983

A5

FIGURE B.5

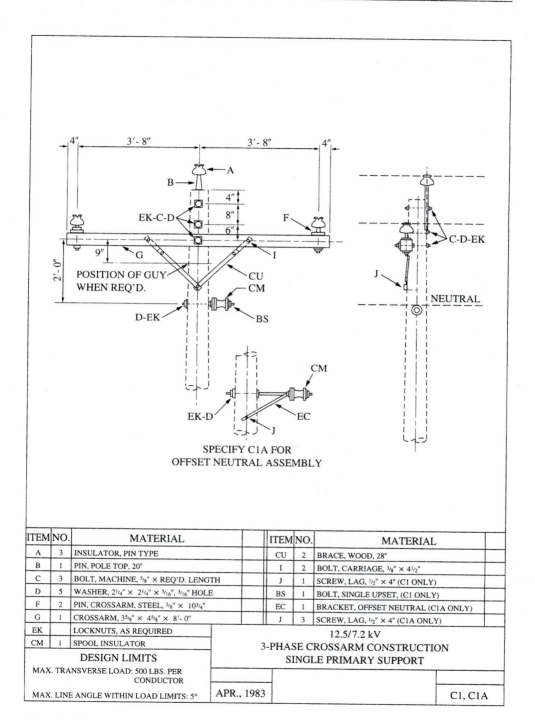

| ITEM | NO. | MATERIAL | | ITEM | NO. | MATERIAL | |
|------|-----|----------|--|------|-----|----------|--|
| A | 3 | INSULATOR, PIN TYPE | | CU | 2 | BRACE, WOOD, 28″ | |
| B | 1 | PIN, POLE TOP, 20″ | | I | 2 | BOLT, CARRIAGE, $3/8″ \times 4^1/2″$ | |
| C | 3 | BOLT, MACHINE, $5/8″ \times$ REQ'D. LENGTH | | J | 1 | SCREW, LAG, $1/2″ \times 4″$ (C1 ONLY) | |
| D | 5 | WASHER, $2^1/4″ \times 2^1/4″ \times 3/16″$, $3/16″$ HOLE | | BS | 1 | BOLT, SINGLE UPSET, (C1 ONLY) | |
| F | 2 | PIN, CROSSARM, STEEL, $5/8″ \times 10^3/4″$ | | EC | 1 | BRACKET, OFFSET NEUTRAL (C1A ONLY) | |
| G | 1 | CROSSARM, $3^5/8″ \times 4^5/8″ \times 8′$-$0″$ | | J | 3 | SCREW, LAG, $1/2″ \times 4″$ (C1A ONLY) | |
| EK | | LOCKNUTS, AS REQUIRED | | | | 12.5/7.2 kV | |
| CM | 1 | SPOOL INSULATOR | | | | 3-PHASE CROSSARM CONSTRUCTION | |
| | | **DESIGN LIMITS** | | | | SINGLE PRIMARY SUPPORT | |
| | | MAX. TRANSVERSE LOAD: 500 LBS. PER CONDUCTOR | | | | | |
| | | MAX. LINE ANGLE WITHIN LOAD LIMITS: 5° | | APR., 1983 | | | C1, C1A |

**FIGURE B.6**

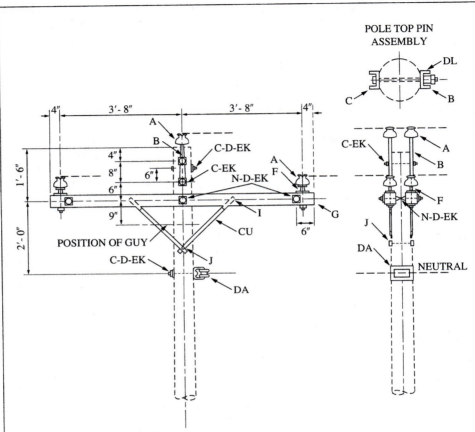

POLE TOP PIN ASSEMBLY

NOTE: WHEN THE TRANSVERSE LOAD IS MORE THAN 1000 POUNDS,
SUBSTITUTE C2-1 OR C2-2 AS REQUIRED

| ITEM | NO. | MATERIAL | ITEM | NO. | MATERIAL |
|---|---|---|---|---|---|
| A | 6 | INSULATOR, PIN TYPE | I | 4 | BOLT, CARRIAGE, 3/8" × 4 1/2" |
| B | 2 | PIN, POLE TOP, 20" | J | 2 | SCREW, LAG, 1/2" × 4" |
| C | 4 | BOLT, MACHINE, 5/8" × REQ'D. LENGTH | N | 3 | BOLT, DOUBLE ARMING, 5/8" × REQ'D. L'GTH |
| D | 13 | WASHER, 2 1/4" × 2 1/4" × 3/16", 13/16" HOLE | DA | 1 | BRACKET, INSULATED |
| F | 4 | PIN, CROSSARM, STEEL, 5/8" × 10 3/4" | DL | 2 | PIPE, SPACER, 3/4" DIA. × 1 1/2" |
| G | 2 | CROSSARM, 3 5/8" × 4 5/8" × 8 - 0" | EK | | LOCKNUTS, AS REQUIRED |
| CU | 4 | BRACE, WOOD, 28" | | | |

**DESIGN LIMITS**

MAX. TRANSVERSE LOAD: 1000 LBS. PER CONDUCTOR

MAX. LINE ANGLE WITHIN LOAD LIMITS: 20°

**12.5/7.2 kV 3-PHASE**

**CROSSARM CONSTR. DOUBLE PRIMARY SUPPORT**

APR., 1983

C2

**FIGURE B.7**

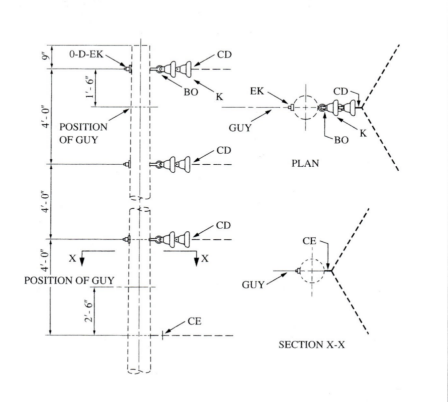

NOTE: ITEMS CD AND CE ARE SHOWN ON ASSEMBLY DRAWINGS M41-1, AND M42-10

| ITEM | NO. | MATERIAL | | ITEM | NO. | MATERIAL | |
|------|-----|----------|---|------|-----|----------|---|
| | | | | BO | 3 | SHACKLE, ANCHOR | |
| D | 3 | WASHER, 2¼" × 2¼" × ³⁄₁₆", ¹³⁄₁₆" HOLE | | CD | 3 | ANGLE ASSEMBLY, PRIMARY | |
| K | 6 | INSULATOR, SUSPENSION | | CE | 1 | ANGLE ASSEMBLY, NEUTRAL | |
| O | 3 | BOLT, EYE, ⅝" × REQ'D. LENGTH | | EK | | LOCKNUTS, AS REQUIRED | |

| DESIGN LIMITS | | | | |
|---|---|---|---|---|
| MAX. TRANSVERSE LOAD: 4000 LBS. PER CONDUCTOR | 12.5/7.2 kV - THREE PHASE VERTICAL CONSTRUCTION | | | |
| ANGLE: 20° - 60° | | | | |
| | APR., 1983 | | | C3 |

**FIGURE B.8**

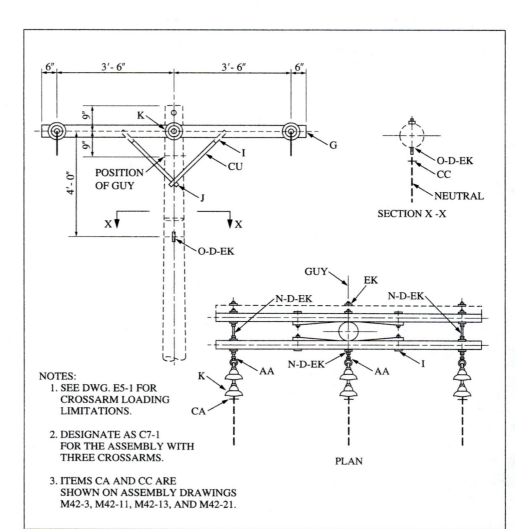

NOTES:
  1. SEE DWG. E5-1 FOR
     CROSSARM LOADING
     LIMITATIONS.

  2. DESIGNATE AS C7-1
     FOR THE ASSEMBLY WITH
     THREE CROSSARMS.

  3. ITEMS CA AND CC ARE
     SHOWN ON ASSEMBLY DRAWINGS
     M42-3, M42-11, M42-13, AND M42-21.

| ITEM | NO. | MATERIAL | | ITEM | NO. | MATERIAL | |
|------|-----|----------|---|------|-----|----------|---|
| D | 11 | WASHER, $2^{1}/_{4}'' \times 2^{1}/_{4}'' \times {}^{3}/_{16}''$ , ${}^{13}/_{16}''$ HOLE | | N | 3 | BOLT, DOUBLE ARMING, ${}^{5}/_{8}'' \times$ REQ'D. L'GTH | |
| G | 2 | CROSSARM, $3^{5}/_{8}'' \times 4^{5}/_{8}'' \times 8'$ - $0''$ | | O | 1 | BOLT, EYE, ${}^{5}/_{8}'' \times$ REQ'D. LENGTH | |
| CU | 4 | BRACE, WOOD, 28" | | AA | 3 | NUT, EYE, ${}^{5}/_{8}''$ | |
| I | 4 | BOLT, CARRIAGE, ${}^{3}/_{8}'' \times 4^{1}/_{2}''$ | | CA | 3 | DEADEND ASSEMBLY, PRIMARY | |
| J | 2 | SCREW, LAG, ${}^{1}/_{2}'' \times 4''$ | | CC | 1 | DEADEND ASSEMBLY, NEUTRAL | |
| K | 6 | INSULATOR, SUSPENSION | | EK | | LOCKNUTS, AS REQUIRED | |

| | |
|---|---|
| 12.5/7.2 kV, 3 - PHASE CROSSARM CONSTRUCTION DEAD END (SINGLE) | |
| APR., 1983 | C7, C7-1 |

**FIGURE B.9**

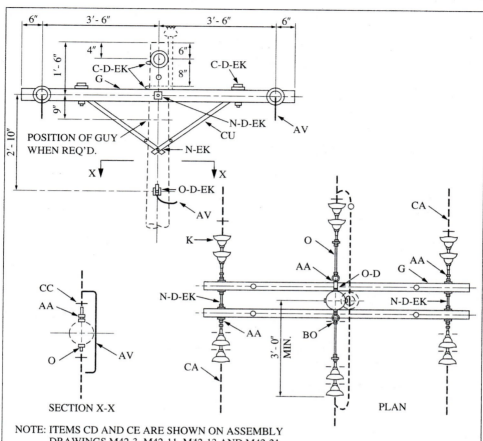

NOTE: ITEMS CD AND CE ARE SHOWN ON ASSEMBLY
DRAWINGS M42-3, M42-11, M42-13 AND M42-21

| ITEM | NO. | MATERIAL | | ITEM | NO. | MATERIAL | |
|------|-----|----------|---|------|-----|----------|---|
| | | | | O | 4 | BOLT, EYE, ⅝" × REQ'D. LENGTH | |
| | | | | P | | CONNECTORS, AS REQ'D. | |
| C | 4 | BOLT, MACHINE, ½" × REQ'D. LENGTH | | AA | 8 | NUT, EYE, ⅝" | |
| D | 14 | WASHER, 2¼" × 2¼" × 3/16", 13/16" HOLE | | AV | | JUMPERS AND LEADS AS REQ'D | |
| D | 4 | WASHER, ROUND, 1⅜" DIA., 9/16" HOLE | | BO | 2 | SHACKLE, ANCHOR | |
| | | | | CA | 6 | DEADEND, ASSEMBLY, PRIMARY | |
| G | 2 | CROSSARM, 3⅝" × 4⅝" × 8 - 0" | | CC | 2 | DEADEND, ASSEMBLY, NEUTRAL | |
| K | 12 | INSULATOR, SUSPENSION | | CU | 2 | BRACE, WOOD, 60" SPAN | |
| N | 4 | BOLT, DOUBLE ARMING, ⅝" × REQ'D. LENGTH | | EK | | LOCKNUTS, AS REQUIRED | |

12.5/7.2 kV, 3-PHASE
CROSSARM CONSTRUCTION, DEADEND (DOUBLE)

| APR., 1983 | | C8 |
|------------|---|-----|

**FIGURE B.10**

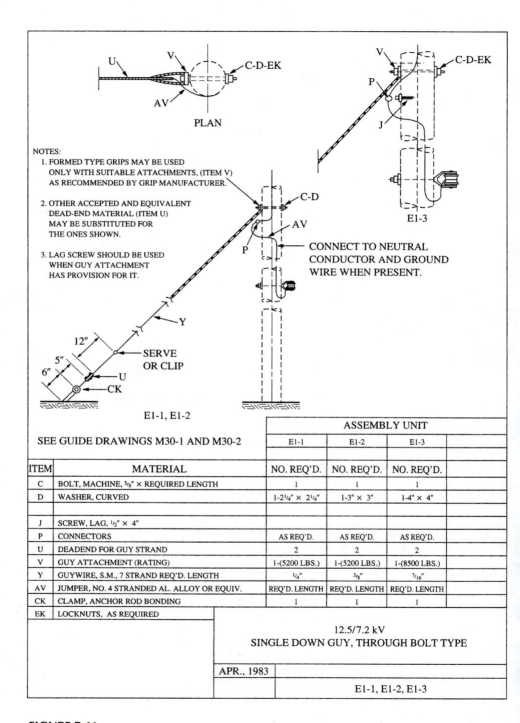

PLAN

NOTES:
1. FORMED TYPE GRIPS MAY BE USED ONLY WITH SUITABLE ATTACHMENTS, (ITEM V) AS RECOMMENDED BY GRIP MANUFACTURER.

2. OTHER ACCEPTED AND EQUIVALENT DEAD-END MATERIAL (ITEM U) MAY BE SUBSTITUTED FOR THE ONES SHOWN.

3. LAG SCREW SHOULD BE USED WHEN GUY ATTACHMENT HAS PROVISION FOR IT.

CONNECT TO NEUTRAL CONDUCTOR AND GROUND WIRE WHEN PRESENT.

SERVE OR CLIP

E1-1, E1-2

E1-3

SEE GUIDE DRAWINGS M30-1 AND M30-2

| | | ASSEMBLY UNIT | | |
|---|---|---|---|---|
| | | E1-1 | E1-2 | E1-3 |
| ITEM | MATERIAL | NO. REQ'D. | NO. REQ'D. | NO. REQ'D. |
| C | BOLT, MACHINE, ⅝″ × REQUIRED LENGTH | 1 | 1 | 1 |
| D | WASHER, CURVED | 1-2¼″ × 2¼″ | 1-3″ × 3″ | 1-4″ × 4″ |
| | | | | |
| J | SCREW, LAG, ½″ × 4″ | | | |
| P | CONNECTORS | AS REQ'D. | AS REQ'D. | AS REQ'D. |
| U | DEADEND FOR GUY STRAND | 2 | 2 | 2 |
| V | GUY ATTACHMENT (RATING) | 1-(5200 LBS.) | 1-(5200 LBS.) | 1-(8500 LBS.) |
| Y | GUYWIRE, S.M., 7 STRAND REQ'D. LENGTH | ¼″ | ⅜″ | ⁷⁄₁₆″ |
| AV | JUMPER, NO. 4 STRANDED AL. ALLOY OR EQUIV. | REQ'D. LENGTH | REQ'D. LENGTH | REQ'D. LENGTH |
| CK | CLAMP, ANCHOR ROD BONDING | 1 | 1 | 1 |
| EK | LOCKNUTS, AS REQUIRED | | | |

12.5/7.2 kV
SINGLE DOWN GUY, THROUGH BOLT TYPE

APR., 1983

E1-1, E1-2, E1-3

**FIGURE B.11**

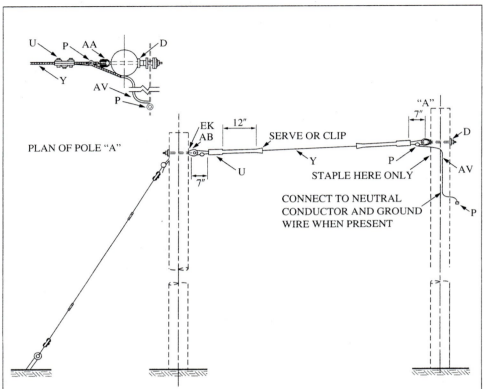

PLAN OF POLE "A"

EK
AB
12"
SERVE OR CLIP
"A"
7"
D
Y
P
AV
STAPLE HERE ONLY

CONNECT TO NEUTRAL
CONDUCTOR AND GROUND
WIRE WHEN PRESENT
P

U
7"

NOTE: OTHER ACCEPTED AND EQUIVALENT ITEMS OF DEADEND MATERIAL MAY BE
SUBSTITUTED FOR THE 3-BOLT CLAMP SHOWN.

| | | ASSEMBLY UNIT | | | |
|---|---|---|---|---|---|
| | | E2-1 | E2-2 | E2-3 | |
| ITEM | MATERIAL | NO. REQ'D. | NO. REQ'D. | NO. REQ'D. | |
| D | WASHER, CURVED | 1-2¼" × 2¼" | 1-3" × 3" | 1-4" × 4" | |
| U | DEADEND FOR GUY STRAND | LIGHT DUTY (2) | HEAVY DUTY (2) | HEAVY DUTY (2) | |
| Y | GUY WIRE, 7 STRAND S.M. REQ'D. LENGTH | ¼" | ⅜" | ⁷⁄₁₆" | |
| AB | NUT, THIMBLE TYPE EYE, ⅝" | 1 | 1 | 1 | |
| AA | BOLT, THIMBLEYE, ⅝" × REQ'D. LENGTH BY | 1 | 1 | 1 | |
| AV | JUMPER, #4 STRANDED AL. ALLOY OR EQUIV. | 1 | 1 | 1 | |
| P | CONNECTORS, AS REQ'D. | | | | |
| EK | LOCKNUTS, AS REQUIRED | | 12.5/7.2 kV SINGLE OVERHEAD GUY, THROUGH BOLT TYPE | | | |
| | | APR., 1983 | | | |
| | | | E2-1, E2-2, E2-3 | | |

**FIGURE B.12**

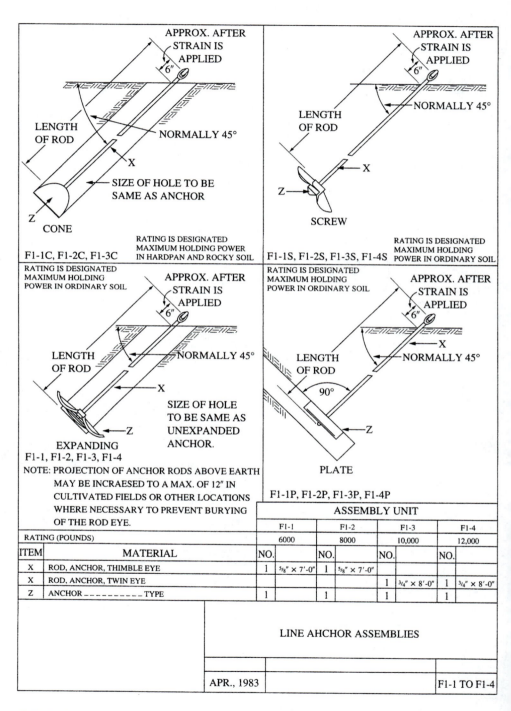

| ASSEMBLY UNIT | | | | | | | | |
|---|---|---|---|---|---|---|---|---|
| | | F1-1 | | F1-2 | | F1-3 | | F1-4 |
| RATING (POUNDS) | | 6000 | | 8000 | | 10,000 | | 12,000 |
| ITEM | MATERIAL | NO. | | NO. | | NO. | | NO. |
| X | ROD, ANCHOR, THIMBLE EYE | 1 | 5/8″ × 7′-0″ | 1 | 5/8″ × 7′-0″ | | | |
| X | ROD, ANCHOR, TWIN EYE | | | | | 1 | 3/4″ × 8′-0″ | 1 | 3/4″ × 8′-0″ |
| Z | ANCHOR _ _ _ _ _ _ _ _ TYPE | 1 | | 1 | | 1 | | 1 |

| | | | | |
|---|---|---|---|---|
| | | LINE AHCHOR ASSEMBLIES | | |
| | APR., 1983 | | | F1-1 TO F1-4 |

**FIGURE B.13**

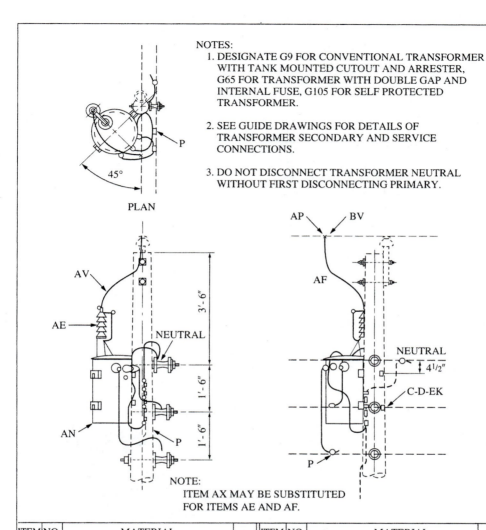

NOTES:
1. DESIGNATE G9 FOR CONVENTIONAL TRANSFORMER WITH TANK MOUNTED CUTOUT AND ARRESTER, G65 FOR TRANSFORMER WITH DOUBLE GAP AND INTERNAL FUSE, G105 FOR SELF PROTECTED TRANSFORMER.

2. SEE GUIDE DRAWINGS FOR DETAILS OF TRANSFORMER SECONDARY AND SERVICE CONNECTIONS.

3. DO NOT DISCONNECT TRANSFORMER NEUTRAL WITHOUT FIRST DISCONNECTING PRIMARY.

PLAN

NOTE:
ITEM AX MAY BE SUBSTITUTED
FOR ITEMS AE AND AF.

| ITEM | NO. | MATERIAL | | ITEM | NO. | MATERIAL | |
|------|-----|----------|---|------|-----|----------|---|
| C | 2 | BOLT, MACHINE, $5/8''$ × REQ'D. LENGTH | | AN | 1 | TRANSFORMER | |
| D | 2 | WASHER, $2^{1}/_{4}''$ × $2^{1}/_{4}''$ × $3/16''$, $13/16''$ HOLE | | AP | 1 | CLAMP, HOT LINE, TAP ASSEMBLY | |
| P | | CONNECTORS, AS REQUIRED | | AV | | JUMPERS, STRANDED AS REQUIRED | |
| AE | | SURGE ARRESTER (G9 ONLY) | | BV | 1 | RODS, ARMOR | |
| AF | | CUTOUT, FUSE, OPEN LINK (G9 ONLY) | | EK | | LOCKNUTS, AS REQ'D. | |

12.5/7.2 kV,
SINGLE PHASE TRANSFORMER
AT 1 - PHASE TANGENT

APR., 1983

G9 -, G65 -,
G105

**FIGURE B.14**

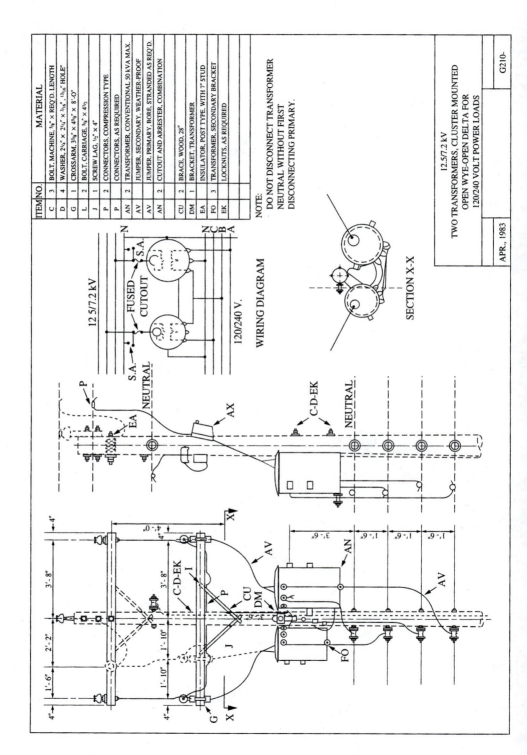

**FIGURE B.15**

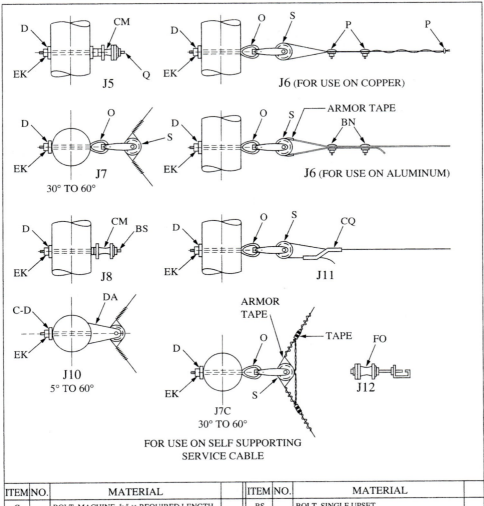

| ITEM NO. | MATERIAL | ITEM NO. | MATERIAL | |
|---|---|---|---|---|
| C | BOLT, MACHINE, $5/8''$ × REQUIRED LENGTH | BS | BOLT, SINGLE UPSET | |
| D | WASHER, $2^{1}/4''$ × $2^{1}/4''$ × $3/16''$, $13/16''$ HOLE | BN | CLAMP, LOOP, DEADEND | |
| O: | BOLT, EYE, $5/8''$ × REQUIRED LENGTH | CQ | SLEEVE, OFFSET, SPLICING | |
| P | CONNECTORS, AS REQUIRED | DA | BRACKET, INSULATED | |
| Q | BOLT, DOUBLE UPSET | FO | TRANSFORMER SECONDARY BRACKET | |
| S | CLEVIS, SECONDARY, SWINGING, INSULATED | EK | LOCKNUTS AS REQUIRED | |
| CM | INSULATOR SPOOL | | | |

SECONDARY ASSEMBLIES

| APR., 1983 | | | J5 TO J12 |
|---|---|---|---|

**FIGURE B.16**

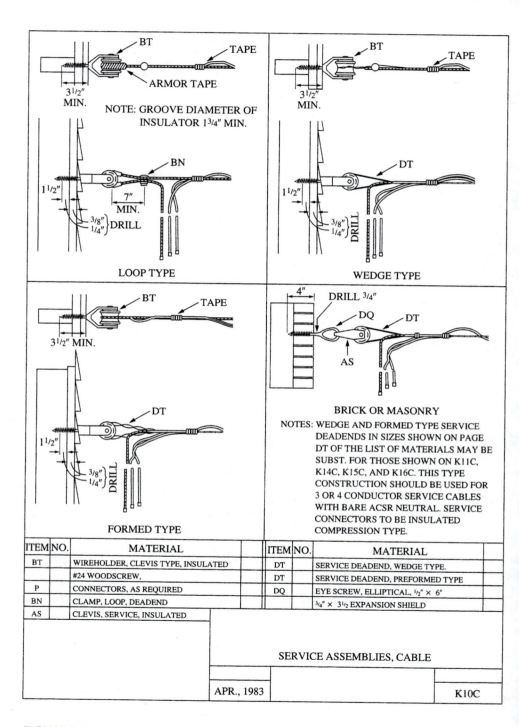

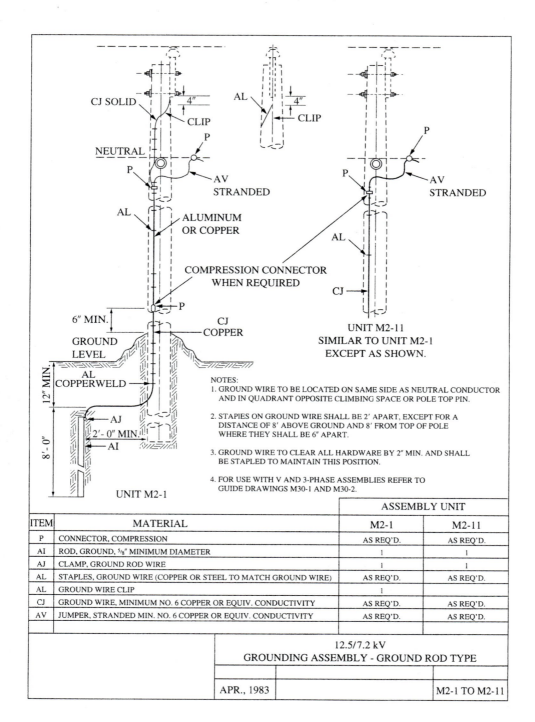

NOTES:
1. GROUND WIRE TO BE LOCATED ON SAME SIDE AS NEUTRAL CONDUCTOR AND IN QUADRANT OPPOSITE CLIMBING SPACE OR POLE TOP PIN.

2. STAPIES ON GROUND WIRE SHALL BE 2' APART, EXCEPT FOR A DISTANCE OF 8' ABOVE GROUND AND 8' FROM TOP OF POLE WHERE THEY SHALL BE 6" APART.

3. GROUND WIRE TO CLEAR ALL HARDWARE BY 2" MIN. AND SHALL BE STAPLED TO MAINTAIN THIS POSITION.

4. FOR USE WITH V AND 3-PHASE ASSEMBLIES REFER TO GUIDE DRAWINGS M30-1 AND M30-2.

| ITEM | MATERIAL | ASSEMBLY UNIT | |
| | | M2-1 | M2-11 |
|---|---|---|---|
| P | CONNECTOR, COMPRESSION | AS REQ'D. | AS REQ'D. |
| AI | ROD, GROUND, ⅝" MINIMUM DIAMETER | 1 | 1 |
| AJ | CLAMP, GROUND ROD WIRE | 1 | 1 |
| AL | STAPLES, GROUND WIRE (COPPER OR STEEL TO MATCH GROUND WIRE) | AS REQ'D. | AS REQ'D. |
| AL | GROUND WIRE CLIP | 1 | |
| CJ | GROUND WIRE, MINIMUM NO. 6 COPPER OR EQUIV. CONDUCTIVITY | AS REQ'D. | AS REQ'D. |
| AV | JUMPER, STRANDED MIN. NO. 6 COPPER OR EQUIV. CONDUCTIVITY | AS REQ'D. | AS REQ'D. |
| | | | |

| | 12.5/7.2 kV GROUNDING ASSEMBLY - GROUND ROD TYPE | |
|---|---|---|
| APR., 1983 | | M2-1 TO M2-11 |

**FIGURE B.18**

# Generalized Circuit Constants

### Series Impedance

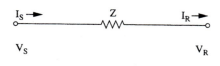

$A = 1 \qquad B = Z$

$C = 0 \qquad D = 1$

### Shunt Admittance

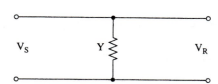

$A = 1 \qquad B = 0$

$C = Y \qquad D = 1$

### General T

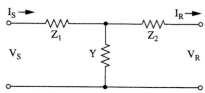

$A = 1 + YZ_1 \qquad B = Z_1 + Z_2 + YZ_1$

$C = Y \qquad\qquad D = 1 + YZ_2$

### General TT

$A = 1 + Y_2 Z \qquad\qquad B = Z$

$C = Y_1 + Y_2 + ZY_1 Y_2 \qquad D = 1 + Y_1 Z$

### Network and series impedance at receiving end

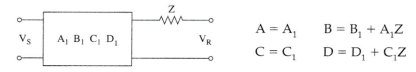

$$A = A_1 \qquad B = B_1 + A_1Z$$
$$C = C_1 \qquad D = D_1 + C_1Z$$

### Network and series impedance at sending end

$$A = A_1 + C_1Z \qquad B = B_1 + D_1Z$$
$$C = C_1 \qquad D = D_1$$

### Network with series impedance at both ends

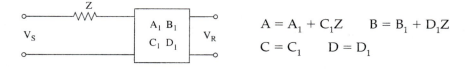

$$A = A_1 + C_1Z_S \qquad B = B_1 + A_1Z_R + D_1Z_S + C_1Z_RZ_S$$
$$C = C_1 \qquad D = D_1 + C_1Z_R$$

### Network with shunt admittance at receiving end

$$A = A_1 + B_1Y \qquad B = B_1$$
$$C = C_1 + D_1Y \qquad D = D_1$$

### Network with shunt admittance at sending end

$$A = A_1 \qquad B = B_1$$
$$C = C_1 + A_1Y \qquad D = D_1 + B_1Y$$

### Network with shunt admittance at both ends

$$A = A_1 + B_1Y_R \qquad B = B_1$$
$$C = C_1 + A_1Y_S + D_1Y_R + B_1Y_RY_S \qquad D = D_1 + B_1Y_S$$

Two networks in series

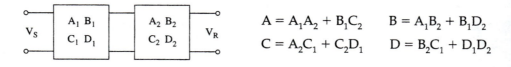

$$A = A_1A_2 + B_1C_2 \qquad B = A_1B_2 + B_1D_2$$
$$C = A_2C_1 + C_2D_1 \qquad D = B_2C_1 + D_1D_2$$

Two networks in series with intermediate impedance

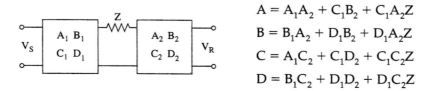

$$A = A_1A_2 + C_1B_2 + C_1A_2Z$$
$$B = B_1A_2 + D_1B_2 + D_1A_2Z$$
$$C = A_1C_2 + C_1D_2 + C_1C_2Z$$
$$D = B_1C_2 + D_1D_2 + D_1C_2Z$$

Two networks in series with intermediate admittance

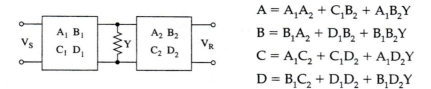

$$A = A_1A_2 + C_1B_2 + A_1B_2Y$$
$$B = B_1A_2 + D_1B_2 + B_1B_2Y$$
$$C = A_1C_2 + C_1D_2 + A_1D_2Y$$
$$D = B_1C_2 + D_1D_2 + B_1D_2Y$$

Two networks in parallel

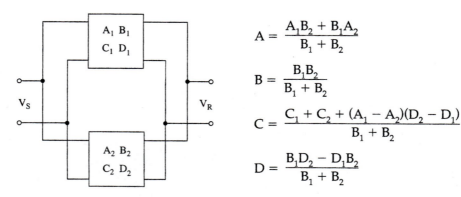

$$A = \frac{A_1B_2 + B_1A_2}{B_1 + B_2}$$

$$B = \frac{B_1B_2}{B_1 + B_2}$$

$$C = \frac{C_1 + C_2 + (A_1 - A_2)(D_2 - D_1)}{B_1 + B_2}$$

$$D = \frac{B_1D_2 - D_1B_2}{B_1 + B_2}$$

# Mathcad Solution to Example 9.11

by
Jeff Gallihugh

Problem Statement: Calculate the line power loss for 50, 100, and 200 mile lines for each line to line voltage 69 kV, 138 kV, 345 kV, and 500 kV using LINNET conductor at 50°C with phase spacing and bundle distances as shown below. The load Current is 100 amps per conductor at .8 PF lagging. The 345 kV is to use a bundle of two conductors per phase and the 500 kV is to use a bundle of three conductors per phase. Use both the medium length and long line equivalent circuits.

Common line spacing in feet:

| | | | | |
|---|---|---|---|---|
| 69 kV | 10 | 10 | 20 | |
| 138 kV | 14 | 14 | 28 | |
| 345 kV | 22 | 22 | 44 | 1.5 bundle of two spacing |
| 500 kV | 30 | 30 | 60 | 1.5 bundle of three spacing |

LINNET Conductor Characteristics

R = 56.932 micro $\Omega$/ft
radius = 0.030 ft
Ds = 0.0243 ft

Calculation of the conductor single phase inductance and single phase capacitance to neutral per unit length for a three phase transmission line requires that the equivalent GMD (Deq) and GMR for the inductance (Dsl) and capacitance

(Dsc) using the bundling distance and the phase spacing distance be determined. The following equations are used.

$$D_{eq} = (D_{ab} \cdot D_{ac} \cdot D_{bc})^{\frac{1}{3}} \qquad D_{sl} = (d^{n-1} \cdot D_s)^{\frac{1}{n}} \qquad D_{sc} = (d^{n-1} \cdot r)^{\frac{1}{n}}$$

Such that:

D = phase distance in feet
Ds = GMR in feet
r = conductor radius in feet
d = bundling distance in feet where d must be = 1 for single conductor
n = number of conductors per bundle
f = frequency of conductor in hertz
L = conductor length in miles

To calculate the per phase inductive reactance per unit length and capacitive reactance to neutral per unit length for bundled conductors up to a bundle of three the following equations are used, and are a function of Deq, Dsl, and Dsc.

For units set $\qquad$ mi := 1

$$x_l = .1213 \cdot \ln\left(\frac{D_{eq}}{D_{sl}}\right) \cdot j \cdot \frac{\Omega}{mi} \qquad\qquad x_{cn} = 2.965 \cdot 10^4 \cdot \ln\left(\frac{D_{eq}}{D_{sc}}\right) \cdot {^-}j \cdot \Omega \cdot mi$$

$$D_{eq}(D_{ab},D_{ac},D_{bc}) := (D_{ab} \cdot D_{ac} \cdot D_{bc})^{\frac{1}{3}} \cdot ft \qquad D_{sl}(d,n,D_s) := [d^{(n-1)} \cdot D_s]^{\frac{1}{n}} \cdot ft$$

$$D_{sc}(d,n,r) := [d^{(n-1)} \cdot r]^{\frac{1}{n}} \cdot ft \qquad\qquad R_{eq}(R,n) := \frac{R}{n} \cdot \frac{\Omega}{mi}$$

To calculate the inductance and capacitance per unit length of the line and the series impedance and shunt admittance per unit length the following equations are used. To avoid units errors the dependencies of the equation are in parentheses. The dependencies are not necessary for the program to work properly.

$$x_l(D_{ab},D_{ac},D_{bc},d,n,D_s) := .1213 \cdot \ln\left(\frac{D_{eq}(D_{ab},D_{ac},D_{bc})}{D_{sl}(d,n,D_s)}\right) \cdot j \cdot \frac{\Omega}{mi}$$

$$x_{cn}(D_{ab},D_{ac},D_{bc},d,n,r) := 2.965 \cdot 10^4 \cdot \ln\left(\frac{D_{eq}(D_{ab},D_{ac},D_{bc})}{D_{sc}(d,n,r)}\right) \cdot {^-}j \cdot \Omega \cdot mi$$

$$y(D_{ab},D_{ac},D_{bc},d,n,r) := \frac{1}{x_{cn}(D_{ab},D_{ac},D_{bc},d,n,r)}$$

$$z(R,D_{ab},D_{ac},D_{bc},d,n,D_s) := (R_{eq}(R,n) + x_l(D_{ab},D_{ac},D_{bc},d,n,D_s))$$

The long line equivalent circuit the generalized circuit constants need to be developed from the unit length impedance and admittance which yields the line characteristic impedance, Zc, and the line propagation constant, γ.

$$Z_c = \sqrt{\frac{z}{y}} \text{ and } \gamma = \sqrt{z \cdot y} \text{ therefore } A_a = D_d = \cosh(\gamma \cdot 1) \quad B_b = Z_c \cdot \sinh(\gamma \cdot 1) \quad C_c = \frac{\sinh(\gamma \cdot 1)}{Z_c}$$

$$Z_c(R,D_{ab},D_{ac},D_{bc},d,n,D_s,R,r) := \sqrt{\frac{z(R,D_{ab},D_{ac},D_{bc},d,n,D_s)}{y(D_{ab},D_{ac},D_{bc},d,n,r)}}$$

$$\gamma(R,D_{ab},D_{ac},D_{bc},d,n,D_s,r) := \sqrt{z(R,D_{ab},D_{ac},D_{bc},d,n,D_s) \cdot y(D_{ab},D_{ac},D_{bc},d,n,r)}$$

$$A_a(1,R,D_{ab},D_{ac},D_{bc},d,n,D_s,r) := \cosh(\gamma(R,D_{ab},D_{ac},D_{bc},d,n,D_s,r) \cdot 1)$$

$$B_b(1,R,D_{ab},D_{ac},D_{bc},d,n,D_s,R,r) := Z_c(R,D_{ab},D_{ac},D_{bc},d,n,D_s,R,r) \cdot \sinh(\gamma(R,D_{ab},D_{ac},D_{bc},d,n,D_s,r) \cdot 1)$$

$$C_c(1,R,D_{ab},D_{ac},D_{bc},d,n,D_s,r) := \frac{\sinh(\gamma(R,D_{ab},D_{ac},D_{bc},d,n,D_s,R,r) \cdot 1)}{Z_c(R,D_{ab},D_{ac},D_{bc},d,n,D_s,r)}$$

$$D_d(1,R,D_{ab},D_{ac},D_{bc},d,n,D_s,r) := \cosh(\gamma(R,D_{ab},D_{ac},D_{bc},d,n,D_s,r) \cdot 1)$$

To develop the medium length lumped parameter equivalent circuit the generalized circuit constants need to be calculated from the unit length impedance and admittance from the conductor length. The equations are:

$$Z = z \cdot 1 \qquad Y = y \cdot 1 \qquad A = D = \frac{Z \cdot Y}{2} + 1 \qquad B = Z \qquad C = Y \cdot \left(1 + \frac{Z \cdot Y}{4}\right)$$

$$Z(1,R,D_{ab},D_{ac},D_{bc},d,n,D_s) := z(R,D_{ab},D_{ac},D_{bc},d,n,D_s) \cdot 1$$

$$Y(1,D_{ab},D_{ac},D_{bc},d,n,r) := y(D_{ab},D_{ac},D_{bc},d,n,r) \cdot 1$$

$$A(1,R,D_{ab},D_{ac},D_{bc},d,n,D_s,r) := \frac{Z(1,R,D_{ab},D_{ac},D_{bc},d,n,D_s) \cdot Y(1,D_{ab},D_{ac},D_{bc},d,n,r)}{2} + 1$$

$$B(1,R,D_{ab},D_{ac},D_{bc},d,n,D_s) := Z(1,R,D_{ab},D_{ac},D_{bc},d,n,D_s)$$

$$C(1,R,D_{ab},D_{ac},D_{bc},d,n,D_s,r) := Y(1,D_{ab},D_{ac},D_{bc},d,n,r) \cdot$$
$$\left(1 + \frac{Z(1,R,D_{ab},D_{ac},D_{bc},d,n,D_s) \cdot Y(1,D_{ab},D_{ac},d,n,r)}{4}\right)$$

$$D(1,R,D_{ab},D_{ac},D_{bc},d,n,D_s,r) := \frac{Z(1,R,D_{ab},D_{ac},D_{bc},d,n,D_s) \cdot Y(1,D_{ab},D_{ac},D_{bc},d,n,r)}{2} + 1$$

To calculate the source voltage and current the following equations are used.

Long Line Equivalent

$$V_s = V_r \cdot A_a + I_r \cdot B_b$$

$$\frac{(V_s - I_r \cdot B_b)}{A_a} = V_r$$

$$I_s = I_r \cdot D_d + V_r \cdot C_c$$

$$\frac{(I_s - V_r \cdot C_c)}{D_d} = I_r$$

Medium Line Equivalent

$$V_s = V_r \cdot A + I_r \cdot B$$

$$\frac{(V_s - I_r \cdot B)}{A} = V_r$$

$$I_s = V_r \cdot C + I_r \cdot D$$

$$\frac{(I_s - V_r \cdot C)}{D} = I_r$$

Enter the conductor variables and phase spacing and bundle spacing without units to calculate conductor physical parameters. An equation test with dummy variables follows.

$$D_{ab} := 1 \quad \text{ft} \qquad D_{ac} := 1 \quad \text{ft} \qquad D_{bc} := 1 \quad \text{ft} \qquad D_s := .1 \quad \text{ft} \qquad r := .1 \quad \text{ft}$$

$$n := 1 \qquad d := 1 \quad \text{ft} \qquad f := 60 \quad \text{Hz} \qquad R := 100 \quad \Omega/\text{mile}$$

The program now calculates the line impedance, characteristic impedance, admittance, and propagation constant.

$$D_{eq}(D_{ab},D_{ac},D_{bc}) = 1.0000 \cdot \text{ft} \qquad D_{sl}(d,n,D_s) = 0.1 \cdot \text{ft} \qquad D_{sc}(d,n,r) = 0.1000 \cdot \text{ft}$$

$$x_l(D_{ab},D_{ac},D_{bc},d,n,D_s) = 0.2793j \cdot \frac{\Omega}{\text{mi}}$$

$$y(D_{ab},D_{ac},D_{bc},d,n,r) = 1.4647 \cdot 10^{-5}j \cdot \frac{\text{siemens}}{\text{mi}}$$

$$x_{cn}(D_{ab},D_{ac},D_{bc},d,n,r) = -6.8272 \cdot 10^4 j \cdot \Omega \cdot \text{mi}$$

$$\gamma(R,D_{ab},D_{ac},D_{bc},d,n,D_s,r) = 0.0270 + 0.0271j \cdot \frac{1}{\text{mi}}$$

$$z(R,D_{ab},D_{ac},D_{bc},d,n,D_s) = 100.0000 + 0.2793j \cdot \frac{\Omega}{\text{mi}}$$

$$Z_c(R,D_{ab},D_{ac},D_{bc},d,n,D_s,R,r) = 1.8502 \cdot 10^3 - 1.8450 \cdot 10^3 j \cdot \Omega$$

Input the load terminal line to line voltage, voltage phase angle, load line current, and power factor angle without units in the values shown:

$$v_{ri} := 1 \quad \text{kV} \qquad \delta_{ri} := 0 \quad \text{deg} \qquad i_{ri} := 1 \quad \text{kA} \qquad \theta_{ri} := 0$$

Calculated load terminal phase voltage

$$V_{ri} := \frac{v_{ri}}{\sqrt{3}} \cdot e^{j \cdot \delta_{ri} \cdot \text{deg}} \cdot \text{kV}$$

$$V_{ri} = 0.5774 \cdot \text{kV}$$

Calculated load phase current

$$I_{ri} := i_{ri} \cdot e^{j \cdot \theta_{ri} \cdot \text{deg}} \cdot \text{kA}$$

$$I_{ri} = 1.0000 \cdot \text{kA}$$

The long line supply voltage and current are calculated as:

$$V_{sLL}(1) := V_{ri} \cdot A_a(1,R,D_{ab},D_{ac},D_{bc},d,n,D_s,r) + I_{ri} \cdot B_b(1,R,D_{ab},D_{ac},D_{bc},d,n,D_s,R,r)$$

$$I_{sLL}(1) := I_{ri} \cdot D_d(1,R,D_{ab},D_{ac},D_{bc},d,n,D_s,r) + V_{ri} \cdot C_c(1,R,D_{ab},D_{ac},D_{bc},d,n,D_s,r)$$

The medium line supply voltage and current are calculated as:

$$V_{sML}(1) := V_{ri} \cdot A(1,R,D_{ab},D_{ac},D_{bc},d,n,D_s r) + I_{ri} \cdot B(1,R,D_{ab},D_{ac},D_{bc},d,n,D_s)$$

$$I_{sML}(1) := I_{ri} \cdot D(1,R,D_{ab},D_{ac},D_{bc},d,n,D_s r) + V_{ri} \cdot C(1,R,D_{ab},D_{ac},D_{bc},d,n,D_s,r)$$

The transmission line power loss is calculated by entering the transmission length in miles without units, and subtracting received power from sending power:

| Long Line Equivalent $\quad l := 1$ | Medium Line Equivalent |
|---|---|
| $V_{sLL}(l \cdot mi) = 100.5772 + 0.3041j \cdot kV$ | $V_{sML}(l \cdot mi) = 100.5773 + 0.2797j \cdot kV$ |
| $|V_{sLL}(l \cdot mi)| = 100.5777 \cdot kV$ | $|V_{sML}(l \cdot mi)| = 100.5777 \cdot kV$ |
| $\arg(V_{sLL}(l \cdot mi)) = 0.1733 \cdot deg$ | $\arg(V_{sML}(l \cdot mi)) = 0.1594 \cdot deg$ |
| $S_{sLL} := V_{sLL}(l \cdot mi) \cdot \overline{I_{sLL}(1)}$ | $S_{sML} := V_{sML}(l \cdot mi) \cdot \overline{I_{sML}(1)}$ |
| $S_{sLL} = 1.0058 \cdot 10^5 + 229.6278j \cdot VA$ | $S_{sML} = 1.0058 \cdot 10^5 + 205.2156j \cdot VA$ |
| $P_{sLL} := 3 \cdot Re(S_{sLL})$ | $P_{sML} := 3 \cdot Re(S_{sML})$ |
| $S_{rLL} := V_{ri} \cdot \overline{I_{ri}}$ | $S_{rML} := V_{ri} \cdot \overline{I_{ri}}$ |
| $P_{rLL} := 3 \cdot Re(S_{rLL})$ | $P_{rML} := 3 \cdot Re(S_{rML})$ |
| $P_{lossLL} := P_{sLL} - P_{rLL}$ | $P_{lossML} := P_{sML} - P_{rML}$ |
| $P_{lossLL} = 3.0000 \cdot 10^5 \cdot kW$ | $P_{lossML} = 3.0000 \cdot 10^5 \cdot kW$ |

Now the line loss will be calculated.

### 69 kV

For 69 kV load and 10 ft, 10 ft, 20 ft phase spacing and ft. bundle spacing for single conductor per phase

Enter the conductor variables and phase spacing and bundle spacing without units to calculate conductor physical parameters:

$D_{ab} := 10 \text{ ft}$ $\qquad D_{ac} := 10 \text{ ft}$ $\qquad D_{bc} := 20 \quad \text{ft}$ $\qquad D_s := 0.0243 \text{ ft}$ $\qquad r := 0.0300 \text{ ft}$

$n := 1$ $\qquad d := 1 \text{ ft}$ $\qquad f := 60 \text{ Hz}$ $\qquad R := .3006 \, \Omega/\text{mile}$

Which calculates the line impedance, characteristic impedance, admittance, and propagation constant.

$D_{eq}(D_{ab},D_{ac},D_{bc}) = 12.5992 \cdot \text{ft}$ $\qquad D_{sl}(d,n,D_s) = 0.0243 \cdot \text{ft}$ $\qquad D_{sc}(d,n,r) = 0.0300 \cdot \text{ft}$

$$x_1(D_{ab},D_{ac},D_{bc},d,n,D_s) = 0.7582j \cdot \frac{\Omega}{mi}$$

$$y(D_{ab},D_{ac},D_{bc},d,n,r) = 5.5837 \cdot 10^{-6}j \cdot \frac{siemens}{mi}$$

$$x_{cn}(D_{ab},D_{ac},D_{bc},d,n,r) = -1.7909 \cdot 10^5 j \cdot \Omega \cdot mi$$

$$\gamma(R,D_{ab},D_{ac},D_{bc},d,n,D_s,r) = 4.0036 \cdot 10^{-4} + 0.0021j \cdot \frac{1}{mi}$$

$$z(R,D_{ab},D_{ac},D_{bc},d,n,D_s) = 0.3006 + 0.7582j \cdot \frac{\Omega}{mi}$$

$$Z_c(R,D_{ab},D_{ac},D_{bc},d,n,D_s,R,r) = 375.4128 - 71.7010j \cdot \Omega$$

Input the load terminal line to line voltage, voltage phase angle, load line current, and power factor angle without units in the values shown:

$v_{ri} := 69$ kV         $\delta_{ri} := 0$ deg                    $i_{ri} := .100$ kA                    $\theta_{ri} := -36.87$ deg

Calculated load terminal phase voltage              Calculated load phase current

$$V_{ri} := \frac{v_{ri}}{\sqrt{3}} \cdot e^{j \cdot \delta_{ri} \cdot deg} \cdot kV \qquad\qquad I_{ri} := i_{ri} \cdot e^{j \cdot \theta_{ri} \cdot deg} \cdot kA$$

$V_{ri} = 39.8372 \cdot$ kV                                  $I_{ri} = 0.0800 - 0.0600j \cdot$ kA

The long line supply voltage and current are calculated as:

$$V_{sLL}(1) := V_{ri} \cdot A_a(1,R,D_{ab},D_{ac},D_{bc},d,n,D_s,r) + I_{ri} \cdot B_b(1,R,D_{ab},D_{ac},D_{bc},d,n,D_s,R,r)$$

$$I_{sLL}(1) := I_{ri} \cdot D_d(1,R,D_{ab},D_{ac},D_{bc},d,n,D_s,r) + V_{ri} \cdot C_c(1,R,D_{ab},D_{ac},D_{bc},d,n,D_s,r)$$

The medium line supply voltage and current are calculated as:

$$V_{sML}(1) := V_{ri} \cdot A(1,R,D_{ab},D_{ac},D_{bc},d,n,D_s,r) + I_{ri} \cdot B(1,R,D_{ab},D_{ac},D_{bc},d,n,D_s)$$

$$I_{sML}(1) := I_{ri} \cdot D(1,R,D_{ab},D_{ac},D_{bc},d,n,D_s,r) + V_{ri} \cdot C(1,R,D_{ab},D_{ac},D_{bc},d,n,D_s,r)$$

The transmission line power loss is calculated by entering the transmission length in miles without units as:

At 50 miles

Long Line Equivalent            $l := 50$          Medium Line Equivalent

$V_{sLL}(l \cdot mi) = 43.0960 + 2.2132j \cdot$ kV        $V_{sML}(l \cdot mi) = 43.1035 + 2.2147j \cdot$ kV

$|V_{sLL}(l \cdot mi)| = 43.1528 \cdot$ kV              $|V_{sML}(l \cdot mi)| = 43.1603 \cdot$ kV

$arg(V_{sLL}(l \cdot mi)) = 2.9399 \cdot$ deg          $arg(V_{sML}(l \cdot mi)) = 2.9414 \cdot$ deg

$S_{sLL} := V_{sLL}(l \cdot mi) \cdot \overline{I_{sLL}(1)}$              $S_{sML} := V_{sML}(l \cdot mi) \cdot \overline{I_{sML}(1)}$

$S_{sLL} = 3.3274 \cdot 10^3 + 2.2628 \cdot 10^3 j \cdot$ VA        $S_{sML} = 3.3277 \cdot 10^3 + 2.2637 \cdot 10^3 j \cdot$ VA

$P_{sLL} := 3 \cdot Re(S_{sLL})$

$S_{rLL} := V_{ri} \cdot \overline{I_{ri}}$

$P_{rLL} := 3 \cdot Re(S_{rLL})$

$P_{lossLL} := P_{sLL} - P_{rLL}$

$P_{lossLL} = 421.2125 \cdot kW$

$P_{sML} := 3 \cdot Re(S_{sML})$

$S_{rML} := V_{ri} \cdot \overline{I_{ri}}$

$P_{rML} := 3 \cdot Re(S_{rML})$

$P_{lossML} := P_{sML} - P_{rML}$

$P_{lossML} = 422.2049 \cdot kW$

### At 100 miles

$$l := 100$$

$V_{sLL}(1 \cdot mi) = 45.8897 + 4.5836j \cdot kV$

$|V_{sLL}(1 \cdot mi)| = 46.1181 \cdot kV$

$arg(V_{sLL}(1 \cdot mi)) = 5.7039 \cdot deg$

$S_{sLL} := V_{sLL}(1 \cdot mi) \cdot \overline{I_{sLL}(1)}$

$S_{sLL} = 3.4489 \cdot 10^3 + 2.0121 \cdot 10^3 j \cdot VA$

$P_{sLL} := 3 \cdot Re(S_{sLL})$

$S_{rLL} := V_{ri} \cdot \overline{I_{ri}}$

$P_{rLL} := 3 \cdot Re(S_{rLL})$

$P_{lossLL} := P_{sLL} - P_{rLL}$

$P_{lossLL} = 785.7140 \cdot kW$

$V_{sML}(1 \cdot mi) = 45.9481 + 4.5966j \cdot kV$

$|V_{sML}(1 \cdot mi)| = 46.1774 \cdot kV$

$arg(V_{sML}(1 \cdot mi)) = 5.7128 \cdot deg$

$S_{sML} := V_{sML}(1 \cdot mi) \cdot \overline{I_{sML}(1)}$

$S_{sML} = 3.4512 \cdot 10^3 + 2.0183 \cdot 10^3 j \cdot VA$

$P_{sML} := 3 \cdot Re(S_{sML})$

$S_{rML} := V_{ri} \cdot \overline{I_{ri}}$

$P_{rML} := 3 \cdot Re(S_{rML})$

$P_{lossML} := P_{sML} - P_{rML}$

$P_{lossML} = 792.5970 \cdot kW$

### At 200 miles

$$l := 200$$

$V_{sLL}(1 \cdot mi) = 49.9288 + 9.7385j \cdot kV$

$|V_{sLL}(1 \cdot mi)| = 50.8697 \cdot kV$

$arg(V_{sLL}(1 \cdot mi) = 11.0369 \cdot deg$

$S_{sLL} := V_{sLL}(1 \cdot mi) \cdot \overline{I_{sLL}(1)}$

$S_{sLL} = 3.6445 \cdot 10^3 + 1.1840 \cdot 10^3 j \cdot VA$

$P_{sLL} := 3 \cdot Re(S_{sLL})$

$S_{rLL} := V_{ri} \cdot \overline{I_{ri}}$

$P_{rLL} := 3 \cdot Re(S_{rLL})$

$P_{lossLL} := P_{sLL} - P_{rLL}$

$P_{lossLL} = 1.3726 \cdot 10^3 \cdot kW$

$V_{sML}(1 \cdot mi) = 50.3724 + 9.8619j \cdot kV$

$|V_{sML}(1 \cdot mi)| = 51.3287 \cdot kV$

$arg(V_{sML}(1 \cdot mi)) = 11.0772 \cdot deg$

$S_{sML} := V_{sML}(1 \cdot mi) \cdot \overline{I_{sML}(1)}$

$S_{sML} = 3.6574 \cdot 10^3 + 1.2197 \cdot 10^3 j \cdot VA$

$P_{sML} := 3 \cdot Re(S_{sML})$

$S_{rML} := V_{ri} \cdot \overline{I_{ri}}$

$P_{rML} := 3 \cdot Re(S_{rML})$

$P_{lossML} := P_{sML} - P_{rML}$

$P_{lossML} = 1.4114 \cdot 10^3 \cdot kW$

## 138 kV

For 138 kV load and 14 ft, 14 ft, 28 ft phase spacing for single conductor per phase

Enter the conductor variables and phase spacing and bundle spacing without units to calculate conductor physical parameters:

$D_{ab} := 14$ ft    $D_{ac} := 14$ ft    $D_{bc} := 28$ ft    $D_s := 0.0243$ ft    $r := 0.0600$ ft

$n := 1$       $d := 1$ ft          $f := 60$ Hz          $R := .3006\ \Omega/\text{mile}$

Which calculates the line impedance, characteristic impedance, admittance, and propagation constant.

$D_{eq}(D_{ab},D_{ac},D_{bc}) = 17.6389 \cdot$ ft    $D_{sl}(d,n,D_s) = 0.0243 \cdot$ ft    $D_{sc}(d,n,r) = 0.0300 \cdot$ ft

$$x_l(D_{ab},D_{ac},D_{bc},d,n,D_s) = 0.7990j \cdot \frac{\Omega}{\text{mi}}$$

$$y(D_{ab},D_{ac},D_{bc},d,n,r) = 5.2891 \cdot 10^{-6}j \cdot \frac{\text{siemens}}{\text{mi}}$$

$$x_{cn}(D_{ab},D_{ac},D_{bc},d,n,r) = -1.8907 \cdot 10^5 j \cdot \Omega \cdot \text{mi}$$

$$\gamma(R,D_{ab},D_{ac},D_{bc},d,n,D_s,r) = 3.8024 \cdot 10^{-4} + 0.0021j \cdot \frac{1}{\text{mi}}$$

$$z(R,D_{ab},D_{ac},D_{bc},d,n,D_s) = 0.3006 + 0.7990j \cdot \frac{\Omega}{\text{mi}}$$

$$Z_c(R,D_{ab},D_{ac},D_{bc},d,n,D_s,R,r) = 395.2761 - 71.8914j \cdot \Omega$$

Input the load terminal line to line voltage, voltage phase angle, load line current, and power factor angle without units in the values shown:

$v_{ri} := 138$ kV          $\delta_{ri} := 0$ deg          $i_{ri} := .100$ kA          $\theta_{ri} := -36.87$ deg

Calculated load terminal phase voltage          Calculated load phase current

$$V_{ri} := \frac{v_{ri}}{\sqrt{3}} \cdot e^{j \cdot \delta_{ri} \cdot \text{deg}} \cdot \text{kV}$$          $$I_{ri} := i_{ri} \cdot e^{j \cdot \theta_{ri} \cdot \text{deg}} \cdot \text{kA}$$

$V_{ri} = 79.6743 \cdot$ kV          $I_{ri} = 0.0800 - 0.0600j \cdot$ kA

The long line supply voltage and current are calculated as:

$$V_{sLL}(1) := V_{ri} \cdot A_a(1,R,D_{ab},D_{ac},D_{bc},d,n,D_s,r) + I_{ri} \cdot B_b(1,R,D_{ab},D_{ac},D_{bc},d,n,D_s,R,r)$$

$$I_{sLL}(1) := I_{ri} \cdot D_d(1,R,D_{ab},D_{ac},D_{bc},d,n,D_s,r) + V_{ri} \cdot C_c(1,R,D_{ab},D_{ac},D_{bc},d,n,D_s,r)$$

The medium line supply voltage and current are calculated as:

$$V_{sML}(1) := V_{ri} \cdot A(1,R,D_{ab},D_{ac},D_{bc},d,n,D_s,r) + I_{ri} \cdot B(1,R,D_{ab},D_{ac},D_{bc},d,n,D_s)$$

$$I_{sML}(1) := I_{ri} \cdot D(1,R,D_{ab},D_{ac},D_{bc},d,n,D_s,r) + V_{ri} \cdot C(1,R,D_{ab},D_{ac},D_{bc},d,n,D_s,r)$$

The transmission line power loss is calculated by entering the transmission length in miles without units as:

At 50 miles

$l := 50$

**Long Line Equivalent**

$V_{sLL}(l \cdot mi) = 82.8454 + 2.4508j \cdot kV$

$|V_{sLL}(l \cdot mi)| = 82.8817 \cdot kV$

$arg(V_{sLL}(l \cdot mi)) = 1.6945 \cdot deg$

$S_{sLL} := V_{sLL}(l \cdot mi) \cdot \overline{I_{sLL}(1)}$

$S_{sLL} = 6.5070 \cdot 10^3 + 3.3841 \cdot 10^3 j \cdot VA$

$P_{sLL} := 3 \cdot Re(S_{sLL})$

$S_{rLL} := V_{ri} \cdot \overline{I_{ri}}$

$P_{rLL} := 3 \cdot Re(S_{rLL})$

$P_{lossLL} := P_{sLL} - P_{rLL}$

$P_{lossLL} = 399.2212 \cdot kW$

**Medium Line Equivalent**

$V_{sML}(l \cdot mi) = 82.8530 + 2.4527j \cdot kV$

$|V_{sML}(l \cdot mi)| = 82.8893 \cdot kV$

$arg(V_{sML}(l \cdot mi)) = 1.6957 \cdot deg$

$S_{sML} := V_{sML}(l \cdot mi) \cdot \overline{I_{sML}(1)}$

$S_{sML} = 6.5069 \cdot 10^3 + 3.3861 \cdot 10^3 j \cdot VA$

$P_{sML} := 3 \cdot Re(S_{sML})$

$S_{rML} := V_{ri} \cdot \overline{I_{ri}}$

$P_{rML} := 3 \cdot Re(S_{rML})$

$P_{lossML} := P_{sML} - P_{rML}$

$P_{lossML} = 398.9008 \cdot kW$

At 100 miles

$l := 100$

$V_{sLL}(l \cdot mi) = 85.1322 + 5.2044j \cdot kV$

$|V_{sLL}(l \cdot mi)| = 85.2911 \cdot kV$

$arg(V_{sLL}(l \cdot mi)) = 3.4983 \cdot deg$

$S_{sLL} := V_{sLL}(l \cdot mi) \cdot \overline{I_{sLL}(1)}$

$S_{sLL} = 6.6133 \cdot 10^3 + 1.7937 \cdot 10^3 j \cdot VA$

$P_{sLL} := 3 \cdot Re(S_{sLL})$

$S_{rLL} := V_{ri} \cdot \overline{I_{ri}}$

$P_{rLL} := 3 \cdot Re(S_{rLL})$

$P_{lossLL} := P_{sLL} - P_{rLL}$

$P_{lossLL} = 718.1223 \cdot kW$

$V_{sML}(l \cdot mi) = 85.1898 + 5.2222j \cdot kV$

$|V_{sML}(l \cdot mi)| = 85.3497 \cdot kV$

$arg(V_{sML}(l \cdot mi)) = 3.5079 \cdot deg$

$S_{sML} := V_{sML}(l \cdot mi) \cdot \overline{I_{sML}(1)}$

$S_{sML} = 6.6119 \cdot 10^3 + 1.8078 \cdot 10^3 j \cdot VA$

$P_{sML} := 3 \cdot Re(S_{sML})$

$S_{rML} := V_{ri} \cdot \overline{I_{ri}}$

$P_{rML} := 3 \cdot Re(S_{rML})$

$P_{lossML} := P_{sML} - P_{rML}$

$P_{lossML} = 713.8214 \cdot kW$

At 200 miles

$l := 200$

$V_{sLL}(1 \cdot mi) = 86.9208 + 11.5336j \cdot kV$     $V_{sML}(1 \cdot mi) = 87.3381 + 11.7111j \cdot kV$

$|V_{sLL}(1 \cdot mi)| = 87.6827 \cdot kV$     $|V_{sML}(1 \cdot mi)| = 88.1197 \cdot kV$

$arg(V_{sLL}(1 \cdot mi)) = 7.5585 \cdot deg$     $arg(V_{sML}(1 \cdot mi)) = 7.6372 \cdot deg$

$S_{sLL} := V_{sLL}(1 \cdot mi) \cdot \overline{I_{sLL}(1)}$     $S_{sML} := V_{sML}(1 \cdot mi) \cdot \overline{I_{sML}(1)}$

$S_{sLL} = 6.7970 \cdot 10^3 + 1.6992 \cdot 10^3 j \cdot VA$     $S_{sML} = 6.7779 \cdot 10^3 - 1.6103 \cdot 10^3 j \cdot VA$

$P_{sLL} := 3 \cdot Re(S_{sLL})$     $P_{sML} := 3 \cdot Re(S_{sML})$

$S_{rLL} := V_{ri} \cdot \overline{I_{ri}}$     $S_{rML} := V_{ri} \cdot \overline{I_{ri}}$

$P_{rLL} := 3 \cdot Re(S_{rLL})$     $P_{rML} := 3 \cdot Re(S_{rML})$

$P_{lossLL} := P_{sLL} - P_{rLL}$     $P_{lossML} := P_{sML} - P_{rML}$

$P_{lossLL} = 1.2692 \cdot 10^3 \cdot kW$     $P_{lossML} = 1.2118 \cdot 10^3 \cdot kW$

### 345 kV

For 345 kV load and 22 ft, 22 ft, 44 ft phase spacing and 1.5 ft bundle spacing for two conductor bundle per phase

Enter the conductor variables and phase spacing and bundle spacing without units to calculate conductor physical parameters:

$D_{ab} := 22$ ft    $D_{ac} := 22$ ft    $D_{bc} := 44$ ft    $D_s := 0.0243$ ft    $r := 0.0300$ ft

     $n := 2$    $d := 1.5$ ft    $f := 60$ Hz    $R := .3006 \, \Omega/mile$

Which calculates the line impedance, characteristic impedance, admittance, and propagation constant.

$D_{eq}(D_{ab},D_{ac},D_{bc}) = 27.7183 \cdot ft$    $D_{sl}(d,n,D_s) = 0.1909 \cdot ft$    $D_{sc}(d,n,r) = 0.2121 \cdot ft$

$x_l(D_{ab},D_{ac},D_{bc},d,n,D_s) = 0.6038j \cdot \dfrac{\Omega}{mi}$

$y(D_{ab},D_{ac},D_{bc},d,n,r) = 6.9217 \cdot 10^{-6} j \cdot \dfrac{siemens}{mi}$

$x_{cn}(D_{ab},D_{ac},D_{bc},d,n,r) = -1.4447 \cdot 10^5 j \cdot \Omega \cdot mi$

$\gamma(R,D_{ab},D_{ac},D_{bc},d,n,D_s,r) = 2.5252 \cdot 10^{-4} + 0.0021j \cdot \dfrac{1}{mi}$

$z(R,D_{ab},D_{ac},D_{bc},d,n,D_s) = 0.1503 + 0.6038j \cdot \dfrac{\Omega}{mi}$

$Z_c(R,D_{ab},D_{ac},D_{bc},d,n,D_s,R,r) = 297.6049 - 36.4819j \cdot \Omega$

Input the load terminal line to line voltage, voltage phase angle, load line current, and power factor angle without units in the values shown:

$v_{ri} := 345 \text{ kV}$          $\delta_{ri} := 0 \text{ deg}$          $i_{ri} := .100 \text{ kA}$          $\theta_{ri} := {}^{-}36.87 \text{ deg}$

Calculated load terminal phase voltage                Calculated load phase current

$$V_{ri} := \frac{v_{ri}}{\sqrt{3}} \cdot e^{j \cdot \delta_{ri} \cdot \text{deg}} \cdot \text{kV} \qquad\qquad I_{ri} := i_{ri} \cdot e^{j \cdot \theta_{ri} \cdot \text{deg}} \cdot \text{kA}$$

$V_{ri} = 199.1858 \cdot \text{kV}$                              $I_{ri} = 0.0800 - 0.0600j \cdot \text{kA}$

The long line supply voltage and current are calculated as:

$$V_{sLL}(1) := V_{ri} \cdot A_a(1,R,D_{ab},D_{ac},D_{bc},d,n,D_s,r) + I_{ri} \cdot B_b(1,R,D_{ab},D_{ac},D_{bc},d,n,D_s,R,r)$$

$$I_{sLL}(1) := I_{ri} \cdot D_d(1,R,D_{ab},D_{ac},D_{bc},d,n,D_s,r) + V_{ri} \cdot C_c(1,R,D_{ab},D_{ac},D_{bc},d,n,D_s,r)$$

The medium line supply voltage and current are calculated as:

$$V_{sML}(1) := V_{ri} \cdot A(1,R,D_{ab},D_{ac},D_{bc},d,n,D_s,r) + I_{ri} \cdot B(1,R,D_{ab},D_{ac},D_{bc},d,n,D_s)$$

$$I_{sML}(1) := I_{ri} \cdot D(1,R,D_{ab},D_{ac},D_{bc},d,n,D_s,r) + V_{ri} \cdot C(1,R,D_{ab},D_{ac},D_{bc},d,n,D_s,r)$$

The transmission line power loss is calculated by entering the transmission length in miles without units as:

### At 50 miles

Long Line Equivalent          $1 := 50$          Medium Line Equivalent

$V_{sLL}(1 \cdot \text{mi}) = 200.5537 + 2.2206j \cdot \text{kV}$          $V_{sML}(1 \cdot \text{mi}) = 200.5579 + 2.2234j \cdot \text{kV}$

$|V_{sLL}(1 \cdot \text{mi})| = 200.5660 \cdot \text{kV}$          $|V_{sML}(1 \cdot \text{mi})| = 200.5702 \cdot \text{kV}$

$\arg(V_{sLL}(1 \cdot \text{mi})) = 0.6344 \cdot \text{deg}$          $\arg(V_{sML}(1 \cdot \text{mi})) = 0.6352 \cdot \text{deg}$

$S_{sLL} := V_{sLL}(1 \cdot \text{mi}) \cdot \overline{I_{sLL}(1)}$          $S_{sML} := V_{sML}(1 \cdot \text{mi}) \cdot \overline{I_{sML}(1)}$

$S_{sLL} = 1.5991 \cdot 10^4 - 1.6747 \cdot 10^3 j \cdot \text{VA}$          $S_{sML} = 1.5988 \cdot 10^4 - 1.6625 \cdot 10^3 j \cdot \text{VA}$

$P_{sLL} := 3 \cdot \text{Re}(S_{sLL})$          $P_{sML} := 3 \cdot \text{Re}(S_{sML})$

$S_{rLL} := V_{ri} \cdot \overline{I_{ri}}$          $S_{rML} := V_{ri} \cdot \overline{I_{ri}}$

$P_{rLL} := 3 \cdot \text{Re}(S_{rLL})$          $P_{rML} := 3 \cdot \text{Re}(S_{rML})$

$P_{lossLL} := P_{sLL} - P_{rLL}$          $P_{lossML} := P_{sML} - P_{rML}$

$P_{lossLL} = 167.4330 \cdot \text{kW}$          $P_{lossML} = 158.9851 \cdot \text{kW}$

### At 100 miles

$$1 := 100$$

$V_{sLL}(1 \cdot \text{mi}) = 199.8220 + 4.9387j \cdot \text{kV}$          $V_{sML}(1 \cdot \text{mi}) = 199.8487 + 4.9649j \cdot \text{kV}$

$|V_{sLL}(1 \cdot \text{mi})| = 199.8830 \cdot \text{kV}$          $|V_{sML}(1 \cdot \text{mi})| = 199.9104 \cdot \text{kV}$

$\arg(V_{sLL}(1 \cdot \text{mi})) = 1.4158 \cdot \text{deg}$          $\arg(V_{sML}(1 \cdot \text{mi})) = 1.4231 \cdot \text{deg}$

$$S_{sLL} := V_{sLL}(1 \cdot mi) \cdot \overline{I_{sLL}(1)}$$
$$S_{sLL} = 1.6055 \cdot 10^4 - 1.5314 \cdot 10^4 j \cdot VA$$
$$P_{sLL} := 3 \cdot Re(S_{sLL})$$
$$S_{rLL} := V_{ri} \cdot \overline{I_{ri}}$$
$$P_{rLL} := 3 \cdot Re(S_{rLL})$$
$$P_{lossLL} := P_{sLL} - P_{rLL}$$
$$P_{lossLL} = 361.0161 \cdot kW$$

$$S_{sML} := V_{sML}(1 \cdot mi) \cdot \overline{I_{sML}(1)}$$
$$S_{sML} = 1.6032 \cdot 10^4 - 1.5219 \cdot 10^4 j \cdot VA$$
$$P_{sML} := 3 \cdot Re(S_{sML})$$
$$S_{rML} := V_{ri} \cdot \overline{I_{ri}}$$
$$P_{rML} := 3 \cdot Re(S_{rML})$$
$$P_{lossML} := P_{sML} - P_{rML}$$
$$P_{lossML} = 292.1748 \cdot kW$$

### At 200 miles

$$l := 200$$

$$V_{sLL}(1 \cdot mi) = 192.0827 + 11.7361j \cdot kV$$
$$|V_{sLL}(1 \cdot mi)| = 192.4409 \cdot kV$$
$$arg(V_{sLL}(1 \cdot mi)) = 3.4964 \cdot deg$$
$$S_{sLL} := V_{sLL}(1 \cdot mi) \cdot \overline{I_{sLL}(1)}$$
$$S_{sLL} = 1.6490 \cdot 10^4 - 4.0384 \cdot 10^4 j \cdot VA$$
$$P_{sLL} := 3 \cdot Re(S_{sLL})$$
$$S_{rLL} := V_{ri} \cdot \overline{I_{ri}}$$
$$P_{rLL} := 3 \cdot Re(S_{rLL})$$
$$P_{lossLL} := P_{sLL} - P_{rLL}$$
$$P_{lossLL} = 1.6664 \cdot 10^3 \cdot kW$$

$$V_{sML}(1 \cdot mi) = 192.1866 + 12.0021j \cdot kV$$
$$|V_{sML}(1 \cdot mi)| = 192.5610 \cdot kV$$
$$arg(V_{sML}(1 \cdot mi)) = 3.5735 \cdot deg$$
$$S_{sML} := V_{sML}(1 \cdot mi) \cdot \overline{I_{sML}(1)}$$
$$S_{sML} = 1.6310 \cdot 10^4 - 3.9671 \cdot 10^4 j \cdot VA$$
$$P_{sML} := 3 \cdot Re(S_{sML})$$
$$S_{rML} := V_{ri} \cdot \overline{I_{ri}}$$
$$P_{rML} := 3 \cdot Re(S_{rML})$$
$$P_{lossML} := P_{sML} - P_{rML}$$
$$P_{lossML} = 1.1240 \cdot 10^3 \cdot kW$$

### 500 kV

For 500 kV load and 30 ft, 30 ft, 60 ft phase spacing and 1.5 ft bundle spacing for a three conductor bundle per phase

Therefore enter the conductor variables and phase spacing and bundle spacing without units to calculate conductor physical parameters:

$$D_{ab} := 30 \text{ ft} \qquad D_{ac} := 30 \text{ ft} \qquad D_{bc} := 60 \text{ ft} \qquad D_s := 0.0243 \text{ ft} \qquad r := 0.0300 \text{ ft}$$
$$n := 3 \qquad\qquad d := 1.5 \text{ ft} \qquad\qquad f := 60 \text{ Hz} \qquad\qquad R := .3006 \ \Omega/mile$$

Which calculates the line impedance, characteristic impedance, admittance, and propagation constant.

$$D_{eq}(D_{ab}, D_{ac}, D_{bc}) = 37.7976 \cdot ft \qquad D_{sl}(d,n,D_s) = 0.3795 \cdot ft \qquad D_{sc}(d,n,r) = 0.4072 \cdot ft$$

$$x_1(D_{ab},D_{ac},D_{bc},d,n,D_s) = 0.5581j \cdot \frac{\Omega}{mi}$$

$$y(D_{ab},D_{ac},D_{bc},d,n,r) = 7.4439 \cdot 10^{-6}j \cdot \frac{siemens}{mi}$$

$$x_{cn}(D_{ab},D_{ac},D_{bc},d,n,r) = -1.3434 \cdot 10^{5}j \cdot \Omega \cdot mi$$

$$\gamma(R,D_{ab},D_{ac},D_{bc},d,n,D_s,r) = 1.8224 \cdot 10^{-4} + 0.0020j \cdot \frac{1}{mi}$$

$$z(R,D_{ab},D_{ac},D_{bc},d,n,D_s) = 0.1002 + 0.5581j \cdot \frac{\Omega}{mi}$$

$$Z_c(R,D_{ab},D_{ac},D_{bc},d,n,D_s,R,r) = 274.9073 - 24.4822j \cdot \Omega$$

Input the load terminal line to line voltage, voltage phase angle, load line current, and power factor angle without units in the values shown:

$v_{ri} := 500$ kV           $\delta_{ri} := 0$ deg           $i_{ri} := .100$ kA           $\theta_{ri} := {}^-36.87$ deg

Calculated load terminal phase voltage          Calculated load phase current

$$V_{ri} := \frac{v_{ri}}{\sqrt{3}} \cdot e^{j \cdot \delta_{ri} \cdot deg} \cdot kV$$                $$I_{ri} := i_{ri} \cdot e^{j \cdot \theta_{ri} \cdot deg} \cdot kA$$

$V_{ri} = 288.6751 \cdot kV$                          $I_{ri} = 0.0800 - 0.0600j \cdot kA$

The long line supply voltage and current are calculated as:

$$V_{sLL}(1) := V_{ri} \cdot A_a(1,R,D_{ab},D_{ac},D_{bc},d,n,D_s,r) + I_{ri} \cdot B_b(1,R,D_{ab},D_{ac},D_{bc},d,n,D_s,R,r)$$

$$I_{sLL}(1) := I_{ri} \cdot D_d(1,R,D_{ab},D_{ac},D_{bc},d,n,D_s,r) + V_{ri} \cdot C_c(1,R,D_{ab},D_{ac},D_{bc},d,n,D_s,r)$$

The medium line supply voltage and current are calculated as:

$$V_{sML}(1) := V_{ri} \cdot A(1,R,D_{ab},D_{ac},D_{bc},d,n,D_s,r) + I_{ri} \cdot B(1,R,D_{ab},D_{ac},D_{bc},d,n,D_s)$$

$$I_{sML}(1) := I_{ri} \cdot D(1,R,D_{ab},D_{ac},D_{bc},d,n,D_s,r) + V_{ri} \cdot C(1,R,D_{ab},D_{ac},D_{bc},d,n,D_s,r)$$

The transmission line power loss is calculated by entering the transmission length in miles without units as:

At 50 miles

Long Line Equivalent           $1 := 50$           Medium Line Equivalent

$V_{sLL}(1 \cdot mi) = 289.2482 + 2.1978j \cdot kV$        $V_{sML}(1 \cdot mi) = 289.2511 + 2.2010j \cdot kV$

$|V_{sLL}(1 \cdot mi)| = 289.2565 \cdot kV$               $|V_{sML}(1 \cdot mi)| = 289.2595 \cdot kV$

$arg(V_{sLL}(1 \cdot mi)) = 0.4353 \cdot deg$              $arg(V_{sML}(1 \cdot mi)) = 0.4360 \cdot deg$

$S_{sLL} := V_{sLL}(1 \cdot mi) \cdot \overline{I_{sLL}(1)}$               $S_{sML} := V_{sML}(1 \cdot mi) \cdot \overline{I_{sML}(1)}$

$S_{sLL} = 2.3131 \cdot 10^4 - 1.3606 \cdot 10^4 j \cdot VA$     $S_{sML} = 2.3126 \cdot 10^4 - 1.3579 \cdot 10^4 j \cdot VA$

$P_{sLL} := 3 \cdot Re(S_{sLL})$                  $P_{sML} := 3 \cdot Re(S_{sML})$

$S_{rLL} := V_{ri} \cdot \overline{I_{ri}}$                      $S_{rML} := V_{ri} \cdot \overline{I_{ri}}$

$P_{rLL} := 3 \cdot Re(S_{rLL})$                  $P_{rML} := 3 \cdot Re(S_{rML})$

$P_{lossLL} := P_{sLL} - P_{rLL}$            $P_{lossML} := P_{sML} - P_{rML}$

$P_{lossLL} = 110.9790 \cdot kW$           $P_{lossML} = 96.7842 \cdot kW$

### At 100 miles

$$1 := 100$$

$V_{sLL}(1 \cdot mi) = 286.8155 + 4.9112j \cdot kV$    $V_{sML}(1 \cdot mi) = 286.8289 + 4.9402j \cdot kV$

$|V_{sLL}(1 \cdot mi)| = 286.8575 \cdot kV$         $|V_{sML}(1 \cdot mi)| = 286.8714 \cdot kV$

$arg(V_{sLL}(1 \cdot mi)) = 0.9810 \cdot deg$      $arg(V_{sML}(1 \cdot mi)) = 0.9867 \cdot deg$

$S_{sLL} := V_{sLL}(1 \cdot mi) \cdot \overline{I_{sLL}(1)}$        $S_{sML} := V_{sML}(1 \cdot mi) \cdot \overline{I_{sML}(1)}$

$S_{sLL} = 2.3219 \cdot 10^4 - 4.4055 \cdot 10^4 j \cdot VA$   $S_{sML} = 2.3181 \cdot 10^4 - 4.3843 \cdot 10^4 j \cdot VA$

$P_{sLL} := 3 \cdot Re(S_{sLL})$                  $P_{sML} := 3 \cdot Re(S_{sML})$

$S_{rLL} := V_{ri} \cdot \overline{I_{ri}}$                      $S_{rML} := V_{ri} \cdot \overline{I_{ri}}$

$P_{rLL} := 3 \cdot Re(S_{rLL})$                  $P_{rML} := 3 \cdot Re(S_{rML})$

$P_{lossLL} := P_{sLL} - P_{rLL}$            $P_{lossML} := P_{sML} - P_{rML}$

$P_{lossLL} = 373.7367 \cdot kW$          $P_{lossML} = 260.0452 \cdot kW$

### At 200 miles

$$1 := 200$$

$V_{sLL}(1 \cdot mi) = 273.0436 + 11.7436j \cdot kV$    $V_{sML}(1 \cdot mi) = 272.9897 + 12.0336j \cdot kV$

$|V_{sLL}(1 \cdot mi)| = 273.2961 \cdot kV$         $|V_{sML}(1 \cdot mi)| = 273.2548 \cdot kV$

$arg(V_{sLL}(1 \cdot mi)) = 2.4628 \cdot deg$      $arg(V_{sML}(1 \cdot mi)) = 2.5240 \cdot deg$

$S_{sLL} := V_{sLL}(1 \cdot mi) \cdot \overline{I_{sLL}(1)}$        $S_{sML} := V_{sML}(1 \cdot mi) \cdot \overline{I_{sML}(1)}$

$S_{sLL} = 2.3992 \cdot 10^4 - 9.8552 \cdot 10^4 j \cdot VA$   $S_{sML} = 2.3703 \cdot 10^4 - 9.6902 \cdot 10^4 j \cdot VA$

$P_{sLL} := 3 \cdot Re(S_{sLL})$                  $P_{sML} := 3 \cdot Re(S_{sML})$

$S_{rLL} := V_{ri} \cdot \overline{I_{ri}}$                      $S_{rML} := V_{ri} \cdot \overline{I_{ri}}$

$P_{rLL} := 3 \cdot Re(S_{rLL})$                  $P_{rML} := 3 \cdot Re(S_{rML})$

$P_{lossLL} := P_{sLL} - P_{rLL}$            $P_{lossML} := P_{sML} - P_{rML}$

$$P_{lossLL} = 2.6932 \cdot 10^3 \cdot kW \qquad\qquad P_{lossML} = 1.8270 \cdot 10^3 \cdot kW$$

Note from authors:

Notice first that as the lines get longer the long line equivalent circuit yields answers that depart significantly from medium length line approximations. Otherwise the long line equivalent circuit would not be necessary.

Notice also that the losses at 200 miles at 345 kV and 500 kV at a receiving current of 100 A are higher than the losses at 200 miles at lower voltages. The additional capacitance at the higher voltages with bundled conductors causes incremental current increases along the line that cause higher losses. Doing the problem over with the receiving current at 300 A (a simple procedure using the Mathcad program, just change the value of $I_{receiving}$) results in losses that are lowest at the highest voltages at any distance in the problem as the table below shows. 300 A is too high for this conductor at the lower voltages, but this illustrates the principle.

Line losses with the receiving current = 300 A

| Voltage | Line length (mil) | Long Line Equivalent (kW) | Medium Line Equivalent (kW) |
|---------|-------------------|---------------------------|-----------------------------|
| 69 kV   | 50                | 3965                      | 3970                        |
|         | 100               | 7663                      | 7666                        |
|         | 200               | 14120                     | 14877                       |
| 138 kV  | 50                | 3880                      | 3892                        |
|         | 100               | 7838                      | 7472                        |
|         | 200               | 13215                     | 13817                       |
| 345 kV  | 50                | 1779                      | 1776                        |
|         | 100               | 3184                      | 3153                        |
|         | 200               | 5698                      | 5354                        |
| 500 kV  | 50                | 1116                      | 1105                        |
|         | 100               | 1982                      | 1890                        |
|         | 200               | 4333                      | 3536                        |

LMF & WC

# APPENDIX E

# Reference List

1. Anderson, P. M. & Fouad, A. A. (1977). *Power System Control and Stability.* Ames, IA: Iowa State University Press.

2. *Applied Protective Relaying.* (1982). Coral Springs, FL: Westinghouse Electric Corporation, Relay-Instrument Division.

3. Brown, D. C. (1980). *Electricity For Rural America, The Fight For the REA.* Westport, CT: Greenwood Press.

4. Buchanan, R. A. (1965). *Technology and Social Progress.* Oxford, England: Pergamon Press.

5. Conn, R. W. (1983, Oct.). The Engineering of Magnetic Fusion Reactors. *Scientific American. 249* (4) pp. 60–71.

6. Cook, V. (1985). *Analysis of Distance Protection.* Letchworth, Herts, England: Research Studies Press.

7. Craxton, S. R., NcCrory, R. L., & Soures, J. M. (1986, Aug.). Progress in Laser Fusion. *Scientific American. 255* (2) pp. 68–79.

8. *Design Guide for Rural Substations, REA Bulletin 65-1.* (1978, June). Rural Electrification Administration, U.S. Department of Agriculture.

9. *Design Guide for High Voltage Transmission Lines, REA Bulletin 62-1.* (1981, Dec.). Rural Electrification Administration, U.S. Department of Agriculture.

10. *Distribution System Protection Manual.* Cannonsburg, PA: McGraw Edison Co.

11. Duboff, R. B. (1979). *Electric Power in American Manufacturing, 1889–1958.* New York, NY: Arno Press.

12. Eaton, R. J. & Cohen, E. (1972). *Electric Power Transmission Systems.* (2nd Ed.) Englewood Cliffs, NJ: Prentice Hall.

13. Editorial Staff of Electrical World. (1949). *The Electric Power Industry, Past, Present, and Future.* New York, NY: McGraw Hill Book Co., Inc.

14. Ekstrom, A. (1975). ASEA's New Thyrister Valve for HVDC Transmission. *ASEA Journal, 3* (48), pp. 55–60.

15. *Electrical Transmission and Distribution Reference Book* (1964). East Pittsburgh, PA: Westinghouse Electric Corporation.

16. Electric Power. (1984). *Encyclopedia Britannica,* (15th ed.).

17. Elgerd, O. I. (1982). *Electric Energy Systems Theory, an Introduction* (2nd Ed.). New York: McGraw-Hill.

18. El-Hawary, M. E. (1983). *Electrical Power Systems: Design and Analysis.* Reston, VA: Reston.

19. Electrical Power Research Institute (EPRI). (1986). Burning Coal More Cleanly and Efficiently. *IEEE Spectrum, 2* (8), pp. 64–69.

20. Eriksson, G. & Hoglof, L. (1975). HVDC Station Design. *ASEA Journal, 3* (48), pp. 61–95.

21. Fishettu, M. A. (1986, May). Electric Utilities: Poised for Deregulation? *IEEE Spectrum, 23* (5), pp. 34–40.

22. Foster, J. A. (1979). *The Coming of the Electrical Age to the United States.* New York: Arno Press.

23. Furfari, C. (1981, Sept.). Criteria for Selecting Electrical System Grounding and Ground Fault Protection. *Plant Engineering,* pp. 117–122.

24. Gonen, T. (1986). *Electrical Power Distribution System Engineering.* New York: McGraw-Hill.

25. Husak, L. (1981). *Compensator Distance Relaying.* Coral Springs, FL: Westinghouse Electric Corporation Relay and Telecommunications Division.

26. Insul, S. (1924). *Public Utilities in Modern Life.* Chicago, IL: Privately printed.

27. Kimbark, E. W. (1971). *Direct Current Transmission, Volume 1.* New York: John Wiley & Sons.

28. Kirtley, J. L., Jr. (1986). Supercool Generation. *IEEE Spectrum, 20* (4), pp. 28–35.

29. Mason, R. C. (1956). *The Art and Science of Protective Relaying.* New York: John Wiley & Sons.

30. Nichols, W. H. (1986, Oct.). *Symmetrical Components, Class 1.* Paper presented at the meeting of the IEEE Industrial Applications Society. Houston, TX.

31. Parkland, C. A. (1945). *Relay Engineering.* Pittman, NJ: Struthers Dunn, Inc.

32. *Power System Communications: Power line Carrier and Insulated Static Wire Systems, REA Bulletin 66-5.* (1978, May). Rural Electrification Administration, U.S. Department of Agriculture.

33. Rice, D. E. (1986, Nov.). *Symmetrical Components, IEEE-IAS Class 2.* Paper presented at the meeting of the IEEE Industrial Applications Society, Houston, TX.

34. Rustebakke, H. M. (Ed.) (1983). *Electric Utility Systems and Practices,* (4th ed.) New York: John Wiley & Sons.

35. Schultz, R. D. & Smith, R. A. (1985). *Introduction to Electric Power Transmission Systems.* New York: Harper & Row.

36. Schwarzschild, B. (1986, Aug.). Committee Reviews DOE Inertial-Confinement Fusion Program. *Physics Today. 39* (8), pp. 19–22.

37. *Specifications and Drawings for 7.2/12.5 kV Line Construction, REA Form 804.* Rural Electrification Administration, U.S. Department of Agriculture.

38. Stevenson, W. D., Jr. (1982). *Elements of Power System Analysis.* (4th Ed.). New York: McGraw-Hill.

39. Weeks, W. L. (1981). *Transmission and Distribution of Electrical Energy.* New York: Harper & Row.

40. Wood, A. J. & Wollenberg, B. F. (1984). *Power Generation, Operation, and Control.* New York: John Wiley & Sons.

41. Zorpetti, G. (1985). HVDC, Wheeling Lots of Power. *IEEE Spectrum, 22* (6), pp. 30–36.

# INDEX

**569**